Philosophy and Foundations of Physics

Series Editors: Dennis Dieks and Miklos Redei

In this series:

Vol. 1: The Ontology of Spacetime
Edited by Dennis Dieks

Vol. 2: The Structure and Interpretation of the Standard Model
By Gordon McCabe

Vol. 3: Symmetry, Structure, and Spacetime
By Dean Rickles

Symmetry, Structure, and Spacetime

By

Dean Rickles

Unit for History and Philosophy of Science
University of Sydney

ELSEVIER

Amsterdam – Boston – Heidelberg – London – New York – Oxford – Paris
San Diego – San Francisco – Singapore – Sydney – Tokyo

Elsevier
Radarweg 29, PO Box 211, 1000 AE Amsterdam, The Netherlands
The Boulevard, Langford Lane, Kidlington, Oxford OX5 1GB, UK

First edition 2008

Library of Congress Cataloging-in-Publication Data
A catalog record for this book is available from the Library of Congress

British Library Cataloguing in Publication Data
A catalogue record for this book is available from the British Library

ISBN-13: 978-0-444-53116-2
Series ISSN: 1871-1774

Printed and bound in The Netherlands

08 09 10 11 12 10 9 8 7 6 5 4 3 2 1

PREFACE

Symmetry is increasingly becoming a central, and indeed 'hot' topic for philosophers of physics: it is linked to various metaphysical issues having to do with space, time, motion, change, identity, modality, ontology and much more besides. This book examines the current interpretive landscape of symmetry in physics with the emphasis on those issues just listed. I consider a number of physical theories for which symmetry poses a particular and *universal* interpretive difficulty: the *same* difficulty can be found across many theories dealing with apparently different subject matter. The problem, roughly, is this: certain symmetries imply 'redundancies,' and this leads to an 'inability' of the theory to 'choose' between states or configurations related by the symmetry.[1] In certain cases this leads to indeterminism, but in general we have an underdetermination of the state by the theory. It is this underdetermination, or, rather, how one deals with it, that is at the root of many debates, new and old, in the philosophy of physics.

The connections of such matters to ontological issues are old news: the debate between Newton (or, rather, Rev. Samuel Clarke) and Leibniz over the nature of space and time was just such an underdetermination problem (as presented in the 'shift argument'). Historical niceties aside, two sides were carved out of this debate: absolutism (or 'absolutist substantivalism'), which accepted the existence of multiple physical states that the theory could not distinguish between (these states differing solely *haecceitistically*), and relationism, which denied such a profusion. As is well known, *mutatis mutandis*, the hole argument, as revived by Σ_i John i (i = Stachel, Earman, and Norton) hangs on the same choice, though here the underdetermination of states by theory is tantamount to indeterminism—the latter being understood as an unpalatable consequence of substantivalism, a mere metaphysical position.

More recent times have seen the emergence of so-called "sophisticated" forms of substantivalism which claim to check certain essential boxes of the core substantivalist position without implying the profusion of indistinguishable states generated by the more naïve substantivalism at work in the shift and hole arguments. However, alongside these modifications come the cries, led chiefly by Robert Rynasiewicz [1996], that the whole debate degenerated into a meaningless farce ever since the notion of *fields* appeared in physics, coming to a head in general relativ-

[1] Strictly speaking we should avoid using "symmetry" to describe these cases because symmetries connect physically *distinct* states, but this is still the common parlance and I shall stick to it in what follows. A better term, which is gaining currency among physicists, would be 'gauge redundancy' (*cf.* [Guilini, 2003]). *Gauge* here refers to a certain class of transformations that leave 'physical aspects' unchanged—where the physical aspects are cashed out in terms of the observables of the theory. The idea is that states or configurations related by such a transformation represent *one and the same* physical state or configuration.

ity where the geometric structure of spacetime is represented by a local dynamical variable (the metric field) not readily distinguishable, ontologically speaking, from 'ordinary material fields'. This sentiment seems to be shared by Carlo Rovelli, a leading quantum gravity researcher, who argues that the distinction between viewing the metric field as matter or spacetime is largely *verbal*—see, for example, [Rovelli, 2004], §2.4.2. But, despite this, Rovelli, and many other quantum gravity researchers, nonetheless argue that the grand old debate between substantivalism and relationalism is playing a crucial and *divisive* role in their field. Gordon Belot and John Earman, and a handful of philosophy-minded physicists, think that there is a close connection between the various theories of quantum gravity and traditional ontological positions concerning spacetime: they think whatever research programme eventually is crowned victor in the race for quantum gravity could *settle* the debate between relationalists and substantivalists! I argue that this is not so; their views stem either from a misunderstanding of the terms of the debate or else from a desire to breathe some life into the dying corpse of the debate. I fear it is too late. Such views involve an alignment between these positions and a pair of other positions that certainly *do* play such a crucial and divisive role, namely background dependence and background independence. The primary aim of this book is to argue against this view, a view that is fast on its way to becoming the received wisdom. The alignment does not hold up to scrutiny. I argue instead that the state of play in modern physics, especially quantum gravity, points towards a novel interpretive stance: a structuralist ontology that does not involve objects (in the sense of *subjects*) at all—hence, both relationalism and substantivalism are out.

It isn't just spacetime theories that face these kinds of problems: the quantum theory of indistinguishable particles and gauge theories in general face similar problems and can be met by the same kinds of interpretive responses. This book seeks to show how these various cases are essentially the same by providing a unified account of the problems and their interpretive options.[2] However, throughout this book my eye is always directed firmly towards quantum gravity, the as yet non-existent theory that will bring together the principles of quantum field theory and general relativity. The position I develop is ultimately intended to provide a satisfactory, and indeed safe and sane, interpretation of quantum gravity. Ultimately, then, this is a defense of structuralism about modern spacetime physics. Structuralism has been 'creeping up' in physics for some time, and with the advent of gauge theories I think it is the obvious interpretive stance to adopt; when we get to background independent theories like general relativity, where even the dynamics is pure gauge, then structuralism is almost *forced* upon us.

The book begins with a fairly informal survey of some background material involving elements from philosophy of science (interpretation and ontology) and some basic concepts from the theories we shall be dealing with. I round Chapter 1 off with a discussion of symmetry where I outline the type of symmetries that I am interested in—this chapter could easily be skipped by those philosophers of science who already understand the basics of symmetry. In Chapter 2 I apply these materials to the debate between substantivalists and relationalists where

[2] Indeed, examining these other cases provides us with a better handle on the interpretation of spacetime physics, and *vice versa*.

I focus on the 'Leibniz-shift' argument. The main task in this chapter is to highlight the philosophical significance of the geometric spaces outlined in the opening chapter and to detach these spaces from the traditional debate concerning spacetime ontology—a running theme. Chapter 3 situates the Leibniz-shift argument in the framework of gauge theories and analyzes the interpretive options of such theories. The main result is that there is a lot of elbow room concerning the relationship between interpretive options and formal methods for dealing with gauge freedom—this is the gauge-theoretical analogue of the detachment of spacetime ontology from geometric spaces.[3] The results from this chapter are then applied to the hole argument in Chapters 4 and 5—I show that the methods for dealing with gauge freedom have counterparts in the various resolutions of the hole argument. Chapter 6 focuses on the concept of *observable* in general relativistic theories and provides what I take to be the most natural resolution to the hole argument, and the true 'lesson' of that argument: the observable content of general relativity (and background independent theories in general) is exhausted by structural (manifold-independent) quantities having the form of correlations between field values. Hence, the view I advocate involves thinking of the observables of general relativity as *correlational* quantities that make no mention of spacetime—this, I say, is best understood as a structuralist position. This view is then put to work against the problems of time and change in classical and quantum gravity in Chapter 7—the idea is that in understanding correlations structurally (so that they are not *decomposable* into independent correlates) we are able to sidestep a number of standard problems and objections that other interpretations face. Chapter 8 ties the other chapters together by focusing squarely on the relationship between the treatment of symmetry and various interpretive positions, and draws some general metaphysical and ontological morals. Finally, in Chapter 9 I defend a 'minimal' structuralist position more explicitly. The key results in this and the previous chapter are: (1) a demonstration of the irrelevance of possibility counting (connected to geometric space descriptions and the treatment of symmetries) to spacetime ontology, and (2) the exposing of an extreme underdetermination of metaphysics by physics that engulfs both subject-based ontologies (such as substantivalism and relationalism) and those structural realist ontologies that *eliminate* objects by reconceptualizing them in terms of structures. The more *deflationary* structuralism I espouse is intended to do justice to both of these factors.

Although the book is written fairly linearly, each chapter making use of the previous, they can nonetheless be read independently of one another. In order to make this possible it was necessary to go over some of the material several times. However, in order to avoid ugly repetition, the treatments generally come at the same material from a different angle.

This book grew out of my doctoral thesis, *Spacetime, Change, and Identity: Classical and Quantum Gravity in Philosophical Focus*, written at the University of Leeds under the supervision of Prof. Steven French (completed in June 2004). My thanks

[3] In essence I want to take Nick Huggett's conclusion about the philosophical implications of statistical physics—that permutation symmetry has no "heavy metaphysical implications" ([1999], p. 24)—and apply it to a variety of areas in which similar symmetries operate.

to Steven for showing me the ropes *vis-à-vis* philosophy of physics; for entrusting me with such an ambitious project; and for being a good friend. My thanks also to the members of my thesis examination committee: Dr Joseph Melia and Prof. Carl Hoefer. I have resisted the very strong temptation to modify the content of the original thesis, leaving the original presentation largely intact (any alterations being mostly stylistic). I have, however, altered my stance with respect to numerous of the arguments presented here, but the main one I still hold fast to: structuralism is the only game in town in background independent (gauge) theories. I did feel the need to change the title though: the present title much better reflects the content and aims of the book.[4]

The finishing touches to this monograph were completed while I was a postdoctoral fellow at the University of Calgary: I would like to thank both the department of philosophy and the Markin Institute (Prof. Penny Hawe in particular from the latter) for their support, financial and otherwise. I thank Dennis Dieks and Miklos Redei for taking this project on as part of their excellent new series. Working with them has been a very smooth experience. I would also like to thank Gordon Belot, Steven Weinstein, and Oliver Pooley for supplying me with copies their doctoral theses: my own thesis (this book) owes a great deal to them. Any errors—and I'm sure there are plenty—are, of course, entirely my own. My *entrée* into philosophy of physics was overseen by Nick Unwin at the University of Bolton, who let me attend his excellent MA philosophy of physics course while still a lowly undergraduate. That course left a great impression on me, and I thank him for that. Finally, I thank Kirsty, Sophie, and Gaia for their constant love, support, and laughs. I dedicate this book to my parents, Muriel and Peter Rickles, for giving me drive and determination.

D.P.R.

[4] I would certainly done a number of things differently as regards the presentation of material had I started this project from scratch: the chapter on gauge theory I would have reframed in the language of fiber bundles; the chapters on the hole argument and observables I would have considered from the perspective of the Kretschmann objection and the ongoing debate concerning general covariance, diffeomorphism invariance, absolute objects, and background independence. However, though these are indeed notable absences, I think the book manages to hold up despite them.

ACKNOWLEDGEMENTS

Parts of Chapter 1 appeared as "Interpreting Quantum Gravity" (*Studies in the History and Philosophy of Modern Physics* 36 (2005) 691–715). Parts of Chapter 6 appeared as "A New Spin on the Hole Argument" (*Studies in the History and Philosophy of Modern Physics* 36 (2005) 415–434). I thank Elsevier for allowing me to reproduce this material here.

Chapter 7 is largely identical to "Time and Structure in Canonical Gravity", which appeared in D. Rickles, S. French, and J. Saatsi (eds.). *The Structuralist Foundations of Quantum Gravity*, Oxford University Press, 2006 (pp. 152–195). My thanks to OUP for allowing me to reproduce this.

CONTENTS

CHAPTER 1

Interpretation and Formalism

1.1. INTERPRETATION AND ONTOLOGY

> *Can philosophers really contribute to the project of reconciling general relativity and quantum field theory? Or is this a technical business best left to the experts? [.] General relativity and quantum field theory are based on some profound insights about the nature of reality. These insights are crystallized in the form of mathematics, but there is a limit to how much progress we can make by just playing around with this mathematics. We need to go back to the insights behind general relativity and quantum field theory, learn to hold them together in our minds, and dare to imagine a world more strange, more beautiful, but ultimately more* reasonable *than our current theories of it. For this daunting task, philosophical reflection is bound to be of help.* ([Baez, 2001], p. 177)

Although one might not guess it from the above quote, in recent times physicists and philosophers of physics have tended to tread very different paths, and they have generally been a little suspicious of one another. As Michael Redhead points out in the first of his Tarner lectures, "many physicists would dismiss the sort of question that philosophers of physics tackle as irrelevant to what they see themselves as doing" while "philosophers generally regard physicists as naive people, who do physics in an uncritical way" ([1996], pp. 1–2). Reichenbach expresses much the same point even more strongly, suggesting that there is a "mutual contempt in which each misunderstands the purposes of the other's endeavours" ([1958], p. xi).

This hasn't always been the case, of course. As Reichenbach notes, "[t]he classical philosophers had a close connection with the science of their times" (*ibid.*), Descartes and Leibniz being fine examples. In addition to this, historically, each time a fundamental revolution has occurred in physics (Newtonian mechanics; the relativity theories; quantum theory, *etc.*) there has generally been an associated shift to a more critical, reflective attitude towards theory construction. Indeed, Kuhn appears to suggest that such a shift is a necessary part of scientific revolutions:

> It is no accident that the emergence of Newtonian physics in the seventeenth century and of relativity and quantum mechanics in the twentieth should have been both preceded and accompanied by fundamental philosophical analyses of the contemporary research tradition. ([Kuhn, 1970], p. 88)

Kuhn also seems to give a very accurate depiction of the present situation in quantum gravity[1] when he writes that:

> Confronted with anomaly or with crisis, scientists take a different attitude toward existing paradigms, and the nature of their research changes accordingly. The proliferation of competing articulations, the willingness to try anything, the expression of explicit discontent, the recourse to philosophy and the debate over fundamentals, all these are symptoms of a transition from normal to extraordinary research. ([1970], pp. 90–91)

One of the main aims of this book is to highlight the extent to which many problems of quantum gravity are predominantly 'philosophical' in their nature and origin—these spring in large part from aspects of the classical theory. But despite this, though physicists have been known to take up more philosophical attitudes in times of crisis, 'the philosophers' have previously remained firmly divided from the *constructive* practice of theory building in physics, waiting in the wings, as it were, until the theories were deemed sufficiently well established to warrant their attention.

The strange case of quantum gravity, as this book will try to show, offers us a place where philosophers might play a role in the more constructive parts of the foundations of physics—though, of course, it is highly doubtful that this will include an involvement in the *technical* foundations. Tian Yu Cao, believing consistency and conceptual clarity to be of the essence in quantum gravity (his point being, there is no experimental basis), makes a similar point:

> this is a rare conjuncture for philosophers to intervene, with a good chance to make some positive contributions, rather than just analysing philosophically what physicists have already established. ([Cao, 2001], p. 183)

There are at least three reasons behind this possibility as I see it: (1) these days philosophers of physics are simply better equipped in terms of their command of the necessary parts of mathematics and physics required—most of them having done their original training and research in physics, and often publishing in physics journals, (2) quantum gravity is an area of physics lacking an experimental basis from which to test the various proposals, thus forcing conceptual and mathematical consistency to take center stage. What's more, (3) it appears that the kinds of conceptual problem that litter the field of quantum gravity are ones that philosophers are already well familiar with, as I have already suggested, and aim to show in more detail.[2]

Philosophers, however, have generally been rather slow to pick up the challenge of quantum gravity, seemingly more content to flog poor old non-relativistic quantum mechanics to death! This is made all the more surprising given that many

[1] We discuss quantum gravity in more detail in later chapters; for now it suffices to think of it as some theory that unites quantum field theory and general relativity, in much the same way that a quantum field theory is some theory that unites quantum mechanics and special relativity. The problem of quantum gravity is that, whereas special relativity provides a nice fixed spacetime structure to define the quantum theory with respect to, general relativity involves a dynamical spacetime structure.

[2] Though as I pointed out in the preface, much of what I have to say will be largely negative, involving the detachment of various metaphysical theses from the physics.

researchers engaged in quantum gravity actively encourage the involvement of philosophers in their discipline. For example, Carlo Rovelli (one of the physicists who created the popular approach known as 'loop quantum gravity'—a genuinely viable alternative to string theory) explicitly voices this opinion:

> As a physicist involved in this effort [quantum gravity], I wish the philosophers who are interested in the scientific description of the world would not confine themselves to commenting and polishing the present fragmentary physical theories, but would take the risk of trying to look *ahead*. ([Rovelli, 1997], p. 182)

Up until very recently the same might have been said of (specially) relativistic quantum field theory, though lately there has been a definite shift of emphasis from non-relativistic quantum mechanics to relativistic quantum field theory—a very welcome move in my opinion. As welcome as this shift is though, methodologically the philosophy of relativistic quantum field theory is the same as non-relativistic quantum mechanics (and likewise subject to Rovelli's gripe): the mathematical and theoretical framework exists (admittedly, modulo certain nasty consistency problems in the interacting theory—*i.e.*, Haag's theorem) and the experimental data is there to confirm this pre-given framework; the job of the philosopher of physics is to examine and interpret this framework and its relation to the world. The uniqueness of quantum gravity as a challenge to philosophers arises precisely from the lack of such an established framework and associated medley of confirming experiments, and philosophers would do well to shift their gaze in its direction because of this feature.

Of course, some might suggest that, quite to the contrary, this in fact gives one more reason to *divert* one's gaze! The worry here, if I understand the objection's direction correctly, is that it is pointless[3] for philosophers to apply their skills to a theory that does not yet exist, and they should, therefore, stand on the sidelines until the theory has sufficiently matured to warrant such attention. I don't think we need take this stance. Although it is true that there is no definitive theory of quantum gravity—nor does it look likely that there will be for some time to come—, there are a number of ways in which one can approach the theories we do have without risking irrelevance. The crucial *philosophical* aspect concerning quantum gravity is how to interpret a background free (*i.e.* diffeomorphism invariant) quantum field theory—the technical aspect is, of course, how to *construct* one![4] This notion can be analyzed independently of specific proposals that aim to implement it; moreover, we can be fairly confident (given what we know about general relativity) that the 'final' theory of quantum gravity will implement this

[3] One might even go so far as to declare that the project is 'meaningless', along the lines suggested by early logical positivists, raising the old spectre of the demarcation problem thus: since it is difficult to conceive of physically possible ways to test quantum gravity, can it rightly be called a science at all? However, the recent efforts to design, albeit mostly hypothetical (but in principle physically possible), experiments to test approach-specific features of quantum gravity would make this position harder to hold nowadays. See Amelino-Camelia [1999] for some of the details of these experiments—[Smolin, in press] presents some "generic" predictions of background independent quantum gravity. Ultimately, of course ('strong programmers' notwithstanding!), the choice of theory/approach will come down to testability and empirical confirmation.

[4] Background dependent theories will not be so conceptually hard to handle—though in string theory the existence of higher-dimensional fundamental objects, compactified spatial dimensions, and duality symmetries will bring their own difficulties.

feature.[5] It is, therefore, to this aspect that I devote my attention and, in particular, to the specific conceptual problems that result from background independence and symmetry (the former involving the latter in a highly non-trivial way).

To understand what background independence is, I first need to say something about the relationship between ontology and interpretation, and the way in which symmetry blackens the already muddy waters. My account of this relationship is based upon models that encode information about a theory's (possible) ontology (or ontologies). Background independence is then seen to be a feature of these models that is caused by a high degree of symmetry. It is a certain class of symmetry that I am interested in: those that preserve all *qualitative* structure of a model or world (see below for more on the notion of 'qualitative'). One of the central themes of this work is that these symmetries cause ontology to be radically underdetermined, making the task of the interpreter very difficult indeed. As I will show, it is always possible to choose a formalism (or an interpretation) *without* these symmetries but this too will be underdetermined by the physics, making the interpretive task harder still.

One of the primary functions of the philosopher of physics is, of course, to *interpret* the mathematical constructions of physicists. Hence, one often speaks of "the interpretation of quantum mechanics" or "the interpretation of statistical mechanics". But what is it to interpret a physical theory? I think that it is fair to say that one faces a greater problem in trying to answer this question in relation to modern physics than in relation to, say, biology or psychology. The reasons are twofold: firstly, it is undeniable that theories of physics—especially recent ones—rely very heavily upon the use of very complex mathematics. From Newton onwards, the kinds of mathematics featuring in physics has become ever more complex. In quantum gravity the mathematics is pushed to its limits of application, even forcing the creation of brand new fields of mathematics in order to express its content.[6] Secondly, the realm of objects with which physics concerns itself is often 'unobservable' (in the philosopher of science's sense) and one therefore has a certain amount of freedom in filling in the details concerning these aspects. I don't say that these point to a difference in *kind* between interpretations of physics and interpretations of other sciences (for clearly psychology and biology deal with 'unobservables' too); I merely say it points to a difference in *magnitude*.

An essential part of interpretation is choosing which parts of a formalism are taken to *represent* something. Then one has to give an account of *what* they represent, *i.e.* one has to provide an account of the ontology. I take this latter aspect to be tantamount to the presentation of a set of *physically* possible worlds that make the theory true[7]—for example, a set of individuals over which properties and re-

[5] Even string theorists appear to be approaching some kind of consensus on the necessity of background independence—hence, the claim often made that background independence is "divisive" (see, *e.g.*, [Smolin, 2006]) might be somewhat overcooked. Perhaps. But so far string theory remains background dependent; the *attitudes* of string theorists towards this state of affairs cannot change this fact. It might well be that a truly background independent formulation of string theory is impossible.

[6] This is perhaps most evident in topological quantum field—that part of quantum gravity that sits at the meeting point between knot theory and physics. See Hu [2001] for a nice introduction to the most important of these developments and Kauffman [2001] for a more general taster. Baez & Munian [1994] still offers the best and quickest route for those completely unfamiliar with these connections.

[7] Note that just because one has a model involving individuals with properties and relations defined over them this does not imply that one is thereby committed to a *direct* correspondence between these individuals with the various properties

lations are defined in such a way so as to make the statements and laws of the theory true. Both aspects prove difficult in physics for the two reasons outlined above. A further problem concerning the latter aspect is that there may be many choices available with regards to which bits represent and what they represent.

Of course, what I have just mentioned is only a small component of interpretation and its difficulties, but even this suffices to demonstrate that it is no simple matter to explain what constitutes an interpretation. Giving any kind of comprehensive answer to this question would take me too far afield from the (interpretive!) task I have set myself in this book. However, I think that limited, though adequate, sense can be made by asking what an interpretation of a somewhat simpler, non-physical theory might amount to. To this end, I consider the case of modal logic and the interpretation (semantics) for its model theory.[8] This brief diversion will come in handy in the later chapters where I make extensive use of notions from modality. It also allows us to understand intuitively what is meant by providing an ontology for a theory via the notion of possible world semantics, and thus the notion of an interpretation of a theory. Further, it will allow me to make clear a division between interpretation and ontology on the one hand and the question of realism/anti-realism on the other. These two categories should be cleanly separated, and if we do so then it makes sense to interpret and provide an ontology for theories that have no empirical basis whatsoever—quantum gravity, for example. Seen in these terms, *realism* is an *extra* claim attached to an interpretation of a theory that effectively says that one of the possible worlds in the ontology of a theory corresponds to the *actual* world, *i.e.* that the interpretation of the theory is *true*, as opposed to merely *possible* (or as opposed to being merely a model in the sense of a structure that makes the theory's axioms or laws true)—underdetermination of a theory by the evidence corresponds to glut of possible interpretations.[9]

Nowadays, propositional modal logic is generally defined as standard propositional logic *plus* the axioms and rules governing the behaviour of the modal operator '$\Box$', read as 'it is necessarily true that ...'.[10] I follow this approach here, using the concise Backus–Naur notation for brevity and clarity (see Marcotty & Ledgard [1986], p. 41 for details). Let Ω be a denumerable set of atomic formula,

and relations and reality. Much of what we have to say in this book turns on the issue of just how we should see the model as representing physical reality.

[8] Chagrov & Zakharyaschev [1997] and Gabbay & Guenthner (eds.) [2002] provide very comprehensive guides to this material. Beall & van Fraassen [2003] offers an excellent recent introduction to the subject.

[9] I should point out that the kind of underdetermination I am concerned with in this book is not of this 'traditional' variety, wherein (usually) radically different but empirically equivalent ontologies compete for the place of the true theory that should be the object of the realist's devotion. Rather, it is an underdetermination *within* a single chosen ontological framework: one settles upon a way of making sense of the theory, in terms of fields say, and then one faces a problem with regards to *which* field configuration is the true field configuration, for example. We might distinguish the two sorts of underdetermination by calling the first sort *trans-ontological* and the second sort *inter-ontological*.

[10] There are several other readings for $\Box$. Notably: 'it ought to be that'; 'it is known that', 'it is believed that', 'it will be the case that', defining the *deontic, epistemic, doxastic,* and *temporal* modes respectively. There are also different *grades* concerning 'logical', 'metaphysical', and 'physical' modality. Our concern in this book is with the latter two grades whose difference we can cash out loosely as 'ways the world could have been (or could be)' and 'ways the world could have been *given our laws of physics*'. Take a fairly typical example: 'there could have been a sphere with a radius of 1000 miles made entirely of Uranium' and 'there could have been a sphere with a radius of 1000 miles made entirely of Gold'. The latter is both physically and metaphysically possible whereas the former is only metaphysically possible: Gold is scarce but it is not unstable like Uranium.

and let $p \in \Omega$ be a member from this set (*i.e.* a propositional variable). Let $f(\Omega)$ be the set of formulae that can be generated from Ω, and let $F \in f(\Omega)$ be a member from this set (*i.e.* a well-formed formula). Finally, let '$\perp$' denote the *falsum* (*i.e.* the constant false proposition), and let '$\rightarrow$' denote implication—these are the logical connectives for our language. From these primitive ingredients the Backus–Naur notation allows us to define the *syntax* of propositional logic recursively as:

$$F ::= p|\perp|F_1 \rightarrow F_2 \tag{1.1}$$

This equation states that a formula of propositional logic is either an atomic formula, the *falsum*, or an implication between two formulae. Propositional *modal* logic simply requires the addition of the operator '$\Box$' to this language, so that the syntax is defined as:

$$F ::= p|\perp|F_1 \rightarrow F_2|\Box F \tag{1.2}$$

Hence, the box of a formula can also be a formula in our language. Now, it would be hard to interpret this 'thin' notion of a logic; thus far, I have taken a *syntactical* approach to logic and I have not been concerned with what the symbols stand for; that is, with what they might be taken to *represent*. Implementing this latter feature is tantamount to giving the logic an *interpretation*. The addition of two more concepts facilitates this latter feature. Crucial to the interpretation of propositional modal logic are the notions of *frame* and *model*. A frame is an ordered pair of the form $\mathrm{F} = \langle \mathrm{D}, R\rangle$, where D is a (non-empty) domain of individuals and R is a binary relation on D (that is, $R \subseteq \mathrm{D} \times \mathrm{D}$). A model is defined on a frame, and is given by an ordered triple of the form $\mathrm{M} = \langle \mathrm{D}, R, V\rangle$, where $V : \Omega \rightarrow 2^{\mathrm{D}}$ is a *valuation* function. Hence, V assigns to each $p \in \Omega$ some subset $V(p) \subset \mathrm{D}$, with the intended reading that $V(p)$ is the set of points in D at which p. Various systems of modal logic are constructed by imposing different conditions on R relating the points of D in different ways; these are the different *modalities* or R-modalities.[11]

As it stands, this is still incomplete: the elements of the model itself are in need of an interpretation. According to, what I shall call, the 'standard view', the semantics (or interpretation) is specified in terms of *possible worlds* and relations between possible worlds. Specifically, the members of the domain $x \in \mathrm{D}$ are known as 'possible worlds' and the relation R is called an 'accessibility relation' (named by Peter Geach; *cf.* Bull & Segerberg [1984], p. 15); V is a valuation picking out a subset of the worlds that are the 'truth-values' of modal propositions $p \in \Omega$. In other words, we can talk about propositions by using sets of worlds (via V) and we can talk about sets of worlds using propositions (using the pull-back V^{-1} of V): a proposition *just is* the set of worlds at which it is true. The intuitive content is that for two worlds $x, y \in \mathrm{D}$, xRy iff y is a 'conceivable world' relative to x.

Of course, this is still rather abstract. There are many ways in which we could fill in the details of the worlds and thus explicate the content of our modal discourse. In modal metaphysics they are taken to be, as the previous paragraph suggests, *ways the world might have been*.[12] But there are alternatives that are equally

11 For a list of these with their translations into propositional (modal) language, see Goldblatt ([1992b], p. 12).

12 Of course, this too still needs fleshing out: how are we to understand these "ways"? As fictional entities? Concrete 'flesh and blood' entities like our own world? Ersatz constructions from actually existing elements? See Loux (ed.) [1979] for a nice collection of articles that deal with this aspect.

compatible with the rules of the logic: the syntax is *applicable* to many other areas of discourse. One might see the models as about the structure of time, for example, and see the worlds as instants of time and the accessibility relation as an ordering on the instants (see Rescher & Urquhart [1971]); or *algebraically* concerning lattices with operators (see Goldblatt [1992b]); the models also have applications in computer science, where the domain is taken to correspond to the states of a computer program (see Goldblatt [1992a]). Hence, the Leibniz-inspired 'possible worlds' semantics is by no means unique.

Thus, logic is largely indifferent to how the semantics is filled in: any specification of objects (any worlds) will do just as well, so long as they satisfy certain functional requirements imposed by the purely syntactic rules of the logic. The models of the logic exhibit these requirements *structurally* (the nature of the elements of the domain is ignored); so long as the objects stand in the right kinds of relations, the structure is preserved. According to van Fraassen, the same goes for interpretations of physics (that is, according to a *semantic conception*[13]):

> The syntactic picture of a theory identifies it with a body of theorems, stated in one particular language chosen for the expression of that theory. This should be contrasted with the alternative of presenting a theory in the first instance by identifying a class of structures as its models. In this second, semantic, approach the language used to express the theory is neither basic nor unique; the same class of structures could well be described in radically different ways, each with its own limitations. The models occupy center stage. ([van Fraassen, 1980], p. 44)

In this case, the functional relationships that must be satisfied come from experiment and observation on the one hand and compatibility with the mathematical formalism on the other. Thus, an interpretation of physics gets an extra constraint imposed from without: it must be empirically adequate.[14] The constraint of empirical adequacy is implemented in the models as a *selection* of a subset of worlds called the *physically possible worlds*; this subset comprises just those worlds that satisfy the laws of the theory—the theory itself can be specified by selecting a subset of worlds from possibility space (*cf.* Lewis [1986a], p. 26). The physically possible worlds can be usefully viewed as 'embedded' in the larger class of *metaphysically* possible worlds. We can view the latter as simply the worlds that can be 'constructed' from the raw mathematical materials of the model without regard to their physical viability.[15] The crucial issue for us is knowing how to delineate these categories; as we shall see, this is an issue that is made all the more pressing

[13] According to the semantic view of theories—along the lines of van Fraassen [1980]—a scientific theory consists of two components: the *theoretical structure* and the *theoretical hypotheses*. The former component comprises a family of mathematical models; each model is given by a domain with a set of relations defined over it. The latter component (the theoretical hypotheses) comprises a set of propositions detailing the kinds of worlds represented by the models and how they should be understood as representing. I shall call the latter component the *ontology*, where this notion should be understood as bracketed from the question of realism as outlined above. One of the primary tasks of interpretation is to extract and unpack the ontology and to elucidate the representation relation.

[14] For van Fraassen, empirical adequacy is *all* scientific theories aim for. His notion of empirical adequacy is spelled out a little differently to how I spell it out here; namely, in terms of empirical sub-models (see §9.4 for further discussion).

[15] This corresponds, roughly, to the distinction between 'kinematics' and 'dynamics'. As Wheeler puts it, "[k]inematics describes conceivable motions without asking whether they are allowed or forbidden. Dynamics analyses the difference between a physically reasonable and a disallowed history" ([Wheeler, 1964], p. 65).

for theories possessing an exceptionally high degree of symmetry. In the latter case (when the symmetries preserve all *observable* structure[16]), we face a further problem in understanding the relation between the models and *genuinely* physically possible worlds since the symmetry introduces 'surplus' elements (representing, if anything, states or worlds that differ 'non-qualitatively'—where I understand *qualitative* in terms of the observables of a theory, so that what is qualitative is theory-relative as well).[17]

For the most part, I skirt the very difficult issue of explicating the nature of qualitative properties in this book, choosing to simply associate a general notion of qualitative with a theory's observables—see [Vallentyne, 1997; Lewis, 1983b; Lewis, 1983a] for more details on the definition of 'qualitative'. This results in a context-dependent, non-general account according to which what counts as 'qualitative' tracks changes in the observables of a theory—with the qualification that the theories of which I am considering this to be applicable are of a 'special' type, possessing a remarkably high degree of symmetry (*i.e.* gauge symmetry). Qualitative predicates and relations will be associated with some, perhaps mind-bogglingly complex, functions of the basic variables of a theory. These will be the observables of theory and they are forced to satisfy certain constraints imposed by the symmetries. Thus, to make sense of the problems, I propose to simply *define* qualitative physical properties and relations as observables.[18] Though this has an implication of *measurability*, it is not necessarily equivalent to the notion of *measurable by us*, for there will most always (in most theories of interest) be observables that could never be observed by beings with our limitations (of size and so on). There are, as I will mention, cases of observables in general relativity that it would be hard to conceive of as measurable by *any* being in practice, save, perhaps, God! Thus, I do not assume a crude empiricism in this definition—for more on this, see §9.4. It is better to understand measurability as being constrained by observables, rather than the other way around: if something is not an observable then it is not measurable by us; if something is an observable, then it might be measurable by us, but this is not necessary. One of the major problems that I am concerned with in this book is cases where symmetries of theories lead to an underdetermination of the qualitative character of a world by non-qualitative (=unobservable) aspects; the observables are the same in multiple worlds that are supposed to differ in some non-qualitative respect. In other words, there are many

16 The notion of "observable" here is theory-relative, and not strictly equivalent to the usual notion of observable, namely as something that can be 'viewed' or experienced. Depending upon the theory, and its way of representing physical states and properties, the concept of observable will shift. However, *invariance* plays a crucial role in determining what qualify as observables.

17 Einstein was aware of this implication soon after completing his theory of general relativity, for he writes that "the connection between *quantities in equations* and *measurable quantities* is far more indirect than in the customary theories of old" ([Einstein, 1918], p. 71). Likewise, speaking of gauge symmetries Bergmann and Komar write that "[u]nder this new symmetry group . . . the formulation of statements about physical reality requires much greater sophistication and care than is required when space and time are assumed to have a built-in rigid structure not subject to the dynamical equations of a physical field" ([Bergmann and Komar, 1980], p. 227).

18 We can, however, give a clear-cut case of a non-qualitative property: 'being identical to oneself', or '$a = a$' (*i.e.* haecceities). In other words, non-qualitative properties are those that involve only reference to particular individuals, rather than qualities of particular individuals. Of course, circularity threatens in spelling out the notion of a quality of an individual here. We will see that such properties as 'the value of the strength of the gravitational field at manifold point x' are non-observable, which shows that the identification between 'qualitative' and 'observable' (in the physicists sense) has some merit.

ways to produce a world with the same qualitative character. If we accept this explosion of worlds as a feature of reality, as occupying a place in our ontology, then we have to countenance indiscernible worlds differing haecceitistically (and in no other way), and we have to countenance 'unmeasurable observables' and, in those cases where the symmetry is *dynamical*, indeterminism, for the laws of physics (the equations of motion) cannot uniquely determine *which* of the qualitatively identical worlds will correspond to the time-evolved world: the equations will at best be able to determine the qualitative character of the world (or 'world-slice').

Before introducing the models that I wish to deal with and showing how interpretation works as applied to them, let me first say something about the relation between this conception of interpretation and the issue of realism. One of the fundamental philosophical problems (for scientific realists) of understanding physical theories is that there are usually 'too many' interpretations equally compatible with the same underlying formal system and the experimental data, or multiple formal systems compatible with the data: this is the well-known problem of underdetermination. This often worries aspiring realists since these interpretations may express very different ontologies—for example, 'many-worlds' *vs* 'collapse' interpretations of quantum mechanics. One way out of this problem for the realist is to focus one's realism upon *structural* aspects of the theory—see [Ladyman, 1998]. Hence, one shrugs off commitment to anything other than the relational and functional patterns expressed by the theory as laid out in the isomorphism-equivalence class of the theory's models. I defend a similar stance later. The important point I want to get across here, to repeat, is that interpreting a theory amounts to providing an ontology for that theory, but it does not imply a *commitment* to that ontology in the sense of realism. Interpretation and ontology are in a distinct category to realism and anti-realism; the latter provide stances with respect to the options laid out in the former. Since there are many interpretations compatible with a single theory (as exemplified by a family of models), and an interpretation is just a specification of a set of possible worlds that make the theory true (a set of worlds satisfying the structural requirements laid down by a theory), it follows that realism will either be directed at the class of interpretations or the *invariant* structure itself (where the latter is given by an equivalence class of models). Clearly, since the interpretations will usually be many and varied, without some extra-theoretical constraint to pick out an interpretation the former type of realism will be in trouble. There is another kind of underdetermination in physics that interests me, the one I mentioned above; this concerns an underdetermination about a *fixed* ontology. Thus, if one has an ontology of, say, indistinguishable particles, then the underdetermination might concern the possible states of a pair of such particles; putatively distinct states, namely permutations, yield the same physical predictions because of the indistinguishability with respect to the particles' observables. The latter structural option looks to be the more successful interpretive move; in committing oneself to the invariant structure one papers over the differences at the level of individualistic interpretations. I return to these issues throughout and then in more depth in the final chapter.

I'll be dealing with two types of model in this book: 'spacetime models' and 'geometric models'—though later on, the latter type will include the former, so that spacetime theories will be represented using geometrical models. In the former case the relations of the model will represent the spatiotemporal features of a world and in the latter case will represent certain geometrical facts about an abstract possibility space (a space of possible configuration, states, or histories of a system). Let us spell these out in a little more detail.[19]

The models of spacetime theories are of the general form: $\mathrm{M} = \langle \mathcal{M}, \mathcal{B}^i, \mathcal{D}^j \rangle$—where the left-to-right ordering of the model represents a 'chain of dependency' such that the entry to the right depends *for its mathematical definition*[20] on its partner to the left. $\mathcal{M}$ is a mathematical manifold whose elements are generally taken (by physicists) to represent the point-events of spacetime; it is equipped with a topology and a differential structure appropriate to the theory. $\mathcal{B}^i$ and $\mathcal{D}^j$ are 'geometric object fields' on $\mathcal{M}$ with the following meanings: the former are 'background fields' that characterize the *fixed*[21] structure of a spacetime while the latter denote the 'contents' of a spacetime. Two relationships are called symmetries in the spacetime theory context: covariance and invariance. For now we can understand these as follows: covariance is the preservation of the form of the laws of the theory when *all* of the fields, $\mathcal{D}$ and $\mathcal{B}$, are transformed by a certain operation associated to the particular theory; invariance is the preservation of the laws of the theory when *the dynamical fields alone* are transformed by the same operation so that the background structure is preserved. We cover these ideas in more detail in the next section.

The geometric models $\mathrm{M} = \langle \Gamma, \Omega \rangle$ are formally similar but have a more abstract meaning; Γ is a manifold whose points represent the possible states of a physical system where Ω gives the manifold a certain geometric structure appropriate to the system. In this case, a symmetry will manifest itself directly on the manifold by means of Ω, such that distinct points of Γ will be connected by a symmetry transformation (a symplectomorphism). Again, we are interested in those mappings (symmetries) that preserve the geometry. These models will function as 'possibility spaces' for the theories we are interested in. The idea is that one starts with a space of metaphysical possibilities, as represented by the 'bare' manifold, and then imposes a geometrical structure (the modalities) on the manifold. Further conditions (arising from dynamical or symmetry considerations) may require that we further *restrict* the space of possibilities in order to get a space of

[19] We will go over this material again, from various angles, at various places in the book, as and when required.

[20] Again, we should be careful about reading an ontological chain of dependency from this—*i.e.* we should heed well the warnings of Einstein, Bergmann, and Komar in footnote 17. Many of the puzzles we meet in later chapters have their origin in just such a reading.

[21] Here "fixed" is roughly taken to mean sameness across all of a theory's models (see Maidens [1998] and Smolin [1991; 2006] for a 'philosopher's' and 'physicist's' analysis respectively of this notion)—there are serious problems with the definition of background independence: for an excellent review see [Giulini, in press]. Thus, the $\mathcal{B}$-fields are *dynamically independent* of $\mathcal{D}$-fields. The converse does not hold, however, and the $\mathcal{D}$-fields themselves will generally satisfy laws directly connected to the $\mathcal{B}$-fields. A *background independent* theory is simply one without any $\mathcal{B}$-fields. However, such a theory may still have background *structure*: for example $\mathcal{M}$ itself is such a structure, along with the topological and differential structure. The symmetries of the manifold, however, mean that the observables, states, and dynamics must be constructed out of the physical degrees of freedom (the $\mathcal{D}$-fields) alone. I discuss what can legitimately be drawn from this in Chapter 6.

genuine physical possibilities. Determining the latter is no trivial matter; much of this book is concerned with showing just how non-trivial it is.

1.2. SYMMETRY AND STRUCTURE

Let us begin by distinguishing between a *structure* and the elements or *objects* of a structure. A structure S may be defined as an ordered tuple of the form $\langle D, \mathcal{R}_i \rangle$, where D is a (non-empty) set of individuals (the domain of S), and $\mathcal{R}_i$ is a (non-empty) set of relations over (or 'on') D (including '1-place' relations; *i.e.* predicates). The objects of a structure are simply the elements of D, and they are *characterized* by the $\mathcal{R}_i$. Now suppose we have two structures, $S = \langle D, \mathcal{R}_i \rangle$ and $S' = \langle D', \mathcal{R}'_i \rangle$, and an injective, surjective map $\phi : D \rightarrow D'$ (with inverse $\phi^{-1} : D' \rightarrow D$), such that $\mathcal{R}_i(x_1, \ldots, x_n) = \mathcal{R}'_i(\phi(x_1, \ldots, x_n))$, for $x_i \in D$ and $\phi(x_i) \in D'$. Then S and S′ are *isomorphic* (with ϕ an isomorphism). We then say that the two structures are related by a symmetry, where ϕ is a symmetry transformation.[22]

All this means in this case is that the relevant structure, here encoded in the relations, is preserved by the mapping (one-to-one and onto). Isomorphic structures, such as those related by a symmetry, are regarded as 'the same' in a certain *restricted* sense. The "sense" is determined by the specific features one is interested in, so that one may be interested in 'group theoretical' properties and relations and thus regard preservation of identities, inverses, and compositions as relevant.[23] When two structures are isomorphic each individual in the domain will have a copy in the codomain, which will be some individual with the same relevant properties and standing in the same relevant relations. In terms of the structures themselves, then, 'isomorphic' means 'indistinguishable' with respect to the relevant properties; two thus related structures will be carbon copies of one another.[24] When $\phi : D \rightarrow D$, then ϕ is an *automorphism* of S, and the group (under composition) of such maps is Aut(S), the group of automorphisms of S onto itself. This too is a symmetry. Hence, we have two separate concepts: (1) a symmetry holding *between* structures, and (2) a symmetry holding *of* a structure. In the first case, $x_1, \ldots, x_n \in D$ and $\phi(x_1, \ldots, x_n) \in D'$; whereas in the second case both

[22] An *abstract* structure is then the isomorphism class of such structures. 'Concrete structure' specifies some particular domain that is replete with the relations.

[23] For example, U(1) (the group of 'phases', $e^{i\theta} : \theta \in \mathbb{R}$) is isomorphic to SO(2) (rotations of $\mathbb{R}^2$) under the homomorphism $\rho(e^{i\theta}) = \begin{pmatrix} \cos\theta & \sin\theta \\ -\sin\theta & \cos\theta \end{pmatrix}$, despite the fact that these groups are very different in other non-group-theoretic ways.

[24] One of the issues that will be dealt with in later chapters is whether the 'relevant' structure covers what it should or what we would like it to in philosophical contexts, for example, when we are considering how and what the structures represent. We might think, in certain cases, that what is deemed 'surplus' to the isomorphism should play a role in determining whether the structures, when they are taken to represent, really are the same, and really do represent the same thing. The answer from physics is that the level of relevant structure is determined by the theory, by its kinematical and dynamical structure; once set, the dynamics will have to be insensitive to changes that amount to an isomorphism, and structures that differ thus will be deemed *physically equivalent*—in other words, the apparently additional structures should be taken to represent one and the same physical state or world, thus indicating the presence of some 'redundancy' or 'surplus' in the theory's formalism. The physical equivalence is often thought to be necessitated by considerations of determinism and measurability: if the dynamics are not insensitive to isomorphic shifts (in the physical variables that are fundamental in the theory) then there will be *physical* differences that make no *empirical* difference and variables that are not measurable. But these non-empirical differences might nevertheless be warranted on other grounds. The main thrust of this book is that how one treads here can not be understood as 'read off the physics', but must be seen as stemming from some other commitments.

$x_1, \ldots, x_n \in \mathrm{D}$ and $\phi(x_1, \ldots, x_n) \in \mathrm{D}$ (or $\mathrm{D} = \mathrm{D}'$). This means that in the first case, the objects are mapped onto a different domain of objects in a structure preserving way, but in the second they are mapped into the same domain in a structure preserving way. I am concerned with the latter type of symmetry, and in particular, with those cases where the structure is a space of possibilities relative to some physical theory, so that the symmetries can be seen as generating physical possibilities (solutions of the equations of motion of the theory) from physical possibilities (assuming, that is, that the symmetries are symmetries of the theory), or with cases where the symmetries are symmetries of space(time), mapping points to points or fields to fields—almost all will be examples, or can be understood as examples, of gauge theories (theories with a gauge symmetry expressing some freedom, and indeed *redundancy*, in the choice of values for dynamical variables).

The crucial point of this section is that whenever we have two objects (from the same domain) that are related by a symmetry, they will play identical roles in the relations of the structure they are part of; a permutation of such objects will not affect the structure—in this sense equivalence implies irrelevance. Thus, if we have a structure containing a domain of objects $\mathrm{D} = \{x_i\}$, including y, and there exists an automorphism $\phi : \mathrm{D} \to \mathrm{D}$, so that $\phi(y) = z$, then $\mathcal{R}(x_1, \ldots, y, \ldots, x_n) = \mathcal{R}(\phi(x_1, \ldots, z, \ldots, x_n))$. This clearly depends upon the nature of the relations, for we can surely conceive of cases where some relation isn't invariant under an automorphism, even though other relations are.[25] If we add these relations to S, then $\phi \notin \mathrm{Aut}(\mathrm{S})$, and we can use the relations to *distinguish* objects that were previously deemed to be indistinguishable. But, if D consists of a set of qualitatively indistinguishable objects (with respect to the observables of the relevant theory), then it is reasonable to expect that ϕ will be a symmetry of S for all $\mathcal{R}$ (such that the $\mathcal{R}$ pertain to qualitative *physical* structure). Still, it is possible to add to S with relations that might serve to *distinguish* objects in D that cannot be distinguished by $\mathcal{R}$. That is to say: indistinguishability between objects is largely dependent upon the relations in which the objects play a role. But, in the class of cases I am concerned with (from physics), the set of relations will generally be fixed by the theory, such that any additions would result in a modification of the theory.

Let us briefly see how these concepts work in the context of spacetime theories. We begin by specifying a theory, by writing down its laws as a set of equations of motion representing relations between the elements of the theory. We get the following schema:

$$\mathfrak{E}[\mathcal{D}, \mathcal{B}] = 0 \tag{1.3}$$

Here $\mathcal{D}$ represents the *dynamical* structures (those that have to be solved for to get their values, such as the electromagnetic field and the metric in general relativity—these represent the physical degrees of freedom of the theory, out which the observables will be constructed) and $\mathcal{B}$ the *background* structures (those whose values

[25] For example, suppose we have two objects, L and R, that are enantiomorphs (left and right hands respectively); enantiomorphy is a relation between L and R, call it E, so that $\mathrm{E}(L, R)$. Now suppose we act on L with a Moebius mapping, that transforms it into a right hand. E is not invariant under the Moebius mapping applied to one of the objects. However, there are other relations, such as having the same *intrinsic* shape or geometry that are preserved by the mapping. It is up to a physical theory to provide us with a relevant set of relations, and the cases we are interested in will also provide symmetries too.

are put in 'by hand', such as the topology and, in pre-general relativistic theories, the metric). Solutions of these equations will be the models that satisfy the equations of motion: $\langle\mathcal{M},\mathcal{B},\mathcal{D}\rangle$. Here the domain is the manifold of points $\mathcal{M}$, and the relations are supplied by the fields $\mathcal{B}$ and $\mathcal{D}$. Hence, we have: $\mathrm{D} = \mathcal{M}$ and $\mathcal{R}_i = \{\mathcal{B},\mathcal{D}\}$. *Prima facie*, this would, most naturally, be taken to correspond to a world in which field values are assigned to spacetime points in such a way so as to satisfy the laws as laid out in the equations of motion—though, as I have said before, there are problems with so direct a reading.

Now let us represent the space of *kinematically* possible histories by $\mathcal{K}$ (an example of one of the geometric models I mentioned above). Then $\mathfrak{E}[\mathcal{D},\mathcal{B}] = 0$ selects a subset $\mathcal{P} \subset \mathcal{K}$ of *dynamically* possible histories (or 'physically' possible worlds) relative to $\mathcal{B}$ (or $\langle\mathcal{M},\mathcal{B}\rangle$). Now, if there are no such $\mathcal{B}$s (or, rather, no $\mathcal{B}$*-fields*) then the physically possible histories (the dynamics) are given by relations between the $\mathcal{D}$s (and, at least fiducially, the manifold, but the manifold's symmetry washes this dependence away—see below). This impacts on the observables of the theory (constructed from the fundamental fields), for the observables must then make no reference to the $\mathcal{B}$s, only to the $\mathcal{D}$s. It also impacts on the dynamics: that too can only refer to the $\mathcal{D}$s. This is the source of the much-repeated claim that general relativity, and background independent theories, are *relational*: it simply means that the states and observables of the theory do not make reference to background fields, only to dynamical fields.[26]

This way of understanding a theory lets us recapitulate in a clearer, more precise way our earlier definitions of covariance and invariance. Let $\mathcal{G}$ be a group of spacetime symmetries that acts on $\mathcal{K}$ as $\mathcal{G} \times \mathcal{K} \rightarrow \mathcal{K}$—*i.e.* elements of $\mathcal{G}$ map kinematically possible solutions onto kinematically possible solutions. We say that $\mathcal{G}$ is a *symmetry group* of the theory whose space of kinematically possible histories is $\mathcal{K}$ just in case $\mathcal{P}$ is left invariant by its action—where, in the general relativistic example, $\mathcal{P}$ represents physically kosher distributions of fields over the points of the spacetime manifold. Alternatively, we can express the distinction between covariance and invariance as:[27]

$$[\mathrm{COV}] \Rightarrow \mathfrak{E}[\mathcal{D},\mathcal{B}] = 0 \quad \text{iff} \quad \mathfrak{E}[g\cdot\mathcal{D}, g\cdot\mathcal{B}] = 0 \quad (\forall g \in \mathcal{G}) \tag{1.4}$$

$$[\mathrm{INV}] \Rightarrow \mathfrak{E}[\mathcal{D},\mathcal{B}] = 0 \quad \text{iff} \quad \mathfrak{E}[g\cdot\mathcal{D}, \mathcal{B}] = 0 \quad (\forall g \in \mathcal{G}) \tag{1.5}$$

Now, in the context of general relativity, there aren't any background fields (*i.e.* $\mathcal{B} = \emptyset$ since the metric is solved for in the equations of motion), therefore we have [COV] = [INV]. Of course, the fact that the manifold appears in the laws—and

[26] Though, again, this does not include the manifold which is required for the (formal) definition of the dynamical fields. The inescapable presence of the manifold, in which dimension, topology, differential structure and signature are fixed independently of the equations of motion, leads Smolin to call general relativity only a "partly relational theory" ([2006], §7.4). However, the absence of background fields coupled with the symmetry of the manifold means that a displacement (*via* a diffeomorphism) of the dynamical fields with respect to it simply produces a gauge-equivalent representation of one and the same physical state. Elimination of these redundant possibilities ("surplus structure" in Redhead's sense [1975]) further reduces the size of $\mathcal{P}$, giving us the reduced space $\overline{\mathcal{P}} = \mathcal{P}/\mathrm{Diff}(\mathcal{M})$. This 'superspace' contains points that are entire orbits of the gauge group, representing *abstract* 'delocalized' structures known as a "geometries"—see [Misner et al., 1973], p. 522. This is supposed to be a space fit only for relationalists; however, as I aim to show throughout this book, this is simply not true.

[27] Here I borrow heavily from Giulini [in press], p. 6. I recommend that all philosophers of physics interested in background independence, and the difficulties in defining absolute objects, read this article: it is an exceptional review.

the absence of symmetry-reducing background fields (*i.e.* additional structure to reduce the effective symmetry group of the theory)—means that the there will be surplus structure: here, the localization on the manifold of the dynamical fields is, as they say, pure gauge. That is, any solutions differing solely with respect to such localization will not differ in any qualitatively discernible way.

Underlying these results is the particular symmetry that underwrites general relativity, namely diffeomorphism invariance. This works as follows. Given the manifold one can define a map, a diffeomorphism ϕ (corresponding to the notion of isomorphism above), from the manifold to some other manifold (we will suppose that this is the *same* manifold, so that ϕ is an auto-diffeomorphism):

$$\phi(\langle \mathcal{M}, g_{\mu\nu}, \mathfrak{F} \rangle) \rightarrow \langle \mathcal{M}', g'_{\mu\nu}, \mathfrak{F}' \rangle \tag{1.6}$$

Diffeomorphisms map points of the manifold onto other points[28] of the manifold and have the effect of 'carrying along' (to the image point) the fields (where $g_{\mu\nu}$ is the spacetime metric and $\mathfrak{F}$ represents any non-gravitational entities) defined at the domain point. Roughly:

$$(\phi \cdot g_{\mu\nu})(x) = g_{\mu\nu}(\phi^{-1} \cdot x), \quad x \in \mathcal{M} \tag{1.7}$$

$$(\phi \cdot \mathfrak{F})(x) = \mathfrak{F}(\phi^{-1} \cdot x), \quad x \in \mathcal{M} \tag{1.8}$$

In other words, when we have such fields on the manifold and then perform a diffeomorphism, we will be comparing the geometric (and topological) structure at the domain and image points (this is what is meant by 'carrying along' the structure): if this structure is the same in both cases then the mapping is a symmetry, an invariance (*i.e.* the structures are isomorphic). The set of all such diffeomorphisms forms a group Diff($\mathcal{M}$), the diffeomorphism group. This is the largest group of transformations possible, the symmetry group of the bare manifold—Anderson calls this the "manifold mapping group" ([Anderson, 1967], p. 8). It is also the symmetry group of general relativity: all topological transformations are admissible. Imposing geometric object fields such as metrics onto a manifold generally reduces the symmetry group (the *invariance* group, or what Anderson calls the "relativity group" [1964]) to a *proper* subgroup of the *covariance* group (which will be the diffeomorphism group), namely the symmetry group of the imposed object. This is not so in the case of general relativity; here the covariance group is identical to the invariance group, and this is understood to be a result of the background independence of the theory.[29]

Hence, the action of the group Diff($\mathcal{M}$) of diffeomorphisms on $\mathcal{M}$ has the effect of acting on the dynamical objects $\mathcal{D} = \{g_{\mu\nu}, \mathfrak{F}\}$. It does so in such a way as not to generate a qualitatively distinct state. Whenever $\phi \cdot \mathcal{D}$ is qualitatively indistinguishable from $\mathcal{D}$, then $\phi \in$ Diff($\mathcal{M}$) is a *gauge transformation*, and Diff($\mathcal{M}$) is a

[28] Mapping points of the manifold to other points—using a diffeomorphism $\phi : \mathcal{M} \rightarrow \mathcal{M}$—demands only the restriction that the manifold not be 'torn'. The transformations have to be smooth in the sense that domain points that are close have image points that are close. A mapping from point x to point $y = \phi(x)$ will carry along the topology in the neighbourhood of x to y, and since the manifold is a patchwork of Euclidean spaces, no properties of the manifold are affected by this.

[29] If we stipulate that our (metric) theory has as its invariance group the symmetry group of the manifold then this would seem to *select* a background independent theory, for there is no single metric that possesses this wide a group of possible motions. Just by supplementing this condition with the further conditions that the field equations for the metric are local and of second-differential order, then we automatically get just one system of equations: Einstein's equations!

gauge group.[30] Since Diff($\mathcal{M}$), the symmetry group of general relativity, is a gauge group, its orbits contain redundancy: if we consider the *elements* of the orbits to stand bijectively to physical states then we face a problem of underdetermination of the 'true physical state' by the generally invariant laws of the theory.[31] The standard response here is to say that the object that corresponds to reality is the quotient space with the manifold and the fields factored by the gauge group. Any fields that can be connected by an element of Diff($\mathcal{M}$) are deemed physically equivalent, or 'gauge-equivalent'.

There are a number of methods available in mathematics and physics that offer (sometimes very clear-cut) ways of dealing with these symmetries and the gauge-related states they generate: our job is to interpret these methods and the end-product of their application. Such methods are usually aligned (albeit tacitly in many cases) with deep metaphysical theses concerning the nature of space and time (*e.g.* in the context of the Leibniz-shift and hole arguments, and the problem of time), the nature of quantum particles (permutation invariance: see below), and the ontology of electromagnetism (and gauge theories in general). Let me quickly show how the simplest of these symmetries, permutation symmetry,[32] looks from the perspective I have outlined and show what kind of deep theses have been ripped off it, and why the implications do not hold water.

1.3. PERMUTATION SYMMETRY AND POSSIBILITY

Let Perm($\mathcal{X}$) be the permutation group of bijective maps of a set $\mathcal{X}$ onto itself.[33] In the context of quantum mechanics, these maps act as linear operators on a vector space representing the space of possible states of a quantum system; in mechanics on phase space they map phase points onto phase points. Permutation symmetry then refers to the following equivalence:

$$S = \langle \mathcal{X}, R \rangle \quad \Longleftrightarrow \quad S' = \langle \mathcal{P}(\mathcal{X}), R' \rangle \quad (\forall \mathcal{P} \in \text{Perm}(\mathcal{X})) \tag{1.9}$$

From the above we see that this involves the following preservation of relational structure (where the $x_i \in \mathcal{X}$ labels a family of particles):

$$\mathcal{R}(x_1, \ldots, x_n) = \mathcal{R}'(\mathcal{P}(x_1, \ldots, x_n)), \quad x_i \in \mathcal{X} \tag{1.10}$$

This preservation of structure and relations clearly implies the indistinguishability of the x_i with respect to the $\mathcal{R}$—here the $\mathcal{R}$s are supplied by the rules of classical

[30] Otherwise ϕ is a symmetry mapping an allowed state onto a physically distinct (allowed) state. Generally, I will use the term 'symmetry' to cover both cases, though I am almost exclusively concerned with gauge type symmetries.

[31] Of course, this underdetermination is turned into indeterminism in the Earman and Norton's hole argument [Earman and Norton, 1987]. However, as we shall see, this is just a general feature of theories with gauge symmetry. Chapters 4 and 5 are entirely devoted to the hole argument.

[32] The presentation I give below is fairly non-standard and very brief; for a more comprehensive guide to the issues, see French & Rickles [2003].

[33] Of course, the fact that the set Perm($\mathcal{X}$) has the structure of a group simply means that: (1) we can combine any two elements ($\mathcal{P}_1, \mathcal{P}_2 \in$ Perm($\mathcal{X}$)) in the set to produce another element ($\mathcal{P}_3 = \mathcal{P}_1 \cdot \mathcal{P}_2$) that is also contained within that set (*i.e.* $\mathcal{P}_3 \in$ Perm($\mathcal{X}$)); (2) there is an identity element $\text{id}_{\mathcal{X}}$ such that $\text{id}_{\mathcal{X}}(\mathcal{P}_1) = \mathcal{P}_1$; and (3) each element $\mathcal{P} \in$ Perm($\mathcal{X}$) also has an inverse $\mathcal{P}^{-1} \in$ Perm($\mathcal{X}$), such that $\mathcal{P}^{-1}\mathcal{P} = \text{id}_{\mathcal{X}}$.

or quantum mechanics.[34] Particles are qualitatively indistinguishable if they have the same values for mass, charge, spin, and all other observables.

Now consider the distribution of a system of two indistinguishable particles, 1 and 2, over two distinct one-particle states, ϕ and ψ. Statistical mechanics is then, very loosely, concerned with the number of ways we can get a distribution of systems over states; essentially, it is concerned with *possibility counting*. According to Maxwell–Boltzmann counting we get four possibilities (where $\phi(1) \cdot \psi(2)$ means that particle 1 is in state ϕ and particle 2 is in state ψ):

1. $\phi(1) \cdot \psi(2)$
2. $\phi(2) \cdot \psi(1)$
3. $\phi(1) \cdot \phi(2)$
4. $\psi(1) \cdot \psi(2)$

This set of possibilities, along with the assumption of indifference (equiprobability), yields the Maxwell–Boltzmann distribution realized by classical systems: each possible state is equally weighted by $1/4$.

There are two ways of counting in quantum statistical mechanics: 'Bose–Einstein' and 'Fermi–Dirac'.[35] The former gives the following possibilities:

1. $2^{-1/2}[\phi(1) \cdot \psi(2) + \phi(2) \cdot \psi(1)]$
2. $\phi(1) \cdot \phi(2)$
3. $\psi(1) \cdot \psi(2)$

The latter gives just one possibility:

1. $2^{-1/2}[\phi(1) \cdot \psi(2) - \phi(2) \cdot \psi(1)]$

Though it is intimately connected to symmetry, this latter difference does not concern us. What concerns us is the apparent 'loss' of a classically possible state and the supposed connection between this difference in possibility counting and the nature and existence of classical and quantum particles.

We can get a better grip on what underlies the permutation invariance of quantum mechanics by considering exactly how the permutations work on states. First, let ω_i be a 'list' of the values of the observables for a particle i, and let $\psi(\omega_i)$ be the wave-function for the particle. Take two particles, 1 and 2, described by the maximal sets ω_1 and ω_2. When 1 and 2 are indistinguishable they share their values for all observables, and since ω_1 and ω_2 are maximal, we have $\omega_1 = \omega_2$. The wave-function for the joint system composed of this pair of particles is $\psi(\omega_1, \omega_2)$. The indistinguishability implies that $\psi(\omega_2, \omega_1)$ is the same state of the system as $\psi(\omega_1, \omega_2)$. What we have, in fact, is $\psi(\omega_1, \omega_2) = c \cdot \psi(\omega_2, \omega_1)$, where $c \in \mathbb{C}$ and $|c| = 1$ (for normalization). Permuting again gives $c^2 = 1$, so that $c = \pm 1$. When $c = 1$ the wave-function is *symmetric* with respect to ω_1 and ω_2; when $c = -1$ it is *anti-symmetric*. When a wave-function is only ever symmetric or anti-symmetric, it

34 Note that this symmetry manifests itself at the level of dynamics so that $\{H, \mathcal{P}\} = 0$ and $[\hat{H}, \mathcal{P}] = 0$—*i.e.* the permutation operators commute with the Hamiltonian. Also, for any observables O, we have $\{O, \mathcal{P}\} = 0$ and $[\hat{O}, \mathcal{P}] = 0$. These are simply different ways of expressing the same results: permutations of particles do not affect qualitative structure.

35 That is, ignoring 'parastatistical' complications that arise when one uses a higher-dimensional group than the permutation group—*e.g.* the braid group. See [Rehren, 1990] for details of such cases.

is said to be *symmetrized*—the claim that wave-functions *are* symmetrized is known as the 'symmetrization postulate'. Let us now return to the issue of possibility counting.

Since the possibilities represented by 1 and 2 in the Maxwell–Boltzmann distribution are permutations, the fact that we count them as distinct means that permutation symmetry is violated; the fact that we don't count them as distinct in quantum theory means that permutation symmetry is satisfied. Or so the story goes. The received view is that this signifies a profound metaphysical difference between classical and quantum particles (*cf.* French & Rickles [2003], p. 221): roughly, the fact that we count permutations as generating distinct possibilities in Maxwell–Boltzmann systems *even though the possibilities are indistinguishable* implies that the particles have some form of individuality that transcends their properties while quantum particles lack this property.[36] I think that some consensus has now been reached that the argument is too quick: the quantum counting can be understood even on the assumption that particles *do* possess transcendental individuality provided one imposes a 'symmetrization postulate' as an initial condition on the quantum state of the composite system formed from the particles.

The argument goes as follows—here I switch to Redhead's notation. Let us assume transcendental individuality, and the way of counting that is supposed to go along with it (namely, Maxwell–Boltzmann counting). We get four wave-functions for our joint system (equaling the number available to classical systems):

1. $\Psi_a(\phi_1) \otimes \Psi_a(\phi_2)$
2. $\Psi_b(\phi_1) \otimes \Psi_b(\phi_2)$
3. $\Psi_a(\phi_1) \otimes \Psi_b(\phi_2)$
4. $\Psi_b(\phi_1) \otimes \Psi_a(\phi_2)$

However, as Redhead points out (*ibid.*, p. 12), the four-dimensional vector space spanned by these wave-functions could just as well be spanned by:

1. $\Psi_a(\phi_1) \otimes \Psi_a(\phi_2)$
2. $\Psi_b(\phi_1) \otimes \Psi_b(\phi_2)$
3. $2^{-1/2}[\Psi_a(\phi_1) \otimes \Psi_b(\phi_2)] + [\Psi_b(\phi_1) \otimes \Psi_a(\phi_2)]$
4. $2^{-1/2}[\Psi_a(\phi_1) \otimes \Psi_b(\phi_2)] - [\Psi_b(\phi_1) \otimes \Psi_a(\phi_2)]$

The first three wave-functions form a *symmetric* subspace [S] and the fourth forms an *anti-symmetric* subspace [A]. Since the indistinguishability of the particles requires that the Hamiltonian be symmetrical, the wave-function will be restricted to one or another subspace (*cf.* Dirac [1958], pp. 207–9). Now, Redhead goes on to argue, "if we impose S or A as an initial condition then only one of the two states [*viz.* $2^{-1/2}[\Psi_a(\phi_1) \otimes \Psi_b(\phi_2)] + [\Psi_b(\phi_1) \otimes \Psi_a(\phi_2)]$ and $2^{-1/2}[\Psi_a(\phi_1) \otimes \Psi_b(\phi_2)] - [\Psi_b(\phi_1) \otimes \Psi_a(\phi_2)]$] is ever available to the system" ([1988], p. 12).

[36] Redhead [1988] calls this (non-qualitative) property "transcendental individuality"; philosophers will be more familiar with the term 'haecceity'. *Haecceitism* is the view that there can be distinct possibilities that differ non-qualitatively, solely in terms of which individuals play which roles. In terms of the previous section's terminology, we can understand haecceitistic differences in terms of an isomorphism between a pair of structures, with the only differences concerning which elements of the domains get which properties and stand in which relations (see §2.4 for more details). Haecceities are often thought to entail haecceitism, and *vice versa*; however, as we shall see, there is no necessary connection between the two notions (*cf.* Lewis [1986a], §4.4).

Hence, at any time the wave-function must be lying in either S or A, and, since this is conserved by the symmetry of the Hamiltonian, one of the two states is thereby rendered *inaccessible* to the system. Hence, we can view the quantum statistics as arising from a restriction on the states available to a system; the state is left as a metaphysically possibility but a physical impossibility—of course, this can be useful in grounding modal talk involving counterfactuals and so.[37] Hence, permutation symmetry does not imply non-individuality (or anti-haecceitism and the absence of haecceity).

Nick Huggett [1994; 1999] has sought to defend the view that the argument works in the other direction too: classical mechanics does not imply that classical particles have transcendental individuality either; classical mechanics is not committed to an haecceitistic way of counting, "strictly speaking it is just compatible with it" ([1994], p. 70). His argument demonstrates that the haecceitistic (extended, full, or unreduced) phase space (generally associated to classical statistical mechanics), with its four element (partition) possibility set, leads to a statistical theory that is empirically equivalent to the anti-haecceitistic (reduced) phase space, with its three element (partition) possibility set (generally associated to quantum statistical mechanics). He's quite correct; the result is guaranteed by the equivalence (at the classical level) of the reduced and extended phase space descriptions (see §8.1 for more details). The conclusion Huggett draws is that classical statistical mechanics does not entail haecceitism, though it is compatible with it: it is compatible with both forms of possibility counting. Of course, this does not show that there is no difference at all between classical statistical mechanics and quantum statistical mechanics (clearly there is), but it does show that the difference must be accounted for in some way other than attributing "a more robust metaphysic of identity" ([1999], p. 70) to the former over the latter. The point can in fact be made rather simply: as Belot notes (using coins and their sides as substitutes for particles and their states), "a better way of describing the difference between the two sorts of statistics is to say that under [classical statistical mechanics] $\{H, T\}$ is twice as likely as either $\{H, H\}$ or $\{T, T\}$, while under the [quantum statistical mechanics] these three alternatives are equiprobable" ([2003a], p. 409). We then have to find a way to account for the difference in frequencies, without involving the identities of the particles. The difference, on a phase space formulation, is to be found in nothing more than the difference in measures (discrete vs continuous) placed on the space—this springs from the fact that we have finitely many particles distributed over infinitely many (micro-) states in the classical context.

Similar lines of reasoning have been applied in the other symmetry arguments I consider in this book. I argue against this sort of reading of metaphysical positions (regarding the nature and existence of individuals and their possible states) from the methods of dealing with symmetries by showing that incompatible metaphysical theses can usually be made compatible with the same resulting formal framework. However, there *are* quite definite ontological implications arising from the methods: the reductive methods that take the whole orbits $[x]_i$ as the funda-

[37] The same argument can be found in French & Redhead [1988]. However, Redhead, in collaboration with Redhead and Teller [1992], later adopted a view whereby the surplus elements (unphysical, metaphysical possibilities) are eradicated from the formalism by shifting to a Fock space description. I return to this in the final chapter.

mental objects rid the original theory of the indistinguishables that result from symmetry; the non-reductive methods retain the indistinguishables (at least formally). There are going to be clear implications for metaphysics here (*i.e.* in terms of 'reductive' versus 'non-reductive' methods and the possibility counting), but these implications are not the ones that are usually supposed to follow. Before we get to these issues, let me first introduce some basic technical material that will be used and expanded upon in the chapters that follow. What follows is a brief survey of the geometric models we shall be dealing with; I also outline the quantization procedure for systems thus modeled.[38]

1.4. A VERY BRIEF PRIMER ON CLASSICAL AND QUANTUM SYSTEMS

Recall that a classical system has a number of possible formulations. We will generally be interested in the Hamiltonian ones, represented by triples of the form $\langle \Gamma, \omega, H \rangle$ known as "Hamiltonian systems", where Γ is a symplectic manifold, ω is a symplectic form (*i.e.* a closed, non-degenerate 2-form) on Γ, $\langle \Gamma, \omega \rangle$ is a symplectic geometry (the "phase space"), and H is a distinguished function on Γ called the Hamiltonian function.[39] States are represented by points in the phase space and observables are simply smooth real-valued functions on phase space: $\mathcal{O} : \Gamma \to \mathbb{R}$. The intended interpretation is that the points of Γ represent physically possible states of some system. The Hamiltonian, being a function of Γ, has an associated vector field that generates a flow on the space. This picks out a family of curves, one through each point of phase space. These curves represent the dynamically possible histories of a system; the fact there is a unique curve through each point corresponds to the theory's being deterministic. If our system possesses constraints on the fundamental canonical variables, $\phi_i(q, p) = 0$, so that not all of the degrees of freedom are independent, then the system is represented by a triple of the form $\langle C, \sigma, H \rangle$, where C is a submanifold of Γ on which the constraints are satisfied, and σ (a *presymplectic form*) is the restriction of ω to C, *i.e.* $\sigma = \omega|_C$. The constraints can be viewed as picking out the physically possible worlds, represented by points on C, from the metaphysically possible worlds represented by points in Γ. However, the weakening of the geometry on C induced by σ means that C is partitioned into orbits corresponding to phase points related by the motions generated by the constraints. These orbits are gauge orbits: their elements consist of states that differ only non-qualitatively. States are represented by points on C and observables are represented by smooth functions, though now on C, but they must commute with the constraints, $\{\phi_i, \mathcal{O}\} = 0$, in order to be measurable. The Hamiltonian on C can no longer determine a unique curve through each phase

[38] The presentation is brief and it is rough; the details are fleshed out in different ways in the different chapters. All we need for now is the basic idea that a geometric space represents a space of possibilities and that the particular geometry of the geometric spaces quite naturally affects the structure of possibility space. This structure is later shown to be deeply entangled with symmetry; it is also (supposedly) deeply entangled with interpretive issues too. However, I will argue that the connections are not quite as direct as is often supposed.

[39] There are a number of possible formulations of Hamiltonian systems too, depending upon the polarization of the phase space (*i.e.* what one chooses as the canonical variables). When it comes to quantizing a theory, this choice can lead to non-trivial differences. I leave these technicalities aside in what follows; for details see Woodhouse [1980].

point but instead can only determine which orbit $[x_t]$ an initial state x_0 will lie in at t. However, given that the elements of such an orbit are gauge-equivalent—they represent qualitatively identical states—the indeterminism is unobservable. Thus, interpretation takes a dive: we have to decide how to deal with this underdetermination of the states by the dynamics and the overdetermination of the states by the observables. Do we take the formal elements as corresponding one-to-one or many-to-one with genuine physical possibilities, or, perhaps, some completely different interpretive route? That there is an empirical equivalence between these options is an indication that there is possible 'surplus structure' operating here (*cf.* Redhead [1975]). Thus a simple difference in geometric spaces results, therefore, in a nasty interpretive headache. We return to these problems repeatedly in the subsequent chapters.

Quantum systems, on the other hand, are represented by quite different structures, triples of the form $\langle \mathcal{H}, \mu, \hat{H} \rangle$, where $\mathcal{H}$ is a Hilbert space, μ is an inner-product on $\mathcal{H}$, and $\hat{H}$ is the Hamiltonian operator, once again governing the dynamics of the system, picking out the physically possible histories. To construct a quantum theory we usually begin with a classical system which is then *quantized*.[40] In this book, where I deal with quantum considerations at all, I am generally concerned with those quantum theories that take as their classical input a Hamiltonian system (generally gauge systems arising as constrained Hamiltonian systems). The hope is that the quantum theory has the input classical theory as its classical limit. It should be emphasized that quantization is not a philosophically benign territory: there are problems concerning which out of a panoply of quantizations to choose from,[41] and concerning how to take the classical limit.

A rough and ready guide to quantization might go as follows. Firstly, we need to choose a specific formulation of the classical theory that we wish to quantize. This in hand, we need to choose a set of fundamental variables out of which the physically measurable quantities are to be constructed. What a quantization should provide is a bijection between this set—a subset $O \supset \mathcal{O}_c$ of the classical observables[42]—and the set of quantum observables $\mathcal{O}_q$, such that $\forall f, g \in O$: $[\hat{f}, \hat{g}] = i\hbar\delta_{ij}\{f, g\}$ holds (where $[f, g]$ is the commutator $fg - gf$ and δ_{ij} is the Kronecker delta giving the value 1 if i and j are the same, and 0 otherwise).[43] The Poisson bracket relations holding between the fundamental variables form an algebra, denoted $\mathcal{A}$. One then seeks a representation $\rho : \mathcal{A} \to O(v, v)$ of this algebra as a set of linear operators O on a vector space V, where ρ is such that the algebraic structure of $\mathcal{A}$ is preserved. Hence, ρ provides the bijective correspondence between classical and

[40] Quantization should not be seen as a 'black box' into which one shoves a classical system and gets out a quantum version; rather, as Baez and Munian point out, "one should think of this prescription as an ideal to strive towards, rather than a simple recipe to follow" ([1994], p. 428).

[41] A serious problem in the context of field theories since the Stone–von Neumann theorem, declaring the unitary (and physical) equivalence of any and all continuous, irreducible representations of the Weyl relations (the Schrödinger and Heisenberg representations, for example), is no longer valid therein (see [Wald, 1994], pp. 19–20).

[42] A subset of classical observables is required since we can't simply assume that a quantum observable corresponds to every classical variable thanks to van Hove's theorem. Generally, one works with position q and momentum p, but there are other options (for example, the Bargmann variables forming the basis of a Fock space).

[43] This set of relations—known as the "canonical commutation relations"—yield the uncertainty relations in the quantum theory.

quantum observables that we were after; that is, between functions f on phase space and operators $\hat{f} = \rho(f)$ on a vector space.

At this stage, for Hamiltonian systems, one would then introduce an inner product on the vector space—transforming the space into a Hilbert space and making sure the operators $\hat{f}_i$ are self-adjoint—and look for a quantum Hamiltonian operator to determine the dynamics. That would be enough for the quantization. However, there is an intermediate step for systems with constraints, for we have to have these working somehow at the quantum level. Firstly, we need to express the classical constraints in terms of the fundamental variables $\phi_i(q,p)$. One then writes down operator versions of these as $\hat{\phi}_i = C_i(\hat{q},\hat{p})$. One then imposes these in the quantum theory by requiring that physical wave-functions are annihilated by the constraints. This condition selects a subspace $V_{\text{phys}} \supset V$ defined by $V_{\text{phys}} = \{\Psi \in V\colon \hat{\phi}_i\Psi = 0,\ \forall i\}$—these are the dynamically (physically) possible states. With the constraints duly dealt with, the inner product can be introduced, this time on V_{phys} rather than V—no easy matter in the context of curved space quantum field theories and an extraordinarily hard matter in canonical quantum gravity.

We might expect that, by analogy with Hamiltonian systems, we simply have to find an expression for the Hamiltonian function to get the dynamics and all is well. However, there is, again, an intermediate difficulty we have to face here: namely, that of defining the *physical* observables. The observables of a gauge theory have to commute with all of the constraints, they have to be 'gauge invariant', in order to be measurable and predictable. But, in general, the fundamental variables of a gauge theory are *not* gauge invariant.[44] At the quantum level, we generally require that the observables commute with the quantum constraints; again, this is to ensure measurability and predictability, though it also follows from the preservation of the algebraic structure of the classical theory. Once we have our observables, we can then deal with the dynamics. In this case, we choose a Hamiltonian H for the classical system, such that it can be written in terms of the fundamental variables, and such that it commutes with the constraints. An operator version $\hat{H} = H(\hat{q},\hat{p})$ is then constructed so that it commutes with the quantum constraints. The dynamics is then determined by the Schrödinger equation, $i\frac{\partial}{\partial t}\Psi = \hat{H}\Psi$.

In the case of general relativity, the Hamiltonian itself is a sum of constraints, so we have $\hat{H}\Psi = 0$! This is the infamous Wheeler–DeWitt equation, and it has been interpreted as showing that quantum gravity is a 'timeless' theory. This is the problem of time that I consider in Chapter 7. As I explain in that chapter, the hole argument has a similar root: both are natural consequences of gauge freedom (itself the subject of Chapter 3). It takes a particularly nasty form in general relativity because, thanks to background independence, there the gauge freedom concerns the points of spacetime and the dynamics of spacetime geometry, rather than fields living on spacetime with the dynamics framed against a fixed background

[44] This is true, for example, in the case of general relativity formulated in terms of the metric variables. However, the approach using loop variables is an intrinsically gauge invariant formulation. However, the diffeomorphism invariance is much harder to implement in this approach.

metrical structure. Methods for dealing with the gauge freedom are generally seen as underwriting specific interpretational choices as regards the status of spacetime points, and time and change. However, an underdetermination as regards which method to use leads to a further underdetermination of the interpretational stances by the physics: incompatible interpretations can invoke the same formal methods for dealing with gauge freedom therefore we shouldn't base interpretative stances on these methods. However, we must leave these problems hanging in the air for now until we have developed a more precise grasp of the concepts involved.

CHAPTER 2

Space and Time in the Leibniz–Clarke Debate

The received view regarding the debate between substantivalists and relationalists is that the former are committed to a larger set of possibilities than the latter are.[45] It is the purpose of this chapter to show this to be false. There are already good arguments to the effect that substantivalists can be deflationists;[46] but I have seen no analogous argument in the literature to the effect that relationalists are not necessarily committed to deflationism.[47] The conclusions of this chapter will be: (1) substantivalists can be deflationists, and (2) relationalists can be inflationists. More generally: differences in possibility counting are not an adequate way of characterizing these interpretations. This will be of increasing importance in the chapters that follow, and I will attempt to deepen the result by focusing on the nature of the geometric spaces used to represent the possibility structures of theories and their interpretations.

I begin by getting clear on the central terms of the debate—'substantivalism' and 'relationalism'—setting them up using the tools of the previous chapter. I then introduce a further distinction that is often taken to be implicated in the debate between substantivalists and relationalists; I call the opposing views 'inflationism' and 'deflationism' (later to be connected to the modal notions of 'haecceitism' and 'anti-haecceitism' respectively). Next, I outline the so-called 'Leibniz-shift' argument, and show how it is supposed to vindicate relationalism and undo substantivalism on the basis of possibility counting. Finally, I present the general argument to show that substantivalists can help themselves to deflationism, and follow this up with an analogous argument demonstrating that relationalists are not necessarily bound to deflationism either. The diagnosis is identical for both cases: an unwarranted assumption about modality slipped in to each thesis about spacetime ontology. Specifically: the connection between inflation and substantivalism depends on an assumption of haecceitism, and that between deflation and rela-

[45] This claim can be found in the Leibniz–Clarke correspondence [Alexander, 1956], Earman and Norton's hole argument [1987], various papers by Belot [1996; 2000; 2001; 2003b], and in Belot and Earman [1999; 2001]. This difference in the number of possibilities, one large the other small, corresponds to what I call *inflation* and *deflation* respectively (see §2.2 below).

[46] See, for example, Butterfield [1989], Maudlin [1988], Maidens [1993], Hoefer [1996], and Pooley [in press]. Belot is the most avid of dissenting voices against these arguments (see especially Belot [1996; 2000], Belot and Earman [2001], and §2.5 below).

[47] Saunders [2003b; 2003a] comes close and in fact claims that his "eliminative relationalism" is a "non-reductive" and "deflationary" variety of relationalism. However, as I intend the term 'deflationism' (meaning the cutting of certain indiscernible *possibilities* out of one's ontology), Saunders' relationalism is as reductive and deflationist as the rest.

tionalism on anti-haecceitism. This dependency can easily be questioned, and is concomitant with neither thesis.

2.1. SUBSTANTIVALISM VERSUS RELATIONALISM

Following Sklar, let us define substantivalism as that view that takes "spacetime to be an entity over and above the material inhabitants of the spacetime ... that could exist even were there no material inhabitants of the spacetime" ([1985], p. 8). Relationalism is the denial of this: what the substantivalist calls 'spacetime' is "nothing but a misleading way of representing the fact that there is ordinary matter and that there are spatiotemporal relations among material happenings" (*ibid.*, p. 10). There are two important things to note about these definitions: (1) there is assumed a straightforward distinction between 'matter' and 'space'; (2) the distinction between the positions is grounded in a basic ontological priority claim involving matter or space.[48] Let us try to firm up these definitions, by putting the previous chapter's notions of models and structures into play. I focus upon (2) next and then deal with (1) in the next subsection.

Let us assume that the structure $S = \langle D, \mathcal{R} \rangle$ adequately models spacetime (or, if you like, just space).[49] We can then get a reasonably clear and sharp distinction between substantivalism and relationalism by simply taking $\langle D, \mathcal{R} \rangle$ to consist of a primitive set of space(time) points with a set of relations defined over them in the former case and to consist of a primitive set of material bodies with a set of relations defined over them in the latter case. Since both descriptions are assumed to model the same *empirical* structure, we can assume that the structures are isomorphic (with respect to purely empirical properties and relations); what differs is the domain of objects: spacetime points vs material objects (which we can assume are 'point-like').[50] However, this equivalence aside, there is an important modal distinction to be made on the basis of these definitions. In the case of substantivalism it is clear that the spatiotemporal structure of the world is not supervenient on the relations exemplified by material objects; it can exist *independently* of material objects thus implying the possibility of empty spaces (without change). By contrast, relationalism requires the existence of material objects in order for the spatiotemporal structure to be instantiated;[51] there can be no actual empty spaces according to relationalism, whenever there are objects there is space and whenever

[48] Auyang notes that "both substantivalism and relationalism presuppose that ... matter is somehow already differentiated into individually identifiable entities" ([2001], p. 206). However, I don't think this is implied by the standard definitions, nor do I think that it is necessary; one can easily extend the definitions to include fields, for example. A problem *does* occur when the field is the metric field of general relativity, for then we have to make sense of its dual role in determining both gravitational and chronogeometric structure.

[49] By "adequate" in this context I simply mean *empirically* adequate, in the sense that the phenomenal, objective facts of the spatiotemporal structure of the world are captured.

[50] There is, of course, an immediate problem concerning regions of spacetime where there are no objects. Clearly, if there are less material objects than there are spacetime points then the structures cannot be isomorphic for the relation will not be one-to-one. If one invokes *possibilia* to function as places where an actual object *might* be, then one can restore the isomorphism, though the distinction between the two positions risks being collapsed. I discuss this matter further in §2.1.2.

[51] Where spatiotemporal relations are *external* in Lewis's sense (*cf.* [1986a], p. 62); namely a relation that does not supervene on the intrinsic natures of relata.

there is space there are objects—let us call this the "material dependency thesis". This brings Sklar's slightly vague definitions in to better focus. Clearly this simple modal distinction between positions that results from the difference in primitive objects paves the way for other conceptual (modal and non-modal) distinctions. In the following sections we examine an important one that has to do with the way in which symmetries are accommodated in the respective interpretations. Before I get to that, I should first like to take a look at how these definitions fit in with spacetime theories in general and show how they face a number of problems in this context.

2.1.1 Connection to spacetime theory

The presentation of §1.1 showed that the most common sorts of spacetime theory represent spacetime by means of a 'bare' differentiable manifold $\mathcal{M}$ over which certain structures, called *geometric-object fields*, are defined. The types of geometric-object fields split into two distinct categories: the absolute objects (or what, following contemporary physicists' parlance, I have been calling "background structures") $\mathcal{B}^i$ and dynamical objects $\mathcal{D}^j$. The interpretation usually given to these objects (*e.g.* [Earman, 1989], p. 45) is that the background structures characterize the *fixed* structure of spacetime and the dynamical structures characterize the *physical contents* of spacetime—a common metaphor is that the background structures form the stage on which the actors (the dynamical fields) perform. A model of a spacetime theory is then given by $\mathrm{M} = \langle \mathcal{M}, \mathcal{B}^i, \mathcal{D}^j \rangle$ (where the left-to-right ordering of the model represents the fact that $\mathcal{M}$ can be viewed as 'prior' to $\mathcal{B}^i$, and $\mathcal{B}^i$ as 'prior' to $\mathcal{D}^i$). What structures are placed on the manifold, what symmetries they admit, what laws they obey, and whether they are background or dynamical is what distinguishes the various types of spacetime.

A natural way to view these structures is as an assignment of geometrical properties to the points of $\mathcal{M}$. The background structures are *fixed* across the models of the theory, so that the points retain any properties determined by these structures independently of whatever processes are going on there or elsewhere. Thus, one can consider a model with twice as much energy as another, and yet the background structure remains insensitive to this. The dynamical structure, on the other hand, is allowed to vary across models, so that the points (or their counterparts in other models) can possess different properties as determined by such structures depending upon what processes are going on there or at other points.[52] Background independence is more than this though, it tells us that the values of the dynamical fields, but not the (absolute) background fields, have to be solved for using the theory's field equations.

This interpretation clearly suits the substantivalist, for the points of the manifold become 'bearers' of properties, and there is a sense in which these points exist independently of dynamical processes played out by material objects (as represented by one or more dynamical objects). However, the relationalist only deals

[52] This already involves some questionable modal assumptions concerning the 'sameness' of points across models. It also involves a questionable assumption concerning identity and individuality since we are supposing that the points are the proper subjects of predication. These assumptions will become increasingly important, and will come under closer scrutiny as we progress.

in material objects and their relations, so, for her, both the background structures and the manifold (if it is ineliminable) have to be reconstructed from these raw materials alone.

So grounded, the modern debate between substantivalists and relationalists concerns the ontological status of the elements of $\mathcal{M}$, interpreted as space(time) points, as structured by $\mathcal{B}^i$—these points are the most natural counterparts of Newton's parts of space in the context of modern spacetime theories. The substantivalist will be committed to the points along with the structure they inherit from $\mathcal{B}^i$, the relationalist will want to claim that they are a fiction (albeit a useful one), and that all there really are are material objects and their properties entering into various relations that define the observed structure of space and time.

Newton is generally taken to have held a substantivalist position with regard to absolute space and time.[53] The reason he believed in a substantivally conceived absolute space and time was because of the *work* that such an interpretation could do: it could provide a physical basis for inertial effects; this was something that was supposed to be impossible for the relationalist.[54] I turn to this issue at the end of this subsection. First I consider the question of what structure this space is taken to possess. In the context of Newtonian mechanics, space is represented by a three dimensional manifold of points equipped with a fixed Euclidean metric determining their distance relations—with this metric functioning as a background structure—so that the structure of space is isomorphic to $\mathbb{E}^3$. Time is represented by a one dimensional manifold, and it too is equipped with a Euclidean metric, so that the structure of time is isomorphic to $\mathbb{E}^1$. Newtonian spacetime has the simple product (topological) structure $\mathcal{M} = \mathbb{E}^3 \times \mathbb{E}^1$, understood as describing an *enduring* 3-space according to which the points of space persist through time.[55] We might then reasonably expect that the substantivalists attributions of robust existence concerns the points of this space and their properties and relations (*cf.* [Earman, 1989] §1.1, for such an account drawn from the *Scholium*). Relationalism will then involve, at the very least, an outright denial of the existence of space points, but may involve an agreement about the Euclidean structure of space—the structure will just be seen to be implemented by relations between material objects rather than between substantival points.

Ontological commitment to the points of this space, including the properties they inherit from the metric, is what underwrites the substantivalist's supposed commitment to inflation. Here is why: The points of the space are indiscernible; space is homogeneous and isotropic. The symmetries of the metric on space allow for a notion of invariance under global, rigid spatial translations and rotations. This means that if a system of matter is globally and rigidly translated some distance in space, so that the parts of the system have different absolute locations

[53] Note that substantivalism does not imply absolutism. Absolutism, in this case, implies 'sameness across models', whereas substantivalism implies a denial of the material dependency thesis. Neither implies the other, so that the substantivalist might deny that space is absolute and yet still deny the material dependency thesis. This will become important in subsequent chapters.

[54] However, Berkeley defended a plausible Machian-type line according to which matter, in the form of the fixed stars, was responsible for the inertial forces. See [Barbour, 2001] for an excellent account of this controversy, and others relating to the debate between absolute and relational theories of motion.

[55] In fact, this is an oversimplification. Strictly speaking, a Newtonian spacetime has a well-defined notion of time-separation, but not spatial separation (unless the time separation vanishes).

(*i.e.* occupy different spatial points), the observable properties of the system are unaffected. This is simply a statement of the invariance of the laws of Newtonian mechanics under spatial translations—we can run a similar line concerning rotations, and time-translations. Now let the system of matter be the total material content of the universe. Once again, the observable properties of the two systems are identical, only this time 'the system' comprises everything material that is contained in the universe! The only difference between the translated and untranslated system concerns the roles played by the points of space: *e.g.* in one system the center of mass is at the point x in the other it is at the point $y = gx$ (where g is some symmetry transformation, a spatial translation say). Inflation arises if we say that the two cases generated by the symmetry represent physically distinct states of affairs. The substantivalist is supposedly bound to say this since (1) the points of space are real (and independent of matter), and (2) the system bears different relations to these points of space in the translated and untranslated scenarios. The relationalist can clearly deflate the number of worlds represented, since any worlds agreeing on their material objects and the relational structure these objects exemplify are identified, for this is all there is to a relationalist's world.[56]

The generation by a symmetry of qualitatively indistinguishable worlds that differ only with respect to which objects (here, points) play which role (here, 'location grounders') is what underlies the Leibniz-shift argument.[57] It arises in the context of Newtonian mechanics because that theory is developed against the background of Euclidean space, and this space possesses a lot of symmetry. We can act on systems of matter with this symmetry to produce isomorphic states. Ontological commitment to the points of this space (and the matter) is supposed to lead the substantivalist into the jaws of inflationism for the reasons given above. The relationalist in not being ontologically committed to the points of space is supposedly thereby rendered immune from inflation, for the worlds match up on those properties that are relevant to the relationalist's conception of space and time.

The translation argument that I presented above is, of course, just a slim-line version of Leibniz's shift argument against Newton's brand of absolutist substantivalism.[58] Leibniz accepts the above implication between substantivalism and inflation (as does Clarke), and argues that since inflation violates the principle of sufficient reason [PSR: nothing (contingent) happens without a reason why it is so rather than otherwise], substantivalism should be rejected. I repeat, then, that in-

[56] I will argue in §2.4, as has been argued by so-called 'sophisticated' substantivalists, that, in fact, a similar deflationary option is available to the substantivalists as well. I will also argue that relationalists are not necessarily committed to deflationism either: §2.4. What leads to these two possibilities concerns the modal semantics one combines one's ontological stance on spacetime with. I argue that nothing internal to the debate on spacetime ontology can decide which theory of modality to use.

[57] Of course, not all symmetries lead to indistinguishable *worlds*. In general symmetries map states to physically distinct ones, leading to distinguishable worlds—*i.e.* they are symmetries of the *theory* rather than symmetries of worlds (*cf.* [Ismael and van Fraassen, 2003]). The issue in those cases were we *do* have indistinguishable worlds is whether they are nonetheless *physically* distinct or not, despite their being qualitatively identical. If we say that the multiple states represent one and the same *physical* state (so that the symmetry does not map physical states to physical states) then we have an example of a *gauge symmetry* which signals a redundancy in the theory (*i.e.* in its description of configurations).

[58] I present the real argument in detail in the next section. In the last section of this chapter, I present the mathematical guts of the argument and show how it connects to the general account of symmetry outlined in the previous chapter.

flation is the real target of the shift argument (and, we will see, the hole argument), and I will argue that inflation is not an integral part of substantivalism.[59]

One might be left wondering why anyone would want to be a substantivalist given the trouble it is supposed to generate. Why not avoid the trouble and be a relationalist? Newton had his reasons of course, and they were reasons of physics not philosophy or theology, as was often believed to be the case (*cf.* Reichenbach [1924] for the canonical development of this 'anti-Newtonian' line). It wasn't until Stein [1967] that the soundness of Newton's methodology became apparent. What Newton provided was essentially an 'inference to the best explanation' argument for the existence of a substantival space, what Teller calls "the argument from inertial effects" ([1991], p. 369). Newton believed that an aggregate of genidentical points (or parts) of space was necessary to explain absolute motion, and that a notion of absolute motion was necessary to explain absolute acceleration; the latter is, moreover, an objective *physical* effect with observable consequences. For example, when an airplane accelerates to take off, the passengers will *feel* their seats being forced into their backs. Relationalism comes unstuck on such effects, since the loved ones of the passengers standing stationary in the airport will also be accelerating relative to the airplane: there is a symmetry between those on the airplane and those in the airport. Yet only those on the airplane feel the effects; there is also an *asymmetry* here that the relationalist apparently cannot accommodate.[60] The substantivalist can, for according to her there is a fact of the matter about who is accelerating relative to absolute space: motion is, then, held to be relative to the 'fixed' points of space—from this, one gets absolute rest (location at the same point as time elapses), velocity (location at distinct points as time elapses), *and* acceleration (any motion distinct from the two previous forms). The connection to physics, then, is clear: Newton believed that substantivalism about absolute space and time was a necessary part of an empirically adequate dynamics. The inflationary aspects brought about by absolute position and velocity were an unfortunate, but unavoidable, consequence: the price to pay for absolute acceleration. As Belot explains, "Newton's laws demand a notion of absolute motion while at the same time implying that there exist states of absolute motion [positions and velocities] which are indistinguishable one from another" ([2000], p. 564).

The shift to (non-dynamical) spacetimes that abolish absolute space and time alters the nature of the debate somewhat—but not so much that inflation no longer

[59] I am certainly not the first to notice that substantivalism *per se* is not the real victim of Leibniz's shift argument and the hole argument. However, opinions differ as to exactly what the victim is. Hoefer [1996] argues that assuming primitive identities for points causes the problem; Maidens [1993] takes a similar line, though she simply *stipulates* her way out of the bind to inflation. Butterfield [1988] and Maudlin [1988] think that the particular brand of modal semantics of the arguments is the target and offer counterpart theory and metrical essentialism respectively as alternatives. Pooley [in press] and French [2001] suggest that it is haecceitism that is the real target; Melia [1999] too appears to adopt a similar line, though his concern is with the relevance of haecceitistic differences to questions of determinism. I examine these options in Chapter 5 when I consider the hole argument (the real context of these interpretive options)—these issues will be a running theme from hereon in.

[60] Sklar ([1974], pp. 230–1) suggests that relationalists might overcome this problem by appealing to the notion that absolute acceleration is a primitive property that certain relatively accelerating objects possess. This is, indeed, possible and consistent, but clearly ad hoc; one might question whether the increase in ontological economy brought about by the avoidance of substantival points of space and time is really worth the attendant decrease in ideological economy. As Teller says, "as food to satisfy one's appetite for a theoretical account, this move satisfies about as well as bread made from sawdust" ([1991], p. 370). Another alternative is to argue along Machian lines that the acceleration is to be judged 'relative to the fixed stars'.

occurs. The most important difference is that the notion of spatial points retaining their identity over time no longer makes sense—there are simply spacetime points whose existence is 'fleeting': different spatial points at different times. Thus, we do away with the view attributed to Newton that spacetime is built out of a stack of instantaneous spaces of Euclidean structure, such that there is a definite way to re-identify spatial points over time.

In the case of 'neo-Newtonian' spacetime, the notion of absolute space is done away with; though absolute time (simultaneity) is retained. Hand in hand with the eradication of absolute space is the eradication of the observationally inert concepts of absolute position and velocity. Absolute acceleration is preserved by effectively encoding it in the structure of the spacetime. We can still consider 'translations' in neo-Newtonian spacetime, but rather than supposing that the instantaneous contents of space are translated, we must suppose that *worldlines* are translated. We simply view the history as taking place over different spacetime points. Special relativity simply extends the damage done by neo-Newtonian spacetime to cover absolute time as well. Once again, absolute acceleration is preserved through the use of inertial frames. And once again, we can consider worlds that differ only as to which points of spacetime are occupied.

With the eradication of background structures, like absolute space, there will be an attendant increase in the symmetry group of the theory, resulting above in the abolishment of absolute positions and velocities, but with absolute acceleration remaining. In both of the above cases, the metric still appears as a background structure. The translation argument can easily be reapplied in such contexts: simply consider translations of events on spacetime. The symmetries will always allow one to generate such possibilities, and will, therefore, allow for inflationary scenarios.

However, the situation changes markedly in general relativity, where the landscape of the substantivalism/relationalism debate alters significantly. There are two aspects that are especially problematic from the point of view of this debate. The first concerns the definitions of 'spacetime' and 'matter' in such a context.[61] The second, related to the first, concerns the fact that the only background structure in general relativity is the 'bare' manifold itself: there are no background *fields*. The symmetry group of general relativity is, accordingly, larger still—the largest possible in fact. However, crucial for our purposes is the fact that the symmetries of general relativity nonetheless allow for the generation of new possibilities, though this time concerning the points of spacetime themselves, or the distribution of fields over these points—inflation arises from permutations of these points. We walk the new terrain in Chapter 4; for now let us stick to the relatively *terra firma* of the Leibniz–Clarke correspondence.

2.1.2 Empty space and fields

Before I move on to the Leibniz-shift argument, I should first deal briefly with a potential difficulty faced by relationalism, which is sometimes glossed over in

[61] I mention this aspect briefly in the next subsection, and consider its bearing on the substantivalism–relationalism debate.

the literature. Namely, that the requirement that there be no spatial and temporal vacua might seem to be too stringent a condition for the relationalist to meet. There are two questions to be asked here: (1) does she need to meet the condition? (2) if the answer to (1) is no, how are such vacua to be constructed relationally? This is closely connected with the question of how fields impact upon the debate between relationalists and substantivalists, for the ability to support fields is thought to provide good reason for adopting substantivalism (*cf.* Earman [1989], p. 173; and [Field, 1980]). Yet, at the same time, field theory is supposed to provide a response to the relationalist's problem concerning empty space since fields are continua.

There appear to be empty regions of space. If there aren't, then the, at least conceptual, possibility of producing a vacuum remains (for example, Hooke's experiments). There are at least two options open to the relationalist: (1) to dig her heels in and deny that such vacua are (metaphysically or physically) possible; or (2) to admit possible but non-actual spatial and/or temporal relations.

In the time period of the Leibniz–Clarke debate, the question of what is *space* and what is *matter* was at least unambiguous. As Earman nicely puts it, "the participants of the debate had the luxury of knowing what they were talking about" ([1989], p. 18). Generally, space (in the sense of vacuum) would simply be defined by *a lack of material or mass*.

Advocating (1) would imply plenism, and could perhaps be achieved nowadays with a field ontology; for fields extend to cover all of space. In Leibniz's day, the former option would have been hard to uphold on physical grounds, since experiments had been conducted to produce vacua and the field picture was unknown. It could only be underwritten by metaphysical principles. The latter option could be cashed out in terms of counterfactuals expressing what relations would be instantiated if objects were placed in the evacuated region. Hence, for a relationalist in the days before field theory, a full characterisation would have to involve plenism or else include *possible* spatiotemporal relations—for more on this issue see Sklar ([1974], p. 170). Earman ([1989] §6.6) has Leibniz down as a plenist. Unless one has an ontology of *possibilia* at one's disposal already, the inclusion of possible spatiotemporal relations would appear to be as bad as the spacetime points that relations were supposed to replace, for the disposal of spacetime points was seen as scoring one over the substantivalist in terms of ontological economy. Perhaps a case could be made for bolstering the modal relationalists position by pointing out that *possibilia* are not *things*? That depends upon one's position in modal metaphysics, for some would argue that they are things. But quite regardless of this, even if we allow *possibilia*, a problem remains concerning what they are possibilities *of*. Presumably objects, and the possibilities concern where they might have been placed, over *there* for example. But then how is "there" defined, for the relationalist, if not by the object that is located at it? There is the prospect of circularity: the possible objects appear to require spatial positions in order fulfill their role, yet spatial positions require objects according to relationalism.

The impact of modern physics radically alters the debate's compass. In particular, the notion of 'vacuum' *qua* 'empty space' *prima facie* makes little sense in such the context of field theory, for the field is defined at *every* point of space(time). Moreover, field theories constitute our best scientific description of the world. In quantum field theory, the situation is intensified, since the vacuum generically possesses mass-energy. For this reason, I will assume that some variant of the former option is the better one for the relationalist in the context of modern physics, though not a lot rests on this in what follows—see [Saunders and Brown, 1991] for a collection of essays on the notion of vacuum in modern physics.

There are other problems too, as Butterfield explains:

> classical physics has introduced the electromagnetic field endowed with energy and momentum; and relativity has identified mass and energy. Furthermore, in general relativity there can be material in the sense of mass-energy in a region where there is not only no material object of an ordinary sort like a chair, but also no field apart from the metric-gravitational field. ([Butterfield, 1984], p. 104)

Coupled with quantum field theory, this rather messes up the nice cleanly formulated debate that was taking place in the time-frame of the Leibniz–Clarke correspondence—I return to this problem again in §4.1.

The concept of a field can be used to the substantivalist's advantage too. For example, Hartry Field ([1980], p. 35) argues that since "a field is usually described as an assignment of some property . . . to each point of space–time, this obviously assumes that there are space–time points" (*cf.* Earman [1989], pp. 154–9). Teller also writes that "[t]he idea of a field enters as the idea that values of physical quantities can be attributed to the space–time points" ([1990], p. 53). The idea is clearly the Quinean one that ineliminable quantification over a type of object entails an ontological commitment to them (*cf.* Butterfield [1984], p. 101). This is indeed one way to describe a field; but it is not the only way. For example, Belot describes an alternative description in which fields are simply "extended objects whose parts stand in determinate spatial relations to one another" ([2000], p. 584). Hence, it is not obvious that the quantification over points *is* ineliminable, though the burden of proof is quite definitely rested upon the relationalist to come up with an empirically adequate theory that dispenses with points and involves only kosher material objects—this has tended to be the main problem with relational accounts.

2.2. INFLATION VERSUS DEFLATION

Call *inflationism* the view that a theory admits possible worlds that differ non-qualitatively, and call the denial of this *deflationism*. However, I intend this usage in a very restricted way, here simply concerning (kinematical) state spaces of physical systems (as given by some theory).[62] The idea is that a theory presents us with a mathematical (configuration) space, with a certain geometrical structure appropriate to the system it is invoked to model (and whose coordinates represent the values of some physical variables, with a dimension equal to the number

[62] The account can easily be generalized to include dynamics; it is also applicable to both Lagrangian and Hamiltonian formulations of physical theories in terms of both states and trajectories. However, the configuration space gets the point I want to make across without any excess technical baggage involving rates of change. I extend the account in the subsequent chapters.

of degrees of freedom of the system), such that the points of this space represent physically possible (instantaneous) states of affairs for a system modeled by the theory. The state space thus functions as a kind of physical *possibility space*.[63] If the state space of a theory admits non-trivial symmetries (so that points of the space can be mapped onto others in a structure preserving way), then it will involve inflation in the following way: distinct points of the space will be related by symmetries; these points will form an equivalence class; and elements of this class will represent qualitatively identical possibilities if the space is given a direct (one-to-one) interpretation.

There is a standard method for eliminating the symmetries from a configuration space, and thus eradicating qualitatively indistinguishable worlds from one's ontology: reduction. The basic idea here is to form the quotient space of the space with symmetries by dividing out by those symmetries. The upshot of this procedure is that whole equivalence classes form the points of a new (smaller) space. The space with symmetries is called an *unreduced* or *extended* configuration space, and the space with these symmetries factored out is called a *reduced* configuration space.[64]

Clearly, the former is inflationistic and the latter is deflationistic: reduced spaces contain less possibilities (*qua* possible *worlds*) than extended ones.[65] The claim under investigation here is whether or not there is a connection between these types of space (with their connection to inflation and deflation) and the substantivalism/relationalism debate. I turn to this latter debate next.

2.3. LEIBNIZ VERSUS CLARKE

In the course of his debate with Clarke, Leibniz presents an argument (a *reductio*) directed against Newtonian substantivalism.[66] In this case, as van Fraassen ([1990], p. 239) nicely expresses it, "Leibniz's God was Buridan's ass magnified" (see also Earman [1989], p. 118): given certain features about the nature of Newtonian space, it is impossible for God to make a reasoned decision regarding the placement of the world within it. Here, in its entirety, is the oft-quoted argument designed "to confute the fancy of those who take space to be a substance, or at least an absolute being":

> [I]f space was an absolute being, there would something happen for which it would be impossible there should be a sufficient reason. . . . Space is something absolutely uniform; and, without the things placed in it, one point of space does not absolutely differ in any respect from another point of

[63] *Cf.* Ismael and van Fraassen [2003] and Belot [2003a].

[64] See Marsden [1992] for a fairly elementary introduction to these ideas.

[65] There are ways of making an extended space compatible with a deflationist *interpretation* of a theory in a certain restricted sense: we can impose conditions on measurement theory so that the symmetries in question are not distinguished by any measurement—*i.e.* observables are made insensitive to non-qualitative differences. However, I ignore this complication here (for details see Belot and Earman ([2001], p. 221) and §3.3 of this book).

[66] As I have already intimated, in fact, it is Clarke—in the course of discussing the principle of sufficient reason—who initially suggests the possibility of different places for a system of matter. The reason he gives as to why a system of matter has the place it has is down to "the mere will of God" (C-II.1, [Alexander, 1956], p. 21).

> space. Now from hence it follows, (supposing space to be something in itself, besides the order of bodies among themselves,) that 'tis impossible there should be a reason, why God, preserving the same situations of bodies among themselves, should have placed them in space after one certain particular manner, and not otherwise; why everything was not placed quite the contrary way, for instance by changing East into West. But if space is nothing else, but that order of relation; and is nothing at all without bodies, but the possibility of placing them; then those two states, the one such as it is now, the other supposed to be the quite contrary way, would not at all differ from one another. Their difference therefore is only to be found in the chimerical supposition of the reality of space in itself. But in truth the one would exactly be the same thing as the other, they being absolutely indiscernible; and consequently there is no room to enquire after a reason of the preference of the one to the other.[67] (L-III.5, [Alexander, 1956], p. 26)

As Belot explains, "Newtonians multiply possibilities in a manner which violates the PSR" ([2001], p. 3):

> In things absolutely indifferent, there is no [foundation for] choice; and consequently no election, nor will; since choice must be founded on some reason, or principle. (L-IV.1, [Alexander, 1956], p. 36)

For a relationalist like Leibniz, there is only one possibility corresponding to the entire set of possibilities brought about by the symmetry transformations of the shift argument. Hence, "PSR can be restored if we can assure ourselves that there are many absolutist possibilities for each genuine possibility. Where Clarke sees many possible arrangements of bits of matter, each placed differently in space but all satisfying the same spatial relations between bits, Leibniz sees only one" ([Belot, 2001], pp. 3–4):

> To suppose two things indiscernible, is to suppose the same thing under two names. And therefore to suppose that the universe could have had at first another position of time and place, than that which it actually had; and yet that all the parts of the universe should have had the same situation among themselves, as that which they actually had; such a supposition, I say, is an impossible fiction. (L-IV.6, [Alexander, 1956], p. 37)

Let us attempt to couch this debate in the terms of the previous section. We see that the shift argument in the Leibniz–Clarke correspondence is constructed by using the symmetries of Newton's theory to generate new configurations for some specified system—in this case the material content of the universe! One starts out with a single instantaneous state representing a possible embedding of matter in space, and then acts on this state (configuration) with a symmetry to produce

[67] Leibniz quickly follows up this argument against substantivalism about space with a similar one about time. The idea is that, on the assumption of absolute, substantival time, God could have created the universe sooner or later than He actually did. Hence, He has no sufficient reason to actualize it at one time as opposed to some other, since the points of time are exactly alike (*i.e.*, indifferent). The details of the spatial and temporal arguments are so alike that we can take them as having the same form, and we may, therefore, restrict our discussion to the spatial scenario.

another *equivalent* state (configuration).[68] The symmetries are such that one can produce infinitely many states in this way; specifically they are elements of the (six dimensional) Euclidean group $\mathcal{E}_3$ of transformations on $\mathbb{R}^3$: translations, rotations, and combinations (under the group composition law).[69] The set of states related by the symmetries describe qualitatively identical situations: all qualitative monadic and relational properties are preserved.

Let us call the space containing all of the possible configurations (worlds) for some system, *including those related by symmetries*, $\mathcal{I}$ (for *inflated*). A crucial fact, for Leibniz, is that the states related by the symmetries form equivalence classes—this accounts for the qualitative identity of the represented states of affairs. Thus, for any state $x \in \mathcal{I}$, there is a set $[x] = \{y\colon y \sim x\} = \{y\colon y = g(x)\}\ (\exists g \in \mathcal{E}_3)$ of isomorphic states (where '$\sim$' denotes isomorphism, and g is a Euclidean symmetry transformation). We can factor out the symmetries of the six dimensional Euclidean group, yielding a reduced (relative) configuration space, with states represented by equivalence classes of states from the extended space. Let us call this space $\mathcal{D}$ (for *deflated*). Clearly $\mathcal{D} = \mathcal{I}/\mathcal{E}_3$.

Putting historical accuracy to one side, the shift argument goes as follows. Suppose we have a very simple Newtonian world $\mathcal{W}$ (considered at a single instant of time) containing three point particles q_i ($i = 1, 2, 3$), each with mass m, living in three dimensional Euclidean space, $\mathbb{E}^3$. If we fix the masses $m_i = 1, 2, 3$ for the particles (so distinguishing them and avoiding any initial identity problems), the (extended) configuration space for this system has nine degrees of freedom corresponding to the positions the particles relative to space (three coordinates per particle). $\mathcal{W}$ can be represented by the structure $\mathrm{W} = \langle \mathbb{E}^3, q_i, \mathcal{R}_i \rangle$ (where $\mathcal{R}_i$ represent the relative distances between the particles, and masses are ignored). A Euclidean symmetry g acts on the q_i, and so on W, so as to generate a new configuration $g(q_i)$ with corresponding structure $\mathrm{W}_g = \langle \mathbb{E}^3, g(q_i), \mathcal{R}_i \rangle$ (representing world $\mathcal{W}_g$). But the fact that g is a symmetry means that the relative distances stay the same: $\forall_{jk}\ \mathcal{R}(q_j, q_k) = \mathcal{R}(g(q_j), g(q_k))$. Consider the case where g is a rigid translation, so that $g(q_i)$ moves all of the particles five meters to the East of q_i. Now, suppose that x is the spatial point lying at the center of the q_i-system; suppose also that y is the spatial point lying five meters to the East of x. Then the center of mass of the system of particles is at point x in $\mathcal{W}$ but at the point y in $\mathcal{W}_g$, where $x, y \in \mathbb{E}^3$.

Now, the substantivalist's position can be characterized by the claim that the points (*parts*, for Newton) of space have their existence and identities fixed *independently* of any material objects occupying them—where, in our example, the material objects are just the three particles and space is three dimensional and Euclidean. Therefore, since any (non-trivial) group action will result in these particles having different positions in space, the substantivalist is committed to each such transformed state representing a distinct possibility: $\mathcal{W}$ and $\mathcal{W}_g$ are distinct

[68] The specific kind of equivalence is determined by the kind of symmetry: some symmetries may lead to physical differences (these are symmetries in the strict sense) and others will lead to no discernible alteration whatsoever (these are the gauge redundancies).

[69] I only consider the *kinematical* shift argument here, using only the symmetries of space. But note that Leibniz considers a *dynamical* shift argument too, where the symmetries of Newtonian space *and* time (*i.e.* the action of the ten dimensional Galilean group $\mathcal{G}$ on $\mathbb{E}^3 \times \mathbb{R}$) are used. See Maudlin [1993] for an explanation of the difference.

possible worlds because the center of mass point is played by x in the first and y in the second. However, since each configuration that results from a group action is equivalent, the possibilities will be indistinguishable. Thus, there could be no reason for one to be actualized over any other g-related one. Leibniz, of course, used this feature as a *reductio* of substantivalism on the grounds that it violates PSR. To rule out such indistinguishable possibilities, Leibniz then invoked the principle of identity of indiscernibles [PII: $\forall F, \forall ab \mid (Fa \equiv Fb) \Rightarrow (a = b)$].[70] The latter move simply corresponds to reduction: each symmetrical configuration corresponds to one and the same physical reality.[71] The folk wisdom is that relationalism and reduction (*i.e.* deflation) go hand in hand, as do substantivalism and inflation. The next section will introduce a number of key concepts from the metaphysics of modality. These concepts are then utilized to show that both alignments are false.

2.4. SOPHISTICATED SUBSTANTIVALISM AND UNSOPHISTICATED RELATIONALISM

The two claims I wish to question are: (1) that the substantivalist is committed to taking $\mathcal{I}$ as representing possibility space; and (2) that the relationalist is committed to taking $\mathcal{D}$ as representing possibility space. This pair of claims are stated quite forcibly by Belot. Thus, he writes (modified to suit the example of the previous section):

> I require substantivalists to maintain that there are a large number of such embeddings [of point particles in Euclidean space, with their relative distances fixed]: place the [three] points as you like; you generate a distinct possible embedding by acting upon this first one by any Euclidean [symmetry]. On the other hand, I require relationalists to maintain that there is a single such embedding. ([Belot, 2000], pp. 276–7)

Thus, Belot is quite explicit that it is possibility counting that distinguishes substantivalists and relationalists; he sees the difference at this level as concomitant with the differences concerning the material dependency thesis that I mentioned earlier. He explicitly ties this to the geometric spaces I presented in the previous section:

> Whenever we compare two spaces of possible worlds, one the quotient of the other, we are contrasting two ways of counting possibilities [namely] a relatively haecceitistic means of counting possibilities and a relatively anti-haecceitistic means of counting possibilities. [.] [T]he existence of spacetime

[70] The property variable F ranges over all qualitative monadic and relational properties, and the arguments a and b over a domain of objects.

[71] At least it does in Leibniz's hands. But see Saunders [2003b; 2003a] for a *non-reductive* (that is, *non-eliminativist*) version of the principle. Indeed, Saunders' main thesis in these two works is that relationalism does not imply reduction. However, I think that the position that Saunders ends up with is more like substantivalism than relationalism, for the claim is that relational properties can serve to *individuate* (absolutely indiscernible) objects; this includes the relations spacetime points bear to one another, hence empty spacetimes are possibilities on this account (see §5.3.3). Also, as I mentioned in the introduction, reduction (elimination) is still implemented at the level of possible worlds, for the simple reason that there are no physical relations between such objects that could serve to individuate them (see §8.3).

> points is closely tied up with questions of counting of possibilities—so they are vulnerable to elimination in the transition from a haecceitistic means of counting to an anti-haecceitistic one. ([Belot, 2003a], p. 410)

Thus, Belot essentially argues that one's choice of geometric space is instrumental in one's spacetime ontology. Let me begin by outlining an argument that has been adduced to disprove (1).[72]

2.4.1 Substantivalists can be deflationists just like relationalists!

Sophisticated substantivalism can be defined simply as substantivalism without inflation: $\mathcal{D}$ functions as possibility space, symmetrical worlds are identified, but space is nonetheless considered as being ontologically independent of (indeed, *prior* to) its material contents.[73] They help themselves to this result as follows: when I said in the previous section that "the substantivalist position can be characterized by the claim that the points of space have their existence and identities fixed *independently* of any material objects occupying them", I went too far. The crucial tenet is that the points *exist* independently of material objects, not that they have their identities fixed independently. By denying that points have *primitive* identities,[74] the world $\mathcal{W}_g$ can not be seen as representing *de re* of $q_i \in \mathcal{W}$ that it has a center of mass situated at y, such that $y \neq x$. To do so would require that the points x and y have identities that do not supervene on qualitative facts. Denying this latter claim does not do away with substantivalism, nor does it entail relationalism. The substantivalist can respect PII by saying that distinct possible worlds must differ in some qualitative way without affecting the basic ontological independence claim. The upshot of this is that by outlawing haecceitism the substantivalist commits himself to the independent existence of a spacetime with an intrinsic structure given by an equivalence class of isomorphic configurations; the shifted worlds can be viewed simply as different (local) representations of this more basic, intrinsic geometry. These representations will generally be much easier to work with both conceptually and computationally, but strictly speaking the shifts do not make metaphysical sense to the sophisticated (anti-haecceitistic) substantivalist.

Thus, substantivalism is not necessarily committed to an inflationary way of counting possibilities; rather, the specified counting depends on a subtle modal

72 All of these types of argument arise in the context of the hole argument, but many of them can be applied with equal force across the board to all spacetime theories. That is to say, the result applies to substantivalism *per se* and not just substantivalism about general relativistic spacetimes.

73 The name 'sophisticated substantivalism' comes from Belot and Earman ([2001], p. 228). See Pooley [in press] for a more detailed characterization and lengthy defense of this position.

74 Primitive identities are non-qualitative properties that serve to individuate objects (see Adams [1979] for the canonical account). Thus, if two individuals possess primitive identities, then, *ceteris paribus*, we can consider swapping *all* of their properties as resulting in a distinct possible world so that their identities can not be seen as supervening on qualitative facts (*cf.* Hoefer [1996] and Kaplan [1975]). The worlds that result from this procedure are said to differ merely *haecceitistically*; that is, they differ solely with respect to which individuals get which properties (with the property distribution itself *invariant* across worlds). It is just such differences that hold between the shifted worlds $\mathcal{W}$ and $\mathcal{W}_g$ and, therefore, that correspond to inflation.

factor; namely, that substantivalism entails haecceitism. Dropping this unnecessary modal baggage entitles substantivalists to $\mathcal{D}$: they can be deflationists too![75]

2.4.2 Relationalists can be inflationists just like substantivalists!

Given that relationalism is often just stated as the denial of substantivalism,[76] it would be rather odd if there weren't some position opposed to sophisticated substantivalism. There is: I call it *unsophisticated relationalism*. The idea is this. We can accept that the relationalist will classify each nominally distinct embedding of q_i in space as merely different ways of describing the same *single* configuration (*i.e.* the same *spatial* structure): when the relations are the same the configurations are the same. However, there is nothing in relationalism that means we cannot permute the identities of the q_i themselves. Thus, we can consider q_1 and q_2 as having masses m_1 and m_2 respectively in $\mathcal{W}$, but as having m_2 and m_1 respectively in some other world $\mathcal{W}_{P_{12}(m_1 m_2)}$ (where $P_{ij}(m_i m_j)$ is a permutation operator shuffling the particles' masses). Applying this permutation procedure in our 'three body'-'three state' Leibnizian (*i.e.* relational) world will give us six distinct, yet qualitatively identical possible worlds (corresponding to each possible permutation). But if we allow this, then we lose out on the parsimonious account that lead Leibniz to choose it in the first place in a bid to avoid indistinguishable possibilities violating PSR. The obvious response is to outlaw haecceitism (*i.e.* to enforce PII so that the action of $P_{ij}(m_i m_j)$ is factored out of possibility space), so that permuting identities just gives us the original world back.[77] Hence, relationalism avoids indistinguishable possibilities only if haecceitism is ruled out, just as substantivalism entails them only if it is accepted. However, neither anti-haecceitism nor haecceitism are necessary corollaries of either substantivalism or relationalism.

Strictly speaking there is no reason why there should not exist a profusion of worlds that have the same relational structure and even the same objects, only differently distributed under the relations—God may have liked a particular configuration so much that he made lots of copies (perhaps infinitely many). Suppose we begin with two worlds that differ with respect to just one relation, say $R(x,y)$ in $\mathcal{W}_1$ but $\neg R(x,y)$ in $\mathcal{W}_2$. Are we to suppose that this single difference is all that prevents $\mathcal{W}_2$ from being snuffed out of existence on pain of violating PII? Anti-haecceitism is not a necessary component of relationalism; it is an additional assumption that can be denied. The reason it can be denied is that relationalism involves a primitive set of (material) objects, which allow us to permute them while keeping the relational structure itself intact. If the existence of this set of objects is denied, or else if we adopt the view that they are individuated by the relational

[75] Note that there is another type of anti-inflationist substantivalism according to which neither $\mathcal{D}$ nor $\mathcal{I}$ represents possibility space. Instead, one configuration is *selected* from each equivalence class (we might call this space $\mathcal{S}$, for *selective*: see §3.3 for a characterization of this and 'direct' and 'indirect' interpretations in the context of gauge theory, where I show that these options carve out an adequate taxonomy for interpretations of gauge freedom). Butterfield gets this interpretation by coupling substantivalism to counterpart theory (see Butterfield [1989]). Thus, *contra* Hoefer ([1996], p. 15), denying PII (*i.e.* deflation or *Leibniz equivalence*) does not entail haecceitism (inflation); there's more than one way to deny it. Therefore, even if the sophisticated substantivalist argument just presented is wrong, it still stands that the substantivalist is not committed to inflation.

[76] See, for example, Hoefer [1998].

[77] See Wilson [1959] for an identical response to a similar permutation argument.

structure then this move won't work: permuting the individuals is either outlawed or else it is idle. Such a view is generally part of structuralist positions. But even on such a structuralist position which appears to definitively rule out haecceitistic differences there is still room to fit such differences in. Given two objects x and y, individuated by the relational structure they are part of, we can simply say that x represents *de re* of y that (*i.e.* y) might have been x (*cf.* Lewis [1983a], p. 395). What we have here is a profusion of possibilities without a profusion of worlds. Naturally, this position—"cheap quasi-Haecceitism"—is open to both substantivalists and relationalists too. Therefore, modal metaphysics cannot dictate spacetime ontology and spacetime ontology cannot dictate modal metaphysics.

More simply, the above argument can be spelt out in terms of the definition of relationalism I gave in §2.1:

1. Relationalism is just that view of spacetime that takes the structure to be determined by the relations holding between a primitive set of material objects.
2. The above definition does not include an analysis of the modal behaviour of this set of objects; in particular is does not involve any claim about whether the objects have haecceities or not.
3. Therefore, on the assumption that the objects *do* have haecceities (and ruling out *e.g.* essentialism, counterparts, *etc.*) we can clearly build up an enlarged possibility space comprising qualitatively indistinguishable worlds that differ solely in terms of which individuals get which properties.[78]

Of course, the argument I have just presented is similar to Clarke's *tu quoque* to Leibniz in the *Correspondence*: granted that space is a relational entity,

> it would be absolutely indifferent, and there could be no other reason but mere will, why three equal particles should be placed or ranged in the order *a*, *b*, *c*, rather than in the contrary order. ([Alexander, 1956] C III-2)[79]

Leibniz simply took this to be a *reductio* of atoms, in much the same way that he took the shift arguments to be a *reductio* of the substantivalist's conception space and time: both violate PSR:

> 'Tis a thing indifferent, to place three bodies, equal and perfectly alike, in any order whatsoever; and consequently they will never be placed in any order, by Him who does nothing without wisdom. But then He being the author of all things, no such things will be produced in nature. (L-IV.3, [Alexander, 1956], p. 36)

However, the notion of *order* in a relationally defined space is rather a weak base on which to construct a permutation argument of the kind Clarke wants. Moreover, there is no reason for Clarke to invoke particles that are equal. Both factors simply serve to supply Leibniz with yet more cannon fodder for PSR to target and blow away with PII. But the shuffling of properties over individuals gets to the anti-haecceitistic core of Leibniz's relationalism. The trouble is, it is nowhere spelt

[78] In fact, there might even be cases where the label "unsophisticated" is a misnomer. For example, the relationalist might require truthmakers for certain modal claims that demand such worlds.

[79] See also Horwich ([1978], p. 409).

out that anti-haecceitism is a necessary part of relationalism. There is nothing internal to relationalism *per se* that implies the identification of those possibilities that differ solely in the redistribution of (non-spatiotemporal) properties over individuals. Relationalism concerns the ontological priority of material objects (the relata) over space and time (the relational structure); this does *not* obviously involve any commitment to any additional modal semantics of either ontological category beyond the basic priority claim: relationalists can be inflationists too! There is no reason why one's views about spacetime should determine one's modal commitments (and *vice versa*).

There is an obvious objection to the argument I have just outlined. Namely, acting on configurations of matter by Euclidean isometries was a part of the Newtonian theory, and therefore is physically well motivated, as are the reductive methods. The shifts are simply elements of the group of symmetries of Newtonian mechanics. Permuting identities is not included in this way. It is true that although we managed to allow the substantivalist to set up camp in what was believed to be the relationalist's site, $\mathcal{D}$, we cannot construct a converse situation wherein the relationalist is allowed to set up camp in $\mathcal{I}$. The reason for this is that the sophisticated substantivalist's occupation of $\mathcal{D}$ can be explicated using resources to be found within the theory and within the mathematical representation. On the other hand, the relationalist cannot similarly occupy $\mathcal{I}$ because the possibility that inflation is compatible with relationalism takes place *outside* of the physical theory and outside of the mathematical representation. Therefore, unless some operation corresponding to $P_{ij}(m_i m_j)$ can be found within the theory and the geometric spaces used to represent the relevant system, $\mathcal{I}$ will be out of bounds, even though inflation *per se* is an option. The objection has some force, but recall that the whole point of going to the reduced space was to enforce the outlawing of indistinguishable possibilities. The fact that such possibilities are compatible with relationalism is enough to show that the enterprise of associating relationalism with deflationism is seriously misguided: possibility counting simply isn't relevant to spacetime ontology.[80]

2.5. LOOKING AHEAD TO THE MODERN DEBATE

I have been speaking liberally about symmetries being at the core of Leibniz's complaints. But I haven't yet properly unpacked this idea. This brief section does just that. The question to be asked is: What allows Leibniz's shift argument to get its grip? Belot suggests this:

[80] Note also that just such an operator occurs in the context of quantum statistical mechanics, where it can be understood precisely as shuffling particles over states. A similar problem to that raised by the shift argument occurs in this context too. Since no observable can distinguish between permuted states (*i.e.* they commute with the permutation operators) the question arises as to whether we should allow worlds corresponding to permuted states into our ontology. The debate gets an extra kick, however, since it appears that the statistical behaviour of quantum systems can be understood as slicing these worlds out of possibility space, and thus as underwriting an anti-haecceitism. Naturally, there is more to the story than this; indeed, as is the case with spacetime ontologies, there are empirically equivalent interpretations that can support both haecceitism and anti-haecceitism in both classical and quantum mechanics. Hence, a general picture is beginning to emerge about the relationship between ontology and symmetry.

> Clarke begins with a single possible configuration of matter in space, and generates many more by allowing the symmetries of Euclidean space (translations, rotations, and compositions thereof) to create new possible dispositions of this system in absolute space, such that the spatial relations between objects are the same in each of the possible configurations, although the disposition of the objects in absolute space will differ. Leibniz retorts that Clarke's many possibilities are in reality one: the bodies, in space, with such and such relations between them. The trick is to allow the absolutist to specify a large space of possibilities which falls into equivalence classes—where two possibilities are considered to be equivalent if and only if they arise from one another by the application of a symmetry. The advocate of PSR can then claim that the true space of possibilities arises by identifying equivalent absolutist possibilities, so that there is exactly one possibility corresponding to each of the absolutist's equivalence classes. ([Belot, 2001], p. 4)

What Leibniz has drawn attention to, albeit in different terms than Belot, is the fact that Newton's theory, formulated on absolute space, possesses symmetries, and these symmetries introduce potential redundancy in the theory in the form of distinct yet indistinguishable states. In reifying absolute space, lifting it from formal to physical, Newton supposedly thereby commits himself to the redundant elements, to their *physical* existence. It is this redundancy that the shift arguments trade on.[81] The symmetries that Leibniz utilizes can be viewed as the Galilean symmetries of a Newtonian spacetime. As Belot points out, these are: symmetry under spatial transformations; symmetry under rotations; symmetry under boosts, and symmetry under time translations. Ten symmetries in all, giving the group of symmetries the structure of a 10-parameter Lie group $\mathcal{G}_N$ (I use the "*N*" to highlight the fact that the group is associated with Newtonian mechanics). Newton's laws of motion are *covariant* with respect to these symmetries, and systems are *invariant* under the symmetries. What this means is that, if we act on a system (*e.g.* the material contents of a world) with one, or some combination, of these ten symmetries, we get a state that is indistinguishable from the original. There is no way of telling whether we are in a universe that is uniformly rotating and moving with a uniform velocity in some direction or in one that is at rest: the laws of physics are indifferent with respect to these cases.

We can connect this up to the general account of symmetry outlined in the previous chapter as follows. We begin with a space $\mathcal{S}$ such that the points $x \in \mathcal{S}$ are taken to represent physically possible states of affairs according to Newton's theory. A set of curves $\gamma \in \mathcal{S}$, as picked out by Newton's equations of motion (as derived from the action), will represent the dynamically possible histories of this theory—in this case there will be one through each point of $\mathcal{S}$. The idea is

[81] Clearly, the claim that a certain structure is redundant depends upon whether or not we can formulate an equally successful, empirically adequate theory without that structure. Newton thought that absolute space was indispensable since he believed that absolute accelerations mandated it. The burden of proof clearly rests on the relationalist to prove him wrong. Leibniz made the assumption that the distinct possibilities were redundant simply because they were indistinguishable and, therefore, in violation of PSR and PII, without actually backing it up with a workable relational alternative. This task wasn't achieved until Barbour and Bertotti formulated their theory of 'intrinsic dynamics' [1982]. Pooley [in press] contains the best discussion of these issues that I have seen.

that the points represent the possible instantaneous states of a system and the histories represent possible sequences of these instantaneous states. Of course, this is simply the notion of a geometric (possibility) space and dynamics that I introduced in the previous chapter. Recall also from that chapter that an interpretation of this setup involves the explication of a representation relation between the formalism (*qua* instantaneous states and dynamical histories) and a set of possible worlds (*qua* ontology). This brings us back full circle to the nature of the representation relation between formal and physical: do we look for an interpretation of the formalism that lines up the states and dynamical trajectories *one-to-one* or *many-to-one* with physically possible states and worlds? Leibniz argues that Newton is committed to the former interpretation and that this leads to inflation via the symmetries of the theory: he is therefore guilty of violating PSR and PII. Leibniz himself appears to be committed to the latter interpretation on account of his commitment to PII which he believes is enforced by potential violations of PSR by indiscernibles.[82]

How exactly does this work? I have already mentioned that Newton's theory possesses certain symmetries, of which there are ten in all. These symmetries are manifested on $\mathcal{S}$ as structure preserving maps from states to states and histories to histories. Consider two states $x, y \in \mathcal{S}$, such that $y = g(x)$ and $g \in \mathcal{G}_N$, so that they are related by a symmetry. Suppose that g corresponds to one of Leibniz's shift operations, the symmetry transformation that shifts the material contents of the world some distance in some direction, 10 meters to the West, say. Then x and y will be isomorphic and so will γ_x and γ_y, the histories containing x and y, since g is structure preserving. If we consider γ_x and γ_y to be parameterized by $t \in \mathbb{R}$, we will find that for any choice of t, $x(t)$ and $y(t)$ will be related by g. The worlds corresponding to γ_x and γ_y will be indistinguishable, differing solely with respect to the location of matter in absolute space. A one-to-one interpretation of the elements of $\mathcal{S}$ will result in an infinity of indistinguishable physically possible states and histories, for there are infinitely many ways to choose g: inflation indeed! We have seen, though, that substantivalists are not necessarily committed to such interpretations, nor are relationalists necessarily committed to many-to-one interpretations.

The same argument can be put in more technical terms, using notions that arise in the context of gauge theory. I deal with this subject in detail in the next chapter. For now let me simply hint at what such a presentation might look like. Newtonian mechanics can be derived from an action that picks out physically possible paths (*i.e.* trajectories γ in the configuration space $\mathcal{S}$). The (inhomogeneous) Galilean group $\mathcal{G}_N$ is a variational symmetry of this action. This implies the conservation of energy, angular momentum, and linear momentum corresponding to invariance under time translations, rotations, and boosts—this is due to Noether's first theorem, on which see Brading & Brown [2003] and §3.2.3. Given that there are 10 such conservation laws, we know that the action admits a 10-parameter Lie group as its group of variational symmetries (*cf.* Earman [2003b], p. 142). These symmetries, as

[82] As I mentioned above, Belot [2001] defends this view whereas I deny it: PSR can be upheld in symmetrical cases without invoking PII. My defense of this claim comes in Chapters 3 and 8. Saunders too offers a way of upholding PSR whilst denying Leibniz's PII; however, he does so by putting another version of PII in its place—this version is *non-reductive* when it comes to interpreting *objects* related by non-trivial symmetries; that is, it renders distinct objects that would be deemed identical (since indiscernible) by Leibniz (see Saunders [2003b; 2003a]).

I showed above, inevitably lead to inflation. Note, however, that Newton's theory is *not* a gauge theory (in the usual sense), for that requires that the action admits an infinite-dimensional Lie group $\mathcal{G}_{\infty}$ as a variational symmetry. However, with the tools of this framework, we can see 'up close' the *source* of inflation: it stems from the variational symmetries of the action. This inflation also arises for 'true' gauge theories whose actions admit $\mathcal{G}_{\infty}$ as a variational symmetry, where the symmetries are 'local' (acting at points) rather than 'global' (acting at all points at once). In such cases we must again ask once about the nature of the representation relation. It arises too in the case of constrained Hamiltonian systems, of which general relativity is an example, and of which gauge theories are a subspecies. In both of these cases a one-to-one interpretation of the representation relation leads to a form of indeterminism that can *always* be eradicated by shifting to a many-to-one interpretation. We are now armed with the knowledge that these choices are not necessarily allied to either of substantivalism or relationalism. Rather, the choice reflects a deeper commitment to a particular understanding of identity and modal semantics.

When I introduce the concept of *gauge* in the next chapter, I show how the shift argument is a fairly natural consequence of interpreting the freedom to transform states that follows from the existence of variational symmetries. In brief, Newton's theory is underdetermined in the sense that absolute position and velocity are arbitrary functions of time. This underdetermination can be best seen in the Hamiltonian framework—in which the configuration space is replaced by the phase space of positions and momenta—as opposed to the Lagrangian one I implicitly assumed above. I show, within this framework, that there are standard methods for dealing with this underdetermination that are generally seen as corresponding, more or less, to 'relationalist' and 'substantivalist' positions. We know better of course!

Belot clings to the old 'Leibniz–Clarke-style' division of substantivalism and relationalism in terms of inflation and deflation. He has his reasons. He wants a clear distinction to show that the correct interpretation of spacetime as drawn from physics is still a live issue—*e.g.* in answer to Rynasiewicz's jibes [1996]. He has his sights set on quantum gravity, and hopes to show, by using an analogous division, that these interpretations underwrite very different approaches to quantum gravity.[83] The idea is that, so long as we can get a nice division set up in this way, quantum gravity might vindicate one or the other position.[84] I strongly disagree with Belot and, indeed, much of this book can be read as charting my disagreement with Belot on this point.[85] Quantum gravity will almost certainly prove to

83 For this reason he is particularly scathing of sophisticated substantivalist positions. He admits that sophisticated substantivalism "is very like relationalism" but "can think of no relevant difference between the two doctrines which would lead to any interesting interplay between serious physics and the (increasingly metaphysical) issue between relationalism and Lockean substantivalism" ([Belot, 1996], p. 183—here, Belot uses the *sobriquet* "Lockean substantivalism" to denote sophisticated substantivalism).

84 The reason being that the extended and reduced spaces that Belot sees as being underwritten by substantivalism and relationalism respectively, though classically equivalent, lead to inequivalent quantizations in general (*cf.* Gotay [1984]). If one of these methods were successful, and if substantivalism and relationalism could be seen to be linked up to them, then quantum gravity physics would support the underwriting interpretation of spacetime.

85 More importantly, however, is the impact the source of this disagreement has on ontology and interpretation. If I am right, then there is a severe underdetermination running through physics (infecting our best (fundamental) physical

be an enormous advancement in our conception of the *structure* of spacetime, but I doubt it can tell us anything about the correct interpretation of this structure *vis-à-vis* relationalism *vs* substantivalism.[86]

theories): distinct (conceptually incompatible; empirically equivalent) interpretations can occupy the same formal frameworks that are usually taken to provide a unique habitat for one or the other interpretation. By analogy with Steven French's arguments on the quantum statistics side (see French [1989; 1998; 2001]; and French & Rickles [2003]), I argue that a structuralist position is shown to be particularly well motivated by this underdetermination. However, the scope of the underdetermination envelops the French & Ladyman 'ontic' version of structural realism [Ladyman, 1998; French and Ladyman, 2003]; therefore I argue for a 'minimalist' version instead.

[86] I defend the preceding claim further in what follows and also in [Rickles, 2005], of which §4 is a technically souped up version of the present chapter, focusing on general relativity and quantum gravity. See also French [2001] and §10.2 of French and Rickles [2003] for a defense of a structuralist conception of spacetime drawn from an analogy of spacetime symmetries with the permutation symmetry of quantum statistical mechanics.

CHAPTER 3

The Interpretation of Gauge Symmetries

The kind of symmetry argument presented in the previous chapter receives its most advanced and most well-motivated expression in the context of gauge theories. Gauge theories are a special class of theory in which symmetry plays the central role in determining physical content.[87] In such theories one finds that a very high degree of symmetry is present in the formalism, leading to more mathematical structure than there is physical structure to be represented in the theory. Gauge theories can deal with these symmetries in such a way as to remove the excess degrees of freedom.[88] In their recent monograph on *Loops, Knots, Gauge Theories, and Quantum Gravity* Gambini and Pullin express exactly this point, and mention several important connected issues:

> Physical theories are not usually described in terms of the minimal possible number of variables. In general, descriptions are made in terms of quantities that present a certain degree of redundancy which results in the fact that the system is invariant under certain symmetries. For instance, one does not usually describe the free electromagnetic field in terms of two helicity components of the electric field, but rather in terms of the vector potential. The resulting formulation is invariant under gauge transformations. What will happen in general is that given a set of initial data the end result of the evolution will not be unique but will lie on a set of equivalent physical configurations related by the symmetries of the theory. ([Gambini and Pullin, 1996], p. 54)

This passage highlights a number of interesting issues. Firstly, they draw attention to the connection between redundancy and invariance under symmetries. This is of course familiar from our discussion of the symmetries of Newtonian spacetime that allowed for the 'shifted world' possibilities suggested by Leibniz and from the permutation symmetry of quantum mechanics. Secondly, they note that what often happens is that an alternative mathematical object is invented to encode the invariant content of the theory. Again, we have walked similar territory

[87] The lectures by Govaerts [2002] and Matschull [1996] are exceptionally clear and very useful for getting to the philosophical problems of such theories. I refer the reader to these for a more general introduction to the technical aspects. Redhead [2003] offers an exceptionally clear and concise overview of interpretive issues concerning gauge theories.

[88] There are a variety of methods for removing the excess, each with their own technical and conceptual benefits and drawbacks. Note, however, that it is not strictly necessary to remove the surplus. I describe the various options in §3.3.

before. The analogy is with Leibniz's notion of a relative configuration of objects or events as opposed to objects and events being defined on Newton's absolute space (spacetime). Thirdly, they mention the fact that in such theories the evolution of an initial data set will be underdetermined, leading to indeterminism. Though, they quickly point out, the symmetries of the theory will connect the evolutions so that they describe one and the same physical state of affairs. This latter feature has a counterpart in Newtonian mechanics and quantum statistical mechanics, though concerning the kinematical rather than the dynamical structure: once the initial conditions have been fixed, and a localization of matter to spacetime and a labeling of particles has been made, the evolutions are uniquely determined by the laws.[89] In these two examples, one does not have a problem concerning the underdetermination of evolutions of states and observables; perhaps if one did have the problem of indeterminism then the postulation of Newtonian absolute space and individual quantum particles would have possibly been viewed as more dubious. However, just this underdetermination/indeterminism and connection of underdetermined elements by symmetries is a characteristic feature of *all* gauge theories. Since our theories of the four forces of nature are all gauge theories, a clear understanding of this underdetermination/indeterminism is a pressing interpretive matter.

The case they mention in support of their claims, Maxwellian electrodynamics, provides a useful starting point for our journey into general relativity and quantum gravity; there are many similarities concerning both the formalisms and the conceptual problems they face. We will see that many of the problems listed above (and described more fully below in §3.3) that apply to electrodynamics, apply also to general relativity and quantum gravity. Many of the problems are simply different expressions of the same problem resulting from the tricky business of interpreting of gauge theories.

3.1. MAXWELLIAN ELECTROMAGNETISM

As traditionally conceived, Maxwell's equations for electromagnetism describe the behaviour of a pair of vector fields—$\vec{E}$ (the electric field) and $\vec{B}$ (the magnetic field)—that are (1) defined at each point of space (taken to possess the structure represented by $\mathbb{R}^3$); (2) functions of time $t \in \mathbb{R}$; and (3) dependent upon the electric charge density ρ and current density $\vec{j}$. Setting the speed of light c to 1, Maxwell's equations are:

$$\nabla \cdot \vec{B} = 0 \tag{3.1}$$

$$\nabla \times \vec{E} + \frac{\partial \vec{B}}{\partial t} = 0 \tag{3.2}$$

[89] In other words, underdetermination is not sufficient for indeterminism; indeterminism requires that no amount of specification of initial data can secure unique future values for some physical quantity or object. In the case of quantum statistical mechanics and Newtonian mechanics, there are enough evolution equations to propagate all of the physical magnitudes once a labeling of particles and localization has been settled on. The indeterminism that lies at the core of gauge theories is an underdetermination of solutions of the equations of motion given an initial data set (*cf.* [Gotay et al., 1998], p. 1, and §3.2.3 below).

$$\nabla \cdot \vec{E} = \rho \tag{3.3}$$

$$\nabla \times \vec{B} - \frac{\partial \vec{E}}{\partial t} = \vec{j} \tag{3.4}$$

A straightforward interpretation of Maxwellian electrodynamics is as a theory describing the behaviour of these fields, where the electromagnetic field can be represented by a pair of potentials[90]: the vector potential A and the scalar potential Φ, where (setting $c = 1$ again):

$$\vec{E} = -\operatorname{grad}\Phi - \frac{\partial A}{\partial t} \tag{3.5}$$

$$\vec{B} = \operatorname{curl} A \tag{3.6}$$

The fields, rather than the potentials, comprise the fundamental ontology of the theory: these are the things we measure. An instantaneous state of the electromagnetic field is given by specifying the values and first derivatives of the electric and magnetic fields at each point of space (at an instant).[91] Maxwell's equations then describe how this state evolves *deterministically* through time. So far there seems to be no problem in representation regarding the ontology of fields, and the match between formalism and system appears to be one-to-one. However, there are two problems: (1) the Aharonov–Bohm effect, and (2) the underdetermination of the potentials by the fields. We deal with both problems in detail in the context of gauge theory in the subsequent sections; in this section we try to make some (intuitive) headway on (2), and simply mention what (1) looks like.

The underdetermination concerns the vector potential A. Since $A = A + \operatorname{grad} f$ (for smooth functions f), $\operatorname{curl} A = \operatorname{curl}(A + \operatorname{grad} f) = \vec{B}$, and so many *formally* distinct vector potentials will represent a physically indistinguishable magnetic field (since the curl of a gradient is zero). As Wigner ([1967], p. 19) points out: "the potentials are not uniquely determined by the field; several potentials (those differing by a gradient) give the same field." However, in this case the underdetermination is a mathematical artifact of the formalism employed, it is gauge. The reason is that since we are dealing with $\vec{B}$, and *not* A, as our basic ontology, this will take the same value on all values of A that differ by a gradient. This gives rise to a *gauge invariant* interpretation of the theory (see §3.3). Classically, this interpretation is fairly unproblematic, though it requires an account of the nature of the gauge invariance; the vector potentials are still not uniquely determined by the magnetic field, nor by the equations of motion, but we are understanding this part of the formalism to be *unphysical*. However, when one considers the behaviour of a charged quantum particle in a classical electromagnetic field (in a non-simply connected space), this gauge invariant interpretation of Maxwell's theory faces a serious problem: maintaining that the magnetic and electric fields exhaust the ontology, and that

[90] An alternative representation is provided by the electromagnetic field tensor F. The equations for this object are: $dF = 0$ and $\star d \star F = J$, where J is the current, and d and $\star$ encode the curls, divergences, and time derivatives in the alternative equations given above. See Part I of Baez & Munian [1994] for an excellent presentation of the various representations of Maxwell's equations. Note that a similar problem to the one I am about to detail applies to this representation too.

[91] The topology of the space turns out to be a crucial factor in assessing the cogency of this interpretation once quantum mechanical effects are taken in to consideration, as is the case in the Aharonov–Bohm effect (see §3.3.2). If the space is not simply connected, the magnetic field must be seen as acting non-locally.

the vector potential is surplus, leads to the conclusion that the magnetic field acts non-locally (*i.e.* where it isn't). This is the content of the Aharonov–Bohm effect: the charged quantum particles undergo a phase shift when the magnetic field is present *even though the field value is zero throughout the regions of space containing the particles' trajectories*. In order to retain locality one must attribute physical reality to the vector potential; this has positive values in the places where the value of the magnetic field is zero, and different vector potentials can give rise to the observable effects of the Aharonov–Bohm effect. This move quite clearly leads right back into the problem of underdetermination of the vector potentials, and the indeterminism which now manifests itself as a *physical* indeterminism since we are now conceiving of the previously *formal* degrees of freedom of the vector potential as *physical* degrees of freedom.

It will be useful to introduce some relevant terminology at this point.[92] A transformation on the potentials that results in the same (*i.e.* physically indistinguishable) fields is known as a *gauge transformation* and such potentials are said to be *gauge equivalent* or *gauge related*. Gauge equivalent potentials are said to lie within the same *gauge orbit*. The freedom to choose from a gauge orbit of vector potentials that represent one and the same field is known as *gauge freedom*. The electromagnetic field vectors (or the electromagnetic field tensor F) are said to be "invariant under gauge transformations", or just *gauge invariant*—the same also applies to Maxwell's equations themselves, with "covariant" replacing "invariant". However, the vector potential is *not* gauge invariant, its value is altered by a gauge transformation.

Suppose that we wish to maintain a one-to-one representation relation between the vector potentials and physically possible worlds (or, simply, possibilities)—because of a desire for a literal, realistic interpretation of the formalism or the desire the accommodate the Aharonov–Bohm effect, for example—then seemingly we must countenance both the haecceitistic differences mentioned in the previous chapters (in this case indiscernible field differences) and a form of indeterminism, since the equations of motion cannot determine the behaviour of the potentials uniquely, but only up to a gradient. As Maudlin explains:

> if the original dynamics implies that a state of the electromagnetic field $E_0 - B_0$ will evolve, after a period of time, into $E_1 - B_1$, then we should expect the new dynamics only to demand that a pair of potentials which yields [by Maxwell's equations] $E_0 - B_0$ ought to evolve into some pair of potentials which yields $E_1 - B_1$. But since many different pairs of potentials yields $E_1 - B_1$, we have no reason to expect the dynamics to pick out one of these pairs over any gauge-equivalent pair. ([Maudlin, 2002], pp. 3–4)

Hence, there is a mismatch between the formalism (in terms of the potentials) and physical reality (in terms of observable field values): many *distinct* potentials give the *same* fields.[93] If we wish to predict how the potentials will evolve into

[92] I give more precise, general definitions of this terminology in the next section. For now I give simple qualitative explanations restricted to the theory of electromagnetism.

[93] Hence, we have an example of what Redhead calls *surplus structure* since "there are more degrees of freedom in the mathematical description than in the physical system itself" ([2003], p. 124).

the future then this mismatch manifests itself as an indeterminism: the dynamics can only determine the evolution of potentials *up to a gauge transformation*. This problem inspires the move to the traditional gauge invariant account involving only an ontology of fields, for we *can* predict which gauge orbit potentials will lie in, and the elements of a gauge orbit give the same field configuration. But anti-realism about the vector potential renders the traditional account unable to deal adequately with the Aharonov–Bohm effect: it leads to a violation of locality. Hence, we have an interpretational dilemma consisting of a trade-off between locality and determinism.

However, it is possible to give a deterministic account using vector potentials, provided we incorporate gauge invariance into the description of the observables. This can be achieved either by using the holonomies of vector potentials (*qua* connections) or the Wilson loops of the holonomies.[94] Both holonomies and Wilson loops contain all of the gauge invariant information contained in the vector potentials. According to the approach in terms of vector potentials, the canonical variables are standardly given by pairs $(A_a(x), E^a(x))$, representing vector potentials and electric fields respectively on a spacelike hypersurface. But, as we saw above, in these variables the phase space contains a lot of surplus structure corresponding to potentials differing by a gradient: good for locality but bad for determinism. However, as I have just noted, all of the gauge invariant content contained in the potentials (where the potential is understood to be a connection on the (fiber bundle) space) is encoded in the holonomies ($H_\gamma(A) = \exp i \oint_a dl^a A_a$, where γ is a loop). (In the case of the electric fields the physical content is encoded in functionals of the form $\int d^3x\, E^a(x) f_a(x)$.) Pairs of holonomies and functionals form a (complete) set of configuration and momentum variables for the theory's phase space. Hence, we have in fact shifted from a description in terms of vector potentials, to one that carries a vector potential around a loop in space (or, rather, a particle is carried around a loop); the holonomy exemplifies the invariant *structure* of the vector potentials in the sense that many vector potentials underlie one and the same holonomy. More precisely, the same holonomy is represented by gauge related vector potentials: $H_\gamma(A) = H_\gamma(A + \text{grad} f)$; the gauge-equivalence class of vector-potentials is, ontologically speaking, what the holonomy is like. Thus, the holonomies are gauge invariant, and vector potentials give the same holonomies iff they are gauge related; moreover, they give a local account. Thus, holonomies give us, interpretively speaking, the best of both worlds: like the magnetic field, they are indifferent to

94 These objects occur in the fiber bundle formalism, the arena for Yang–Mills theories (in which Maxwell's theory belongs) and theories of connections (*e.g.* vector potentials) in general. Connections in such a formalism permit the (frame independent, invariant) comparison of points in neighbouring fibers. This notion of parallel transport then allows for the definition of the directional derivative of some mathematical object, a vector, for example. The parallel transport of an object around a closed curve γ is, in general, dependent upon the choice of curve. The parallel transport itself may be seen as a map (a homomorphism) from closed curves (living in the base space) to the Lie group of the bundle: so, to each curve there is associated a group element $H \in \mathcal{G}$. The action of H induces parallel transportation. We know that the result of parallel transportation is path dependent, so we write $H(\gamma)$. This object is known as the *holonomy*. We can connect this to general relativity by noting that a notion of curvature is obtained by considering the failure of some element of the fiber to return to its pre-parallel transported (around some closed curve) value. This curvature is equivalent to the holonomy when the latter is evaluated on an infinitesimal closed curve. As Gambini and Pullin point out ([1996], p. 2), "[k]nowledge of the holonomy for any closed curve ... allows one ... to reconstruct the connection at any point of the base manifold up to a gauge transformation".

certain smooth changes in the vector potential (namely, the gauge transformations); and, since the vector potential resides outside of the solenoid as well, they provide an explanation of the Aharonov–Bohm effect that does not result in non-local action.

Setting Maxwell's theory up as a theory of connections (and 1-forms) allows for two known representations: (1) the connection representation using holonomies (as outlined above), in which states are represented by holomorphic functionals[95] of the complex connections A^i_a; and (2) a loop representation, according to which the states are represented by functionals of closed loops on a 3-manifold.[96] The gauge invariant content contained in the connections is encoded in the holonomies, and this is in turn reflected in the traces of holonomies of the connections (*i.e.* Wilson loops $W_\gamma(A)$: maps from loops to complex numbers). The configuration variables are then given by functionals of the form $q[\alpha](A) := \mathrm{Tr}\,\mathcal{P}\exp G \oint_a dl^a A_a \equiv \mathrm{Tr}\, U_a$ (where $\mathcal{P}$ means 'path ordered', U_α is a group element given by the holonomy around the loop α, and the index α ranges over closed loops). The momentum variables are given by functionals $P[s](A, E) := \int_s dS^{ab}\, \eta_{abc}\, \mathrm{Tr}\, U_v E^c$ (where S is a closed 'strip' (with topology $S^1 \times \mathbb{R}$), U_v is a group element representing the holonomy of the connection A^i_a, and v labels a specific loop).

Although determinism is restored in these two representations—since the canonical variables are gauge invariant—non-locality re-enters as a result. However, the non-locality is of a rather curious kind, for it isn't of the 'action-at-a-distance' variety that plagued the traditional account. The non-locality concerns the 'spread outness' of the variables: they are not-localized at points. In the case of holonomies and Wilson loops it is better to understand them as living in *loop space* (*cf.* Redhead [2003], p. 138). I return to this issue in §3.3.2.

3.2. ASPECTS OF GAUGE THEORIES

All of the four forces we know to occur in nature are described by gauge theories.[97] The strong, electroweak, and electromagnetic forces are such theories, and the nature of each interaction is characterized by a gauge group: SU(2), SU(2) × U(1), and U(1) respectively—we deal with general relativity in the next chapter. The idea is that invariance under these groups of transformations leads one to the form of the equations for the relevant field (interacting with charges). Let me begin this section with a simple example showing this.

[95] In other words, a functional with a complex derivative at every point in some domain in the complex numbers.

[96] Wilson loops are needed in the formulation of gauge invariant quantities in non-Abelian Yang–Mills theories such as general relativity. The holonomies are not gauge invariant in such cases.

[97] Steven Weinstein [1999] does not agree that general relativity is a gauge theory since "the diffeomorphism group is not the automorphism group of a principle fiber bundle" (p. S154). However, in the sense in which we understand 'gauge theory' in this book, general relativity is the example *par excellence*: the symmetry group of general relativity is a gauge group in precisely the sense that it generates redundancy in the description of physical configurations. See §7.1 for further details. (See also [Bergmann and Komar, 1980], p. 229 for an earlier statement of this viewpoint.)

3.2.1 The gauge argument

For simplicity, consider the example of a free non-relativistic particle with wave-function $\psi(x)$. Invariance under U(1) means that if we act (by multiplication) on this wave-function by an element of $e^{i\theta} \in U(1)$ (*i.e.*, a phase factor), the resulting wave-function is physically equivalent[98] to the original, *i.e.*

$$\psi'(x) = e^{i\theta}\psi(x) \tag{3.7}$$

In this case a *global* U(1) transformation was applied to the wave-function. What this means is that the same operation is applied at every point in space at some time (*i.e.* θ does not depend upon x). Quantum mechanics is thus invariant under global U(1) transformations. However, it is by *demanding* that wave-functions be invariant under *local* (*i.e.* position dependent) U(1) transformations that Maxwell's theory results.[99] These are transformations of the form

$$\psi'(x) = e^{i\theta(x)}\psi(x) \tag{3.8}$$

Such transformations do not result in physically equivalent wave-functions for they differ in terms of their momenta. The charge density *is* invariant, but the current density is not. In order to achieve local invariance, modifications to the Hamiltonian have to made (or the Lagrangian if one is working in a spacetime description). These modifications require that (1) interactions be turned on between the system and charge;[100] and (2) a new field be introduced (the 'gauge potential': A). The system is coupled to this new field, and the modified Hamiltonian (respectively, Lagrangian) is invariant under local U(1) transformations *whenever the field is likewise transformed.* The form of the transformations then matches those of Maxwell's theory, where A is identified as the vector potential for the magnetic field.[101]

The gauge argument highlights philosophically interesting issues regarding the *application* of gauge theories, but it is not the aspect I wish to focus on. I am concerned with the nature of the gauge symmetries, and in particular with the notion of gauge freedom. I think that such a foundational study is required in order to properly understand how the gauge argument works.

One of the characteristic properties of a gauge theory is the conceptual problems faced when setting up a direct, one-to-one correspondence between the mathematical formalism of the theory and the physical system that the theory represents. As Henneaux and Teitelboim put it in their canonical text on gauge systems, "the physical system being dealt with is described by more variables than there are physically independent degrees of freedom" ([1992], p. 1). One extracts

98 Physical equivalence is cashed out in this case in terms of the expectation values of the system's observables—*e.g.* charge density, $\rho = e\psi^*\psi$, and current density, $j = \frac{1}{2}i\exp(\psi^*\nabla\psi - \psi\nabla\psi^*)$.

99 The generic name for the argument that demanding local gauge invariance produces an interacting field theory is known as "the gauge argument" or "gauge principle". It is discussed by Martin, and its "profundity" questioned, in [2003].

100 The charge is the conserved current that results from global U(1) invariance. See §3.2.3 below.

101 I should point out that the argument properly requires fields, rather than a single particle. The imposition of *local* gauge invariance is then underwritten by the fact that field values at spacelike separated points can be specified independently of one another. However, the present example makes the basic idea plain enough without recourse to the complications field theory introduces.

the physical degrees of freedom through the use of a gauge symmetry connecting certain of the variables: either one can *factor* out the action of the symmetry group giving a formalism without gauge freedom, or one can 'ignore' the gauge freedom by focusing on gauge invariant observables.[102] Gauge theories thus supply the machinery for both generating and eradicating (or 'dealing with'—in a way that preserves the 'physically relevant' structure) the 'surplus structure' created by the gauge freedom of the theory.

I think it is appropriate to say something about the case of the permutation symmetry of quantum mechanics at this point. The permutation operation can be construed as a kind of gauge transformation such that states connected by such a transformation are to understood as representing the same physical situation. The quantum statistics are then explained by something akin to gauge freedom in the theory; namely, the freedom to permute indistinguishable particles' labels without thereby altering the structure of the physically measurable quantities (*i.e.* the expectation values of the observables) of the theory: permuted particles will simply differ by a constant phase factor, and this does not allow for the distinction of wave-functions, as we saw earlier. Hence, the theory is *permutation invariant*. According to the received view this was supposed to hold some metaphysical significance concerning the individuality of the particles. But I argued that no such view was forced upon us by permutation invariance, and that permutation invariance will be a feature of *any* reasonable physical theory possessing indistinguishable elements (related to the kinematical and/or dynamical structure of the theory) and a way of generating permutations on these: permutation invariance can be accommodated by 'individualists' and 'non-individualists' alike. Likewise, the symmetries of Newtonian space used in the shift argument can be loosely construed as a kind of gauge transformation too. The generation of haecceitistically differing shifted worlds corresponds to something like gauge freedom: we can globally shift the material contents of a world without altering the physically observable properties. Hence, any physical system defined on this space will be invariant under such shifts. This was supposed to hold some metaphysical significance concerning the ontological status of space (and, *mutatis mutandis*, time). I argued that this wasn't forced upon us by the invariance: the invariance can be accommodated by substantivalists and relationalists alike.

I argued, and will argue, that with respect to these two cases one should view the underdetermination of metaphysical (interpretive) options by the physics as pointing to a structuralist understanding of quantum states and spacetime: all parties will agree about the structure that is exemplified in these cases. Accepting this does not imply that one should be realist *only* about this structure, or even about this structure itself, for I prefer to detach the realism/anti-realism issue from the interpretation of physics. What I do wish to defend is the claim that any interpretive options that go beyond this basic structure are straying from physics, and once this happens, the underdetermination is ineluctable. Since I am concerned with the interpretation of physics, I hold that the invariant structures exhaust what phys-

102 We shall see in §3.2.3 (and, in more detail, in §4.3) that this gauge freedom manifests itself in the form of *constraints* on the variables when the theory is cast into Hamiltonian form. In general relativity these constraints lead to problems with the status of space, time, and change in both the classical and quantum theories.

ical theories can sensibly talk about with definiteness as regards ontology. There might be individuals, there might be objects with strange identities, there might be entities with bizarre modal behaviour, but a physical theory cannot tell us much about these aspects. At best they can function as a *constraint* on these metaphysical positions; but there will be cases where mutually incompatible metaphysical positions are both compatible with some theory. Because of this, what I propose is that we should view physical theories as imposing a version of PSR with respect to such 'meta-physical' issues, such that only those aspects of (mathematical) structure that can be distinguished by the observables count as ontologically relevant in the physical description of the world-structure to be represented. In cases where a mathematical formalism admits more degrees of freedom than are strictly necessary in the representation of a physical system, the theory (*qua* states, observables, and laws) should be understood as indifferent to these. Again, I do *not* rule out that there is more to the world than this invariant structure, only that physical theories cannot 'see' beneath such structure.[103] With this claim out in the open, I can now introduce the details of gauge theories and show how it squares up.

3.2.2 From Hamiltonian to gauge systems

In the first chapter of this book I sketched the bare outlines of the formalism of gauge theory. I noted there that they constituted a *weakening* of a Hamiltonian system. Recall that a Hamiltonian system was given by a triple $\langle \Gamma, \omega, H \rangle$ consisting of a space Γ (the cotangent bundle $T^* \mathcal{Q}$, where $\mathcal{Q}$ is the configuration space of a system), a tensor ω (a symplectic [closed, non-degenerate] 2-form), and a function H (the Hamiltonian $H : \Gamma \to \mathbb{R}$). These elements interact to give the kinematical and dynamical structure of a classical theory. The space inherits its structure from the tensor, making it into a phase space with a symplectic geometry. The points of this space are taken to represent physically possible states of some classical system (*i.e.*, set of particles, a system of fields, a fluid, strings, branes, *etc.*). Finally the Hamiltonian function selects a class of curves from the phase space that are taken to represent physically possible histories of the system (given the symplectic structure of the space). Any system represented by such a triple will be deterministic in the sense that, knowing which phase point represents the state of the system at an initial time, there will be a *unique* curve through that point whose points represent the past and future states of the system. The physical interpretation of this framework is as follows. Recall that the phase space is given by the cotangent bundle of the configuration space, where points of the configuration space represent possible instantaneous configurations of some system (relative to an inertial frame). The cotangent bundle is the set of pairs (q, p), where q is an element of the configuration space and p is a covector at q. Thinking of q as representing the position of a system leads to the view that p represents that system's momentum. The value of the Hamiltonian at a point of phase space is the energy of the system whose state is represented by that point.

[103] Of course, if we think that physics 'tells it how it is', and if we are of a realist disposition, then we are likely to adopt the view that this structure is all there is.

Now, since there is a unique curve through each point of phase space, one can interpret the phase space as *directly* representing the physically possible states of a system, and the curves as *directly* representing the physically possible histories of a system. A simple one-to-one understanding of the representation relation is possible that does not lead to indeterminism or underdetermination as regards the canonical variables q and p and quantities built from them. Weakening the geometry of the phase space, and moving to gauge systems, however, puts pressure on this simple direct interpretation,[104] precisely because indeterminism breaks down and the canonical variables are underdetermined. Let us see how this occurs.

The first change to note is that the symplectic form is replaced by a *presymplectic form* σ, so that the phase space $\tilde{\Gamma}$ of a gauge system inherits its geometrical structure from this. The presymplectic form induces a partitioning of the phase space into subspaces (not necessarily manifolds) known as *gauge orbits*, such that each point x in the phase space lies in exactly one orbit $[x]$. Once again we choose a Hamiltonian function on phase space, such that the value at a phase point represents the energy. However, in this case, given the weaker presymplectic form, the Hamiltonian is not able to determine a unique curve through the phase points. Instead, there are infinitely many curves through the points. However, the presymplectic form does supply the phase space with sufficient structure to determine which gauge orbit a point representing the past or future state will lie in. Hence, for two curves $t \to x(t)$ and $t \to x'(t)$ intersecting the same initial phase point $x(0)$, we find that the gauge orbit containing $x(t)$ is the same as that containing $x'(t)$: $[x(t)] = [x'(t)]$.

Each classical observable is represented by a function $f : \tilde{\Gamma} \to \mathbb{R}$ on the phase space. But given that the future phase point of an initial phase point is underdetermined, it will be impossible to uniquely predict the future value of the so-defined observables. Hence, there appears to be a breakdown of determinism; the initial-value problem does not appear to be well posed, as it is for standard Hamiltonian systems. The reason is clear: there is a unique curve through each phase point in a Hamiltonian system but infinitely many curves through the phase points of a gauge system.

Yet there are many theories that are gauge theories and that are evidently *not* indeterministic in any pathological sense. The trick for restoring determinism and recovering a well posed initial-value problem is to be restrictive about what one takes the observables to be. Rather than allowing *any* real valued functions on the phase space to represent physical observables, one simply chooses those that are constant on gauge orbits, such that if $[x] = [y]$ then $f(x) = f(y)$. Such quantities are said to be *gauge invariant*. The initial-value problem is well posed for such quantities since for an initial state $x(0)$, and curves $x(t)$ and $x'(t)$ through $x(0)$, $f[x(t_1)] = f[x'(t_1)]$.

From what I have said so far we can see that there are two competing interpretations of a gauge theory: on the one hand there is a one-to-one interpretation of the phase points, such that each point represents a distinct possible state of a system; on the other hand there is a many-to-one interpretation according to which

[104] Note that I don't say that such an interpretation isn't possible. It is, provided one either accepts the consequence of indeterminism and underdetermination, or else finds another way to deal with them.

many phase points (those within the same gauge orbit) represent a single possible state of a system.[105] The former leads to indeterminism and (if not supplemented by a gauge invariant account of the observables) an ill-posed initial-value problem, while the latter involves surplus structure that can be eradicated, but only in a way that violates locality. Hence, the choice of interpretation is a highly non-trivial matter. There are other interpretive options, including variations on those just mentioned, but before I consider these, let me first introduce another aspect of gauge theories: the connection to the Lagrangian formalism and Noether's theorems.

3.2.3 Noether's theorems

Noether's theorems primarily concern the relationship between *invariances* and *conservation laws*.[106] The setting is Lagrangian field theory, according to which equations of motion are derived by means of a variational procedure from the action integral of the theory. In particular, the theorems are concerned with the various invariances of the action integral, $S = \int \mathfrak{L}\, dx$, under the continuous groups of transformations of the fields. As I have just shown, the mathematics of gauge theory, broadly construed, takes place within the framework of Hamiltonian systems (with constraints). However, we generally begin with the Lagrangian framework, and (Legendre) transform to the Hamiltonian one: certain features apparent in the former are manifested as constraints in the latter—the *first class* constraints are responsible for the gauge freedom (see §4.3).

The basic object of the Lagrangian approach is the configuration space $\mathcal{Q}$, on which is defined a set of coordinates q^i representing the instantaneous states of some system. A motion in this space is given by a path q(t) in $\mathcal{Q}$ (where q(t) is a function, with t the time and q a point in $\mathcal{Q}$), giving the evolution in time of a system. The dynamics is given by an action functional that assigns a number to each path.

The first theorem says that if the action is invariant (*i.e.*, retains the same form) under the action of an r-parameter Lie group $\mathcal{G}_r$ ($r < \infty$) then there are r 'conserved currents' when the equations of motion (the Euler–Lagrange equations) are satisfied. When the group is the group of spatial and temporal translations and rotations the system described by the equations will exhibit conservation of linear momentum, energy, and angular momentum respectively. Such conservation laws are ultimately responsible for the possibility of the Leibniz-shift examples.

The second theorem says that if the action is invariant under an infinite dimensional group $\mathcal{G}_{\infty r}$ of transformations (given by r functions of the variables) then there will exist r identities (Bianchi identities) holding between the equations derived from the action. When the group is the group of electromagnetic gauge transformations, the Bianchi identities are simply the Maxwell field equations. The appearance of Bianchi identities points to an underdetermination in the dynamical

[105] This option is available because the phase points lying within the same gauge orbit are related by a gauge transformation: if they represent physical possibilities, they represent indistinguishable possibilities. Hence, the one-to-one interpretation of the representation relation if interpreted simplistically will involve haecceitistic differences between the worlds represented by the solutions.

[106] See Brading & Brown [2003] for an excellent philosophical account of these theorems (note that they consider *three* theorems—here, I consider just two). They too connect their account up to the structure of gauge theories, as I do here.

evolution of the fields: there are more dynamical variables than there are (independent) equations of motion.[107] For this reason a *gauge symmetry* is introduced, so that a gauge transformation connects those parts of the formalism that represent physically indistinguishable states of affairs. This symmetry can be seen to eradicate the underdetermination responsible for indeterminism in the evolution of the (gauge) fields.[108]

Thus, the idea of gauge freedom manifests itself at the level of the Lagrangian formalism. In more detail: the action principle $\delta \int \mathcal{L}(q, \dot{q})\, dt = 0$ allows us to derive Euler–Lagrange equations. Sometimes these equations will be non-hyperbolic, they can't be solved for all accelerations. This results in a *singular* Lagrangian, revealing itself in the singularity of the Hessian $\partial^2\mathcal{L}/\partial\dot{q}^k\partial\dot{q}^h$. This implies that when we Legendre transform to the Hamiltonian formulation, the canonical momenta are not independent, but will satisfy a set of relations called primary constraints, related to the identities of the Lagrange formalism.[109] What this functional dependence means from a physical point of view is that not all points of the phase space of the theory represent physical states; if they represent anything it is (physically) *impossible* worlds. The constraints define a submanifold of phase space were each point *does* represent a physical state—this is the constraint surface. The constraint surface itself is partitioned into gauge orbits containing points that differ by a gauge transformation—(with our philosophical hats on) we can view them as representing physically indistinguishable states, worlds differing haecceitistically; more standardly they are viewed as representations of one and the same state or world.

3.3. INTERPRETIVE PROBLEMS OF GAUGE THEORIES

The key problem in trying to interpret gauge theories, is knowing what to do with the gauge freedom. There are multiple options, and hence, multiple ways of interpreting gauge theories. These differences are interpretively non-trivial in the classical theory, and can also lead to technically non-trivial differences when it comes to quantization of a gauge theory.

Let us call an interpretation that takes each phase point as representing a distinct physically possible state of a system a *direct* interpretation. Hence, each point x^i in a gauge orbit $[x]$ represents a distinct possibility. However, such a direct interpretation leads to a form of indeterminism for the reasons outlined in §3.2.2. But, since each of the phase points represents a distinct physical possibility, there is (strictly speaking) no surplus structure according to such an interpretation. Recall also that the indeterminism is of a very peculiar kind: the multiple futures that are

107 In other words, there's more mathematical structure than is strictly necessary to represent the physical system. Thus, we have a direct connection to Redhead's notion of surplus structure even within the Lagrangian framework.

108 As I show below in §3.3, there are a number of ways of understanding how this works, some involving a 'reductive' move (analogous to Leibniz's in the space/time context) and some involving 'non-reductive' moves.

109 Preserving these under evolution may require the imposition of higher-order constraints. Once one has a situation where all the constraints are preserved by the motion, one will have defined a submanifold where all of the constraints are satisfied—this is the "constraint surface". See Earman ([2003b], pp. 144–5) for a clear explanation of these constraints and their relation to the singularity of the Hessian. I offer a slightly more elementary presentation in [Rickles, 2006].

compatible with an initial state are physically indistinguishable, for they are represented by points lying within the same gauge orbit. Hence, the indeterminism concerns haecceitistic differences.[110] However, for realists the indeterminism will still constitute a problem, though it is not insurmountable. As Belot notes:

> if we supplement this account of the ontology of the theory with an account of measurement which implies that its observable quantities are gauge-invariant, then the indeterminism will not interfere with our ability to derive deterministic predictions from the theory. ([1998], p. 538)

I defend an account along similar lines in the following chapters based loosely upon the idea that a form of PSR should be seen as operating on observables, so that the observables are 'indifferent' as to the roles played by particular individuals. I think that Saunders "non-reductive relationalism" ([2003b; 2003a]) can be seen as implementing much the same idea—a claim I defend in §6.2. Using this method one can help oneself to gauge invariance at the level of observable ontology and remain neutral about the rest (spacetime points, quantum particles, shifted worlds, vector potentials, *etc.*). Saunders differs on this point; he does not wish to remain neutral, he wants to retain a notion of an individual, regardless of how thin this notion happens to be. This is a consequence of the fact that his proposal flows from his own version of PII.

Let us call an interpretation that takes many phase points (from within the same gauge orbit) as representing a single physically possible state of a system an *indirect* interpretation. There are two ways of achieving such an interpretation. According to the first method one simply takes the representation relation between phase points from within the same gauge orbit and physically possible states to be many-to-one. Since the points of a gauge orbit represent physically indistinguishable possibilities, there is no indeterminism on this approach. Redhead suggests that "the 'physical' degrees of freedom [*i.e.* the fields] at [a future] time t are being multiply represented by points on the gauge orbit ... in terms of the 'unphysical' degrees of freedom" ([2003], p. 130).[111] The gauge freedom is simply an artifact of the formalism. There are superficial similarities between this approach and the modified direct approach given above. However, the stance taken on this approach is that not all of the phase points represent distinct possibilities. Even on the modified direct approach this is false. The latter approach simply says that the question of whether or not all of the phase points represent distinct possibilities is irrelevant to the observable content of the theory, the observables are indifferent as to what state underlies them provided the states are physically indistinguishable.

The second method involves treating the gauge orbits rather than phase points as the fundamental objects of one's theory. By taking the set of gauge orbits as the points of a new space, and endowing this set with a symplectic structure, one can construct a phase space for a Hamiltonian system—this new space is known as

110 I review this problem of indeterminism below in §3.3.1.

111 Redhead's analysis seems to suggest that this is the *only* way to interpret the direct formulation (in terms of vector potentials)—though he mentions that a gauge invariant or gauge-fixing account can resolve the indeterminism. But clearly, it is open to us to give a direct interpretation and accept the qualitatively indistinguishable worlds that are represented by the isomorphic futures (points within the gauge orbit).

the *reduced phase space*,[112] and the original is then called the *extended phase space*.[113] Hence, the procedure amounts to giving a *direct* interpretation of the reduced phase space—*i.e.* one that takes each gauge orbit as representing a distinct physically possible state—but an *indirect* interpretation of the extended phase space. The resulting system is deterministic since real-valued functions on the reduced space correspond to gauge invariant functions on the extended space. In effect, the structure of the reduced space *encodes* all of the gauge invariant information of the extended space even though no gauge symmetry remains (*i.e.* there is no gauge freedom). Note, however, that complications can arise in reduced space methods: the reduced space might not have the structure of a manifold, and so will not be able to play the role of a phase space; or some phenomenon might arise that requires the gauge freedom to be retained, such as the Aharonov–Bohm effect (*cf.* Earman [2003b], pp. 158–9 and Redhead [2003], p. 132). If these complications do arise, one can nonetheless stick to the claim that complete gauge orbits represent single possibilities, as *per* the above method.

There is another method that involves taking only a single phase point from each gauge orbit as representing a physically possible state of a system. To do this one must introduce *gauge fixing* conditions that pick out a subset of phase points (a *gauge slice*) such that each element of this subset is a unique representative from each gauge orbit (*cf.* Govaerts [2002], p. 63). Gauge fixing thus 'freezes out' the gauge freedom of the extended phase space. In more detail, for a constrained system with constraints $\phi_i(q,p) = 0$ ($i = 1,\ldots,N$), one must impose M further conditions $F_j(q,p) = 0$ ($j = 1,\ldots,M$, where $M = N$). This defines another surface in the phase space that should, if successful, intersect each gauge orbit just once; for all practical purposes, this is the same as the reduced phase space, it is composed of just those points that satisfy both the constraints and the gauge fixing conditions.[114] This method leads to an interpretation that is neither direct nor indirect, I shall call it a *selective* interpretation. There is a serious problem—known as a *Gribov obstruction* (*ibid.*, p. 64)—facing certain gauge fixing procedures. The obstruction implies that the gauge conditions do not result in a unique 'slicing' of phase space, but may result in the selection of two or more points from within the same gauge orbit.[115] Thus, suppose we impose the conditions $F_a(p) = 0$, as best we can; then the Gribov obstruction will manifest itself as there being numerous p^i, all of which are related by gauge-transformations, and all of which satisfy

[112] In order to distinguish this approach from the previous one, let us call it a *reductive* interpretation from now on. Note that this matches Leibniz's form of relationalism since it can be seen as enforcing PII on phase points within the same gauge orbit. Thus, to complete the analogy, an extended phase space Γ would correspond to that containing phase points related by the symmetries associated with $\mathcal{G}_N$ (representing indistinguishable shifted, rotated, and boosted worlds) and the reduced phase space $\tilde{\Gamma}$ would correspond to the space with symmetries removed: $\tilde{\Gamma} = \Gamma/\mathcal{G}_N$.

[113] Thus the points of the reduced space correspond to gauge orbits of the original extended space. Curves in the reduced space contain information about which gauge orbits the system (as represented by the extended space) passes through—*i.e.* about which *qualitatively distinct* states it passes through.

[114] Referring back to the Leibniz–Clarke dispute, the present interpretive move would correspond to keeping absolute spacetime, but imposing a condition such that exactly one localization of the matter relative to it was chosen—*i.e.* where the point particle p_1 is at point x, point particle p_2 is at point y, and so on. However, in this case, it is difficult to see what could be gained by such a move; there is no symmetry or geometrical structure available to explain the various invariance principles and conservation laws.

[115] As Redhead notes ([2003], p. 132), in the case of non-Abelian gauge theories, the application of the gauge fixing method leads to a breakdown of unitarity (in perturbative field theory) that has to be dealt with by the ad hoc introduction of "fictitious" ghost fields—thus replacing one type of surplus structure with another.

the gauge fixings. Even in spite of the Gribov obstruction—which is very serious for any non-Abelian gauge theory (*cf.* Singer [1978])—there is an obvious problem with using this formal method of dealing with gauge freedom as a way of interpreting gauge theories. The method involves the singling out of a point from each gauge-equivalence class, giving a gauge-fixed submanifold, and carries out the physics on that. But since the elements of the equivalence class are deemed to be physically (qualitatively) indistinguishable, it is difficult to know what could *ground* this choice of gauge, other than perhaps the ease with which physics can be done with respect to it, at least in the general case.

Finally, there is an approach, named a "coarse-grained gauge-invariant" interpretation by Belot ([1998], p. 538), according to which the gauge orbits themselves stand many-to-one with physically possible states.[116] This would class as an indirect interpretation of an indirect (or, possibly, reductive) interpretation according to my definitions above. I don't have reason to consider such interpretations in what follows. Presumably, though, we could construe the many-to-one nature of the representation relation as pointing to gauge freedom again, and so apply one of the above interpretations here as well.

Each of these interpretive options is seen to be applicable in both general relativity and quantum gravity; indeed, they are seen to play a crucial role in both their technical and philosophical foundations—though I am skeptical about the scope of their philosophical import (*cf.* [Rickles, 2005]). In the next chapter, I turn to a specific argument (the "hole argument") based upon a direct, local interpretation of general relativity. The argument is connected to the nature of spacetime since the gauge freedom is given by (active) diffeomorphisms of spacetime points (or by 'drag-alongs' of fields over spacetime points). What we appear to have in the hole argument, is an expression of the Leibniz-shift argument couched in the language of the models of general relativity, with diffeomorphisms playing the role of the translations. Earman and Norton [1987] see a direct, local interpretation as being implied by spacetime (manifold) substantivalism (*i.e.* the view that spacetime points, as represented by a differentiable manifold, exist independently of material objects). Clearly, this view is then going to be analogous to the interpretation of Maxwell's theory that takes the vector potential as a physically real field. Such an interpretation is indeterministic: the time-evolution of the potential can only be specified up to a gauge transformation. Earman and Norton extract a similar indeterminism from the direct interpretation in the spacetime case, and use this conclusion to argue against substantivalism.[117] The "problem of time" (the subject of Chapter 7) applies the reasoning of the hole argument (as broadly catalogued in the direct, indirect, reductive, and selective interpretations) to the evolution of data off an initial spatial slice. One's interpretation of the gauge freedom then has an impact on the question of whether or not time and change exist![118]

116 He calls "simply gauge-invariant" those interpretations that take whole gauge orbits as representing a single physical possibility (*ibid.*). As is evident from the many-to-one and reduced space methods, there is more than one way to understand such interpretations.

117 In Chapter 5 I show how both the claim that substantivalism implies a direct interpretation and that the direct interpretation leads to indeterminism can be questioned in a variety of ways. These "ways" are more or less on all fours with the general interpretive options outlined above.

118 Or so the received wisdom goes. I argue that the problems of time and change affect *any* approach that treats general relativity as a gauge theory. All interpretations will face timelessness whether they get rid of the gauge freedom or retain

Let us next consider the two basic problems that face direct and indirect interpretations respectively: indeterminism and non-locality. (I don't consider the problems with selective interpretations here, since they can be lumped with reduced phase space methods for the purposes of the following investigations. I have said enough about the problems in my comments above, to show why I don't think they are good options interpretively speaking. I return to the idea of a selective interpretation in §5, where I argue that Butterfield's application of counterpart theory to spacetime points is the philosophical analogue of such an interpretation of the gauge freedom.)

3.3.1 Indeterminism

The main interpretive problem facing direct interpretations of gauge theories is the underdetermination that results from the gauge freedom. How one chooses to deal with this problem leads quickly into many distinct interpretive problems: indeterminism, non-locality and frozen dynamics (to name the most important for our purposes). I save the latter to a detailed examination in Chapter 7; the non-locality problem that results from a particular attitude to this problem is reviewed in the next subsection, §3.3.2; for now I restrict my attention to the indeterminism problem itself.

I have already briefly mentioned how the gauge freedom in Maxwell's equations gives rise to a form of indeterminism or, more accurately, underdetermination in the evolution of certain kinds of data off an initial surface. Let me expand on this idea some more, for it has a direct bearing on the following chapters. In particular, I argue that the hole argument and the problem of time in canonical quantum gravity are simply special cases of this problem.

The indeterminism flows from a direct (one-to-one) interpretation of the formalism I outlined in the previous section. Here, each point of the phase space is taken to represent a distinct physically possible state of the theory, even for those points occupying the same gauge orbit and which, therefore, describe qualitatively identical physical states. In theories with gauge freedom, there will be cases where a completely specified initial state (even an entire history up to some instant) is not sufficient to uniquely determine the evolution of data. Multiple evolutions are equally compatible with the initial state and the laws of the theory. The best that one can do, is to predict which gauge orbit the data will lie in, the points of which represent isomorphic structures. Hence, for two futures that are compatible with an initial state $x(0)$, there will be two corresponding states $x(t')$ and $x(t'')$ respectively. The physical structures (fields, in the case of the electromagnetism) that $x(t')$ and $x(t'')$ represent will be physically indistinguishable since the states are isomorphic: there is a structure preserving map $\phi : x(t') \rightarrow x(t'')$ (a gauge transformation) connecting the evolutes; in other words $x(t'), x(t'') \in [x]$, where $[x]$ is a gauge orbit 'to the future' of $x(0)$.

it as surplus or not. However, the notion of what counts as observable in such theories is crucial here, for denying that observables are *fully* gauge invariant (*i.e.* with respect to all constraints) can escape the problem, but only at the price of reintroducing indeterminism.

There are a variety of 'tried and tested' methods for dealing with this apparent violation of determinism that results from the gauge freedom of Maxwell's equations; they apply more generally to any gauge theory, and we will meet them again in the chapters that follow. They are as follows:

- **Gauge-invariance: 'Ostrich Style'.** If one sticks to the view that the electric and magnetic fields comprise the basic ontology of the theory, then the indeterminism that results is simply understood to be a by-product of the freedom to choose from a gauge equivalence class of potentials: it is not a *physical* form of indeterminism. As Earman nicely sums it up:

  > The fix for determinism is to blame the apparent failure [of determinism] on redundancy of the descriptive apparatus, which is simply another way of saying that the variational symmetries containing arbitrary functions of the independent variables connect equivalent descriptions of the same physical situation, *i.e.* are gauge transformations. ([Earman, 2006], p. 450)

  Thus, in terms of the electric and magnetic fields (and, *mutatis mutandis*, for fields associated with other gauge theories), the theory is completely deterministic. This state of affairs might lead to one to simply ignore the indeterminism, to bury one's head in the sand. As Maudlin puts it, "if one regards the fields as the real ontology, then one knows that the dynamics of that ontology is deterministic" ([2002], p. 5). Pragmatically, perhaps, this move is admissible; and, in fact, I have some sympathy with this view. Clearly, however, more needs to be said about the origin and nature of the mismatch between the formal representation (with its 'mathematical indeterminism') and the deterministic physical system.
- **Gauge fixing: Restrict the constraint surface.** In this case the indeterminism is removed by imposing a certain condition on the potentials so as to select a single member from each gauge equivalence class. This is tantamount to selecting a submanifold of the constraint surface that contains just one point from each gauge orbit. Fixing a gauge in this way implies that the potentials can be understood as being in a one-to-one correspondence with the fields without any surplus structure: since the field dynamics is deterministic, so is the dynamics for potentials. Hence, in addition to fixing the indeterminism problem, another interpretive problem is resolved; namely, the representation relation between the formalism (potentials) and reality (fields) can be read as direct once again (though such a conception required an original *selective* step). Aside from the technical difficulties I mentioned earlier, this method also loses all of the information contained in the gauge freedom: information about invariances, covariance, conservation laws, *etc.* In many cases this loss will be too high a price to pay for eradicating indeterminism.
- **Reducing phase space.** This method eliminates the indeterminism by forming a new phase space out of equivalence classes of gauge equivalent phase points (*i.e.* the gauge orbits). (Recall that the elements of gauge orbits are isomorphic, representing indistinguishable physical states.) A new space, the *quotient space*, is formed so that the gauge orbits of the extended space are the phase points

for this smaller space. The gauge freedom is thus "quotiented out", the new 'reduced' phase space being the quotient of the original space by the gauge orbits. With no gauge freedom remaining, the problem of indeterminism is eradicated. Unlike the previous method, this method retains information about the symmetries of the extended phase space in its geometrical structure.[119] I mentioned earlier that the reduced phase space method can face severe technical difficulties. For example, the set of gauge orbits does not necessarily have the structure of a manifold. In such cases it may be impossible to construct the reduced space. There are obvious similarities between this strategy and the gauge-fixing strategy: in both cases one gets rid of the gauge freedom, thereby getting rid of the symmetry. The difference: gauge-fixation methods *select* a single state from a gauge orbit while reduced space methods *identify* all states from a gauge orbit.

- Gauge invariance: 'Giraffe style'. The final method involves utilizing the presymplectic structure of the phase space by restricting the observables to those that are gauge invariant. I mentioned that the presymplectic structure ensures that any two trajectories passing through a single phase point on the constraint surface will intersect the same gauge orbits to the future of that point. On this approach, each gauge orbit can nominally be taken to represent a single physically possible world of the theory, but there is no reduction and no 'ostriching': the focus is simply shifted from the states to the observables. By choosing only gauge invariant functions we eradicate the indeterminism, since the initial-value problem is well posed for these functions. (There are two common classes of gauge invariant functions for the theories we deal with, these are holonomies and Wilson loops. Such objects are essentially 'blind' to the differences within gauge orbits, which are retained but treated as whole objects (*i.e.* one does not have to worry about their *individual* elements). There is, then, a large amount of surplus structure that is left largely unexplained. But, as I argue in Chapter 8, this is not necessarily a bad thing.)

To summarize. Any direct interpretation of a gauge theory will lead to indeterminism with regard to the states: fixing the initial state (given the equations of motion) will not fix a possible future uniquely, it will do so only *up to gauge*. The clear advantage of these approaches is that they are, interpretatively speaking, non-arbitrary. Each element of the phase space is assigned the same ontological weight; each represents a physically distinct possibility. One can restore determinism, moreover, by modifying the notion of observables so that one's measurement theory only concerns gauge invariant magnitudes. However, in addition to the indeterminism, there is the problem that the physically distinct states remain, and these physically distinct possibilities are qualitatively indistinguishable: those

119 As Maudlin [2002] points out, this method has an further advantage over the previous method, for the gauge condition may be too stringent on two counts: (1) if more than one point in the extended ("inflated", in Maudlin's terminology) phase space satisfies the condition then determinism may be lost, since the dynamics will not determine which of the points the trajectory will pass through. (2) if some physical state does not meet the gauge condition then certain physical possibilities cannot be represented in the formalism. In the case of quotienting, however, "one is automatically guaranteed that each physical state will correspond to exactly one gauge orbit, since the orbits by definition contain all the points in the phase space that represent the state" (p. 5).

with respect to which the theory was initially indeterministic. They differ *haecceitistically*, in terms of which individual elements of the ontology play which roles (which vector potential it is that determines the fields, for example). Regardless of one's theory of measurement, this problem will remain on a direct account of the representation relation. This might be enough to call the interpretation into question, but I think it can be dealt with (as I try to show in Chapter 8).

Indirect gauge invariant approaches come in two types: reductive and non-reductive. The former remove the gauge freedom by quotienting it out, avoiding both the indeterminism and the qualitatively indistinguishable possibilities—such reductive interpretations are the analogue of Leibnizian relationalism. Each gauge orbit represents a *single* physical possibility, and they are conceived of as phase points in a new space. Non-reductive approaches likewise treat gauge orbits as representing (from the perspective of the observables) a single physical possibility—thus avoiding indeterminism and qualitatively indistinguishable possibilities—but nonetheless treat the elements of gauge orbits as distinct: they are seen simply as different modes of presentation of the same state of affairs.

The selective interpretations will eradicate the indeterminism (and qualitatively indistinguishable worlds), but at the high price of losing the utility that gauge freedom affords (including not being able to deal with the Aharonov–Bohm effect). The technical problems are also severe when compared to the problems of the other approaches. We see in §5.2.2 that Butterfield's One [Butterfield, 1989] as a response to the indeterminism of the hole argument corresponds to a selective interpretation, and must therefore shoulder the burden of these problems. He seeks to avoid the loss of gauge freedom by distinguishing between the 'models' of a theory and the 'worlds' the models represent. His selectivism then applies to worlds but not models. In the case of electromagnetism, this would amount to retaining a formalism with surplus structure given by the vector potentials, but insisting that only one of these represents a physical possibility (*i.e.* only one corresponds to the magnetic induction). This move would also avoid the Gribov obstructions, since that problem applies to the models. I will discuss and argue against this idea when we reach the responses to the hole argument in Chapter 5. The next subsection looks at how the various responses to the indeterminism (corresponding to the various interpretations of a gauge theory) fare with regard to the Aharonov–Bohm effect.

3.3.2 Non-locality

I mentioned in various places above that indirect interpretations of the relation between gauge potentials and a physical system lead to non-locality. In the case of Maxwell's theory, this afflicts the approach that takes the electric and magnetic fields as the ontology of the theory, and views the underdetermination of the vector potential by the magnetic field as highlighting the fact that the field is being multiply represented by the potentials. The non-locality enters the picture when we consider the behaviour of electrons in the neighbourhood of a classical electromagnetic field confined within a solenoid. This is the arena for the Aharonov–Bohm effect.[120]

The setup consists of a two-slit apparatus with a solenoid (long—ideally, infinitely so—and thin) sitting beyond and in between the slits (1 and 2, standing distance d apart). This produces a magnetic field confined within the solenoid when it is turned on. Outside of the solenoid the value of the magnetic field is zero. Electrons are fired through the slits at the screen (separated by length L), and when the solenoid is turned on, they undergo a phase shift—this is manifested by the shift in the interference pattern on the detection screen. This phase shifting is known as the "Aharonov–Bohm effect" (see [Aharonov and Bohm, 1959] for the original presentation). Formally, the (relative) phase shift $\delta = \Phi_1 - \Phi_2$ (for amplitudes $C_1 e^{i\Phi_1}$ and $C_2 e^{i\Phi_2}$ from each slit) is computed as follows.[121] Firstly we note that $\vec{B} = 0$ outside of the solenoid, but $A \neq 0$ in such regions. The presence of a magnetic field in the solenoid alters the phase of an electron by the integral of the vector potential along the path taken by the electron, multiplied by the charge e of the electron divided by Planck's constant $\hbar$.

Let Φ_1 be the phase of the wave that travels through slit one (let [1] denote the whole trajectory of the wave from source to screen), and let $\Phi_1(\vec{B} = 0)$ be the phase when the solenoid is turned off. When the solenoid is turned on, and a magnetic field is produced, confined within the solenoid, the phase is given by:

$$\Phi_1 = \Phi_1(\vec{B} = 0) + \frac{e}{\hbar} \int_{[1]} A \cdot dl \tag{3.9}$$

Replacing instances of '1' with '2' gives us the expression for the phase of the wave that passes through slit 2 (again, let [2] denote the trajectory of the wave passing through slit 2). The interference of waves is then given by the phase difference:

$$\delta = \Phi_1(\vec{B} = 0) - \Phi_2(\vec{B} = 0) + \frac{e}{\hbar} \int_{[1]} A \cdot dl - \frac{e}{\hbar} \int_{[2]} A \cdot dl \tag{3.10}$$

We can now write this as a loop integral going along [1] to the screen, and back along [2] (where $\delta(\vec{B} = 0)$ is the phase difference when the solenoid is off, and γ labels the loop formed by stringing [1] and [2] together):

$$\delta = \delta(\vec{B} = 0) + \frac{e}{\hbar} \oint_{\gamma} A \cdot dl \tag{3.11}$$

Hence, the electrons undergo a phase shift, yet the magnetic field does not extend out beyond the boundary of the solenoid and the electrons cannot penetrate the boundary of the solenoid. As Drieschner *et al.* put it,

> What makes the well known AB effect so astonishing is that there seems to exist "nothing" in the region of space where the electron's possible paths

[120] See Nounou [2003] for a clear exposition of the effect followed by a nice philosophical analysis utilizing the fiber-bundle approach to gauge theories. Leeds [1999] offers an in depth and general discussion of a variety of interpretive options that have been suggested in response to the effect.

[121] The following presentation is derived from Feynman ([1962], Ch. 15).

extend, yet the electron is discernibly influenced by "something". ([2002], p. 305)

Of course, the vector potential *does* extend out beyond the solenoid, there is lots of A circulating around it; if it were a physically real field then we *would* have "something" living in the region of space through which the electrons travel, and this would explain the phase shifts. Classically of course, A has no discernible effect, since the forces, given by the Lorentz force law $F = q(\vec{E} + v \times \vec{B})$, determine the motion of particles inside and outside of the solenoid. Quantum mechanically, however, the force law is not sufficient to determine the shift; the electrons are subject to some force in the regions where the electromagnetic field is not. But note that we still have the problems of gauge freedom associated with the vector potential. The expression for the phase shift is itself invariant under changes of the form $A' = A + \text{grad} f$: both A and A' give the same phase difference, and therefore the same interference pattern. As Feynman puts it "it is only the curl of A that matters; any choice of the function of A which has the correct curl gives the correct physics" ([1962], §15-10).

Hence, thus far we seem to have two options regarding the interpretation of the AB-effect: Either the magnetic field acts non-locally or the vector potential acts locally. The latter option would grant physical status to the vector potential, thus introducing indeterminism at the level of ontology; the former option would introduce action-at-a-distance into a field theory. We thus have an appearance of the trade off between locality and determinism again. Let me develop these rough views some more.

Now, the magnetic field is confined within the solenoid, so if we view it as being responsible for the phase shift (and, hence, accept an indirect, gauge invariant view), then we must accept that it acts non-locally. The vector potential is not so confined, when the current flows, it extends outside of the solenoid. If we take the potential as responsible then we can see how it affects electron phases locally. This clearly requires that we view the vector potential as a physically real field. It seems that we must do this if we don't wish to have the magnetic field acting at a distance. It is, however, going to be an indeterministic account unless we apply one of the above methods for removing gauge freedom. The fact that the vector potential (a gauge potential) is *not* gauge invariant is enough, on gauge invariance accounts of observables (such as Dirac's [1958]) to show that it is not physically real. The problem is that gauge potentials are not measurable; there is no way to determine *which* gauge potential one has (*cf.* Nounou [2003], pp. 176–7). No realist is likely to be happy with this. In the present case, the fields and the AB-effect underdetermine the value of the vector potential. However, in spite of this problem, the view of many physicists, including Feynman, Bohm, and Aharonov, is that the vector potential is a physically real field *more fundamental* that the magnetic field, and the reason seems to boil down to the fact that it provides an explanation of the AB-effect that does not violate the local action principle. Weighed against the indeterminism of the vector potential ontology, the non-locality of the magnetic field loses out: indeterminism is traded in for locality.

Where the traditional accounts given above fail, on pain of violating non-locality and determinism, is where alternative gauge invariant accounts come into their own. The idea is, rather than taking the vector potential field as being responsible for the effect (resulting in a local but non-gauge invariant and, therefore, indeterministic account), one takes certain gauge invariant quantities *constructed* from the potential as responsible. Hence, one can work in terms of the holonomy of the vector potential around a loop encircling the solenoid, or the corresponding Wilson loops (traces of holonomies). These quantities are gauge invariant, and hence determinism is secured, *and* they extend beyond the solenoid (*i.e.* they are defined on paths outside of the solenoid), encircling it, and so can be called upon to explain the phase shifts—indeed, the description of the holonomy is more or less identical to Eq. (3.11).

This is, it is true, non-local too, but in a very different sense from the gauge invariant account involving the magnetic induction, which involved action-at-a-distance. In the holonomy case, the non-locality is simply a manifestation of the fact that holonomies are given as functions on a space of loops rather than a point-manifold (*cf.* Belot [2003b], p. 204). They are therefore not localized to manifold points as the magnetic field and vector potential are. This does not make the holonomy interpretation non-separable (*cf.* [Healey, 2004], pp. 646–7—and see below), for that requires that a physical process in a certain region of spacetime is not supervenient on assignment of qualitative intrinsic properties at the spacetime points in that region; if spacetime points make a showing at all, it is as elements of a loop that wraps around the solenoid. The physical processes are not grounded in spacetime regions in the first place. Hence, strictly speaking, *pace* Healey ([1997]—and see below), there is neither non-locality (*qua* action-at-a-distance) nor non-separability in the holonomy interpretation. Redhead's otherwise superb account of the interpretation of gauge symmetry is marred by his conflation of these two forms of non-locality (see Redhead [2003], pp. 132 and 138).[122]

Redhead ([2003], p. 138) suggests that in adopting a gauge invariant account using holonomies, one thereby waves goodbye to the principle of gauge invariance, thus tarring such an account with the same brush as I tarred the selective interpretation with. His reasoning is that in shifting attention to holonomies one eschews non-gauge invariant quantities, such as the gauge potentials. But gauge transformations are only defined on non-gauge invariant quantities, so gauge invariance cannot be accommodated. Some might think that this is a good thing since such quantities are indeterministic. But even if one doesn't, Redhead's claim does not hold water. Firstly, there are reductive accounts that can incorporate the principle of gauge invariance in the very structure of the phase space. Secondly, if we don't wish to reduce, then we can still adopt a direct interpretation, but modified so that only gauge invariant quantities are measured.

There is a sizeable philosophical literature on the interpretation of the Aharonov–Bohm effect. It is fair to say that no consensus has been reached. This work is largely tangential to my brief though, which has been to examine the general subject of gauge freedom and symmetries and their relation to metaphysical issues. However, I briefly mention some of the most important proposals. Richard

[122] The same conflation is made in the accounts of Nounou ([2003], p. 178), Lyre [2001] (corrected in his [2004]), and Healey [1997].

Healey [1999] argues for a view whereby the effect shows, by analogy with quantum non-locality, that there is non-separability in the world; he uses this to defend *holism*.[123] As I noted above, the non-locality of the holonomies concerns the fact that they are not localized at spacetime points, rather than any violation of local action or separability. Healey defines these two notions as follows:

> *Local Action*
> If A and B are spatially distant things, then an external influence on A has no immediate effect on B. ([Healey, 1997], p. 23)

> *Separability*
> Any physical process occurring in spacetime region R is supervenient upon an assignment of qualitative intrinsic physical properties at spacetime points in R. ([Healey, 1997], p. 24)

Healey claims that "no interpretation gives a completely local account of the effect" ([1997], p. 32), some violate the local action principle, some violate the principle of separability, and others violate both. Now, I agree with Healey that a completely local account cannot be given, if by that is meant an account in terms of interactions at spacetime points, or in infinitesimally close regions of spacetime points. But this is not what Healey means; his idea of locality is exhausted in the two principles above. However, if we view the holonomies as defined with respect to a space of loops, *not* spacetime points, then we find that the account of the AB-effect given in terms of them will trivially satisfy both of Healey's conditions for locality, for the physical processes do not happen in spacetime regions and they are local in loop space. Indeed, I think this is the right way to view the holonomies: spacetime points do not enter into their description; this is precisely why they are so useful in general relativity, where we have the freedom to arbitrarily permute points of the manifold. But the account *is* non-local in the additional sense I mentioned above. Holger Lyre formulates the principle of separability slightly differently to Healey, as follows:

> *Lyre-Separability*
> Given any physical system S and its exhaustive, disjoint decomposition into spatiotemporally divided subsystems, it is possible to retrieve the properties of S from the properties of these subsystems. ([Lyre, 2004], p. 608)

This has the quite definite advantage of avoiding the issue of supervenience (though it isn't clear that the notion of 'property retrieval' is any less of a problem), and of assuming an initial region of spacetime in which the physical occurs. Again, the notion of *spatiotemporally* decomposing a holonomy defined in loop space does not make much sense, so that Lyre's definition is not applicable. However, what *is* violated is something very close to both Healey's and Lyre's notions of separability, though more general; namely, Lewis's thesis of *Humean Supervenience* (*cf., e.g.*, [1986b], pp. IX–X). This is the thesis that *all there is* supervenes on the arrangement of intrinsic properties localized to spacetime points. Thus, the entailment

[123] The type of holism that results is close in many respects to Teller's "relational holism", argued for in the context of the Bell inequalities (see Teller [1989]).

would be that the magnetic interaction, and the Aharonov–Bohm effect, supervenes on some local matters of fact, on the distribution of intrinsic properties over a set of spacetime points. It seems that even an action-at-a-distance account that restores separability will violate Humean supervenience, since the causal connection doesn't admit a reduction of the kind that is required by that thesis; we seem to have something over and above a 'mosaic' of properties attached to points of spacetime. Healey's notion of separability is most commonly associated to many systems bearing relations to one another, and the issue is whether or not the whole (the composite system) is greater than the sum of its parts. This is put to the test by quantum entanglement since that implies that the joint state of the composite system is what determines the states of the components, rather than the other way around. This naturally suggests a holism, and that is where Teller's notion of relational holism finds a home, since it denies the thesis he calls "particularism", the idea that individuals are completely characterized by the intrinsic properties, such that any relations that there are supervene on the intrinsic properties of the relata (see Teller [1989], p. 213). But this is not anything like what we have with holonomies; they are simply *singular* beasts, not composites. For this reason, Lyre's further notion of holism does not apply either. It is as follows:

> *Lyre-Holism*
> Given any physical system S and its exhaustive, disjoint decomposition into subsystems, it is impossible to retrieve the properties of S from the properties of these subsystems. ([Lyre, 2004], p. 609)

This definition evades both supervenience (though with the same proviso as for Lyre-Separability) and any explicit dependence on spacetime points or regions. But the holonomies do not class as Lyre-holistic either, for the simple reason that it does not make sense to speak of decomposing a single holonomy into subsystems; it has no subsystems. However, Lyre is right in saying that "[t]raditional ontology, which thinks of objects as being spatiotemporally fixed [localized], is clearly undermined" ([Lyre, 2004], p. 620); we have something wholly more abstract than magnetic fields or vector potentials with holonomies, yet I think that the interpretive benefits outweigh their *prima facie* ontologically puzzling nature.

Maudlin [1998] argues that a local (in both mine and Healey's senses) and separable interpretation can be given if we take the vector potential to be gauge fixed, described by "one true gauge".[124] It is doubtful that Maudlin wishes to take this view seriously, he mentions it as a counterexample to Healey's claim that nonlocality afflicts any approach to the AB-effect. Leeds [1999] adopts a view similar to Aharonov and Bohm, such that the effect is understood as an interaction between matter and a gauge potential. Such a view has, I think, been superseded by holonomy interpretations.[125] If one is willing to give up on the idea that the ontology has to involve localization to spacetime points, then such an account gives

[124] As evidence for my claim that Butterfield's counterpart theoretic substantivalist should be seen as adopting a selective interpretation, note that an objection of Martin's against Maudlin's account exactly parallels one given against Butterfield by Norton. Martin argues that "nothing in the physics can reveal this one true gauge" ([2003], p. 49); Norton argues that there is no way to distinguish the one true model that represents a general relativistic spacetime from the "imposters" ([1988], p. 63).

[125] See Belot ([1998]; also §7 of Belot [2003b]) for the best account and defense of such a view. Belot claims that the holonomies of A also "provide a good set of coordinates" ([2003b], p. 204) for the reduced phase space of Maxwell's the-

a local (in Healey's sense) and deterministic account of the effect. This view also sits best with the claim that observables should be indifferent to isomorphic states: holonomies, you will recall, contain all of the gauge invariant information of A.

It is somewhat curious that the interpretive issues surrounding the substantivalism/relationalism debate are not mentioned in connection with gauge theories.[126] True, the translations of the Leibniz-shift argument are not gauge in the strict sense, but the philosophical issues are pretty much of a kind nonetheless. The PII wielding relationalist response to the translation invariance by modding out by the action of the group of translations would seem to offer a similar move to the realist about gauge potentials. The obvious move is to regard those states with vector potentials related by a gradient as the same physical state. The physical ontology is then given once one mods out by the action of the gauge group. The (Newtonian) substantivalist position would then correspond to treating the states as distinct though indistinguishable. (I have already mentioned that the idea of 'one true gauge' corresponds to Butterfield's idea that only one model out of a gauge equivalence class represents a physical possibility for the system.)

Gauge or no gauge, the central issue is a representational one: how should we understand the relation between a piece of mathematical formalism and physical reality? Gauge theory makes this issue more urgent, but it is really a feature of any theory with non-trivial symmetries. In gauge theory, too, the interpretive options are more closely allied to particular formal choices (reducing, gauge fixing, *etc.*). This is how the connection to general relativity and the hole argument becomes apparent, for general relativity is a spacetime theory *with gauge freedom*. We then find that the substantivalism/relationalism is *explicitly* connected to the interpretive moves taken with respect to gauge freedom in exactly the way I suggested above.[127] For this reason, I reserve discussion of the relationship between these interpretive moves and issues in the metaphysics of identity and modality until Chapters 4 and 5.

3.4. WHY GAUGE?

Gauge theories clearly constitute a considerable advance in out understanding of the world. But it is not exactly clear what role the gauge freedom of such theories plays. Consider the following passage form a well known textbook on gauge theories:

> [Gauge theories] are theories in which the physical system being dealt with is described by more variables that there are physically independent degrees of freedom. The physically meaningful degrees of freedom then

ory. I agree that they can function as such, but note that the reduced phase space is not a necessary part of the holonomy interpretation. I prefer a view that retains all of the gauge freedom in the formalism, but remains neutral about its ontological status.

126 A notable exception is Belot's masterly survey of the relation between symmetry and gauge freedom: [Belot, 2003b].

127 Even this connection has only recently come to be appreciated by philosophers; in most presentations, the gauge theoretic aspects of the hole argument were either not noticed at all, or merely hinted at. Not until Belot & Earman [1999] did this situation change; largely, it has to be said, as a result of the connections noticed between the hole argument and the problem of time in quantum gravity.

> reemerge as being those invariant under a transformation connecting the variables (gauge transformation). Thus, one introduces extra variables to make the description more transparent, and brings in at the same time a symmetry to extract the physically relevant content. ([Henneaux and Teitelboim, 1992], p. 1)

We have seen this idea in action in the context of Maxwell's theory, quantum statistical mechanics, and Newtonian mechanics. The claim here is that the introduction of gauge freedom is vindicated by an increase in *transparency*. They don't further explain the nature of this transparency, but I expect that they are referring to the fact that gauge theories with gauge freedom can make manifest and explicit various properties of physical systems to do with symmetries: invariance, covariance, conservation laws, *etc*. One can think of other more philosophical reasons too, connected with the previous ones. The idea is that the excess degrees of freedom allow one to understand certain modal claims about the behaviour of physical systems. These will be like the counterfactuals of the Leibniz–Clarke debate, such as 'if we move this system of matter a distance of 5 feet to the West its qualitative properties will be the same'. In formalisms without the excess this doesn't seem to be possible. Hence, the transparency is both technical and philosophical, and the two types are intimately related.

Clearly though, as interpreters of physics, we are faced with the question of how we are to understand the gauge freedom. However useful the apparatus of gauge theory, it does not resolve the problem of where the surplus structure comes from, and how we are to interpret it. At best, it gives us a more precise medium for exposing and investigating surplus structure and symmetries.

Here is how Martin describes the same problem:

> The received way of characterizing the domain of gauge symmetries is that gauge symmetry concerns the covariance of the fundamental equations of motion for specific interactions, and that the covariance is tied to a certain descriptive freedom related to the presence of non-physical and, therefore, redundant or 'surplus' quantities in the theory. The basic idea is that in describing the physics we introduce too much, and the symmetry under the covariance group effectively rids the theory of the non-physical excess. ([Martin, 2003], p. 49)

Now, as I have argued already, the symmetries do not imply reduction. We can reduce, and get rid of the "non-physical excess", or we can deny that it really is excess and either accept the indeterminism or retain the symmetry as effectively imposing a version of PSR such that non-observable differences are not counted as physically relevant (that is, as far as 'the physics' goes). The idea is that, as far as the non-gauge degrees of freedom are concerned, it doesn't matter *which* gauge potential plays the role, providing that the gauge potentials are gauge-equivalent.[128] Hence, there is the possibility of retaining all of the gauge freedom, giving it a direct interpretation, and nonetheless having an account of the physical structure that is not interfered with by the gauge freedom. I suggested above that one might

[128] In Chapters 8 and 9 I shall return to the issue, where I defend the claim that such symmetries fit a structural ontology.

even consider this structure as representing possibilities (or potential states) for the physical structure.

Before I leave this topic, I should point out that what I had to say in the previous chapters, as regards inflation and deflation, passes over into the present context too. The gauge symmetry group generates an inflated possibility set, and direct approaches face the problems concerning commitment to indistinguishable worlds: haecceitism, indeterminism, *etc.* I argued that supposedly direct approaches like substantivalism and individualistic packages are not so committed. The point was that such interpretations are not necessarily direct: one can choose a selective interpretation; a (non-reductive) indirect interpretation, and thereby deflate the inflated possibility set; or, a direct interpretation modified so that the observables do not constitute an inflated possibility set. It is true, though, that Leibniz's version of relationalism (a reductive interpretation) *is* deflationary: PII implies that the whole orbit of gauge-related points itself represents a single possibility. But accepting PII in this sense is not a demand enforced by relationalism—a point stressed by Saunders [2003b].

CHAPTER 4

Spacetime in General Relativity

The hole argument is an attempt to show that a manifold substantivalist conception of the spacetime implies that general relativity is indeterministic. This argument is now a standard piece of equipment in the philosopher of physics' toolbox. For this reason, a detailed analysis might seem like a bit of a waste of ink. However, although there are indeed many excellent presentations and analyses of this argument,[130] there is, I think, still much more to be said, both about the argument itself, about the various responses to it that have been suggested in its aftermath, and about its connections to the issues presented in the previous chapters (especially the connection to gauge theory). Most importantly, however, the connections between this argument and the problems of time and change that crop up in canonical quantum gravity remain to be adequately explicated.[131] I postpone this task until Chapter 7. The purpose of the present chapter is to get clear on exactly what allows the hole argument to get its grip, how the various responses work, and how both the argument and the responses relate to the issues of previous chapters.

My response matches up with what I had to say in those chapters: the indeterminism of the hole argument highlights the fact that general relativity is 'indifferent' to which points of spacetime the metric field is spread out over. This does *not* imply that spacetime points do not exist, nor that they do; nor does it rule out haecceitism, nor does it imply it. It does not mandate relationalism nor does it mandate substantivalism. In fact, the indeterminism of general relativity is not relevant to these conceptions of spacetime. I argue that a structuralist conception works best—though, again, general relativity does not mandate it, part of my reason for opting for structuralism is because many *prima facie* incompatible positions can be made compatible with general relativity (philosophically: with the appropriate tweaks to modality and identity; or technically: with the appropriate tweaks in one's conception of observables).

The plan of this chapter is as follows. In §4.1 I begin by introducing some background material that is essential in what follows. This will involve a discussion of manifold substantivalism, and the relation between the models of general relativity and physically possible worlds. In §4.3 I then go on to introduce the hole argument in the language of gauge theory, and catalogue what the various inter-

[130] For 'six of the best' try Earman & Norton [1987] (the original presentation), Norton [1988], Butterfield [1989], Hoefer [1996], Maudlin [1988], or Stachel [1993]. Einstein's own rendition can be found in [Einstein, 1916].

[131] Though Gordon Belot and John Earman [1999; 2001] have done much to remedy this unfortunate state of affairs. My debt to this pair of articles should be readily apparent.

pretive options for such theories look like in the context of general relativity. In Chapter 5 I then give a taxonomy of responses, split into three classes: (1) determinist; (2) modalist; and (3) relationalist. I show that each response can be seen more or less as on all fours with some method for dealing with gauge freedom. I find fault with each response and then in Chapter 6 I provide a response that is in line with the general account of symmetries that I have been suggesting in the previous chapters.

The target of the hole argument is a certain form of substantivalism known as "manifold substantivalism". The idea is to show that such a conception of spacetime leads to an indeterminism concerning how the dynamical fields are spread over the points. This indeterminism is taken to flow from a direct interpretation of the representation relation between the models of general relativity and physically possible worlds; this is seen by Earman and Norton as concomitant with manifold substantivalism. In the next section I examine Earman and Norton's reasons for believing that manifold substantivalism is the best form of substantivalism available, and then, in §4.2, show how this view connects with the understanding of the relation between models and worlds. I argue both that (1) manifold substantivalism isn't the most defensible form of substantivalism; and (2) that substantivalism is not committed to a one-to-one interpretation of the representation relation.

4.1. MANIFOLD SUBSTANTIVALISM

It is the manifold that Earman and Norton claim is the best candidate for what represents spacetime for a substantivalist. Hence, it is the $\mathcal{M}$ component of the spacetime models that we discussed in §1.1 that the substantivalist should be committed to. This commitment to the manifold amounts to a realism about the points of spacetime along with their topological and differential properties. Note that this part of the model classes as a background structure in that it is fixed across the physically admissible models of the theory (though general relativity can be formulated on different manifolds)—this matches the definition I gave in §1.1).

Why do they choose this structure as being the correct representation of spacetime? They claim that such a view is naturally extractable from the local formulation of spacetime theories:

> We take all the geometric structure, such as the metric and derivative operator, as fields determined by partial differential equations. Thus we look upon the bare manifold—the 'container' of these fields—as spacetime. ([Earman and Norton, 1987], pp. 518–9)

This is most evident, say Earman and Norton, in the context of general relativity. They back this up by pointing out that the metric in general relativity "now incorporates the gravitational field"; "carries energy and momentum[132]", and is

[132] Compare this to the following remark of Feynman's about the status of the electromagnetic field: "The fact that the electro-magnetic field can possess momentum and energy makes that field very real..." ([1962], Vol. 1, Ch. 10, 9). Hence, the carrying of energy and momentum are taken to signify a robust form of reality of the carrier.

such that "a gravitational wave propagating though space [could have] its energy ... collected and converted into other types of energy, such as heat or light, or even massive particles".[133] They say: "If we do not classify such energy bearing structures as the wave as contained within spacetime, then we do not see how we can consistently divide between container and contained" ([Earman and Norton, 1987], p. 519). So presumably, Earman and Norton see the most defensible form of spacetime substantivalism as involving an entity without the usual spatiotemporal properties: the manifold merely has dimension and a notion of betweenness for points. The distance relations, and spatial and temporal relations come from a material field *in* spacetime!

Robert Rynasiewicz [1996] essentially says that we *can't* consistently (unambiguously) divide between container and contained in this way in the context of general relativity as we can in the context of pre-GR (non-field) theories. For this reason he claims that the substantivalist/relationalist debate is "outmoded". I remarked in §2.1.1 that a crucial feature of the relationalist/substantivalist debate was the availability of a clear-cut distinction between 'matter' and 'space(time)'. This was required in order that the fundaments of the debate, concerning the relative ontological priority of matter and space(time), have a clear expression. For space, for a substantivalist, is defined to be that which functions as an independent *container* for everything else; whereas for a relationalist space is defined by matter: the distinction between container and contained is collapsed. Hence Rynasiewicz's rejection of the debate in the context of general relativity.[134]

Clearly, the substantivalism on offer in the context of general relativity is a very different beast to the ones considered earlier. Substantivalism about Newtonian, neo-Newtonian, and Minkowskian spacetimes incorporated the metric structure; they could do this because it was a background structure, dynamically decoupled from matter-energy. But, as I just mentioned, the only available background structure in general relativity is the manifold. In their discussion, Earman and Norton do in fact consider the metric as representing spacetime, only to reject it on the grounds that:

> classifying the metric as part of the container spacetime leads to trivialisation of the substantivalist view in unified field theories of the type developed by Einstein, in which all matter is represented by a generalised metric tensor. For there would no longer be anything contained in spacetime, so that the substantivalist view would in essence just assert the independent existence of the entire universe. ([Earman and Norton, 1987], p. 519)

In any case, according to Rynasiewicz, the metric field in general relativity does not definitively admit an interpretation as contents or container; neither substantival structure nor relative structure. Hoefer [1998] disagrees, and argues that at least one side of the debate, substantivalism, can be given a clear definition and,

[133] This belief is amplified by Kuchař, who writes: "[t]he ripples of gravitational radiation can travel around, interfere, attract each other, and amplify. They can hold themselves together in a gravitational geon. Part of the gravitational radiation can leak out, part of it may collapse and form a black hole" ([1993], p. 4).

[134] Rynasiewicz also extends his thesis to the classical theory of electromagnetism. I do not discuss this aspect of his argument here; for a nice discussion see Hoefer ([1998], pp. 454–7).

given that relationalism is just the denial of substantivalism, the full debate can be formulated: "for better or worse the debate still goes on". Hoefer disregards the importance of the 'container/contained' distinction, and instead focuses the dispute on the ontological status accorded to the metric field. I agree with Hoefer that the debate can be formulated as he suggests, but the debate also admits a formulation in terms of the manifold, as Earman and Norton suggest. Which is correct? I have to agree with Hoefer, for the reason that the metric carries all the significant aspects concerning spacetime. The manifold only has topological and differential structure, and, as I intimated above, these are hardly sufficient to ground spatiotemporal properties and relations. It simply isn't equipped with enough structure to represent spacetime (*cf.* [Maudlin, 1988], p. 87).

Many, however, have thought that the dynamical nature of the metric field in general relativity, coupled with its role as determiner of chronogeometrical properties and relations pushes strongly towards a relationalist conception. Hence, Rynasiewicz's point can simply be reinstated at this level: is the metric to be conceived of as spacetime or matter? The reasons for choosing the latter option are to do with the features mentioned by Earman and Norton above, that the metric is dynamical, has energy, and so on.[135] This relationalist understanding of the metric field certainly seems to be the prevalent one amongst physicists (*cf.* [Rovelli, 1997] and [Smolin, 2006], for example), but the view is that the metric field also represents spacetime. However, I agree here with Howard Stein on the falsity of the view that the metric field's having these properties leads automatically to relationalism if we view it as representing spacetime. Let me quote him at length:

> It is often claimed that the general theory of relativity has demonstrated the correctness of Leibniz's view. This is a drastic oversimplification. It is no more true in the general theory than it in Newtonian dynamics that the geometry of space–time is determined by *relations among bodies*. If the general theory does in a sense conform better to Leibniz's views than classical mechanics does, this is not because it relegates "space" to the ideal status ascribed to it by Leibniz, but rather because the space—or rather the space–time structure—that Newton requires to be real, appears in the general theory with attributes that might allow Leibniz to accept it as real. The general theory does not deny the *existence* of something that corresponds to Newton's "immobile being"; but it denies the rigid *immobility* of this "being," and represents it as interacting with the other constituents of physical reality. ([Stein, 1967], p. 271)

Hence, Newton was committed to the existence of absolute substantival space and time; but the absoluteness was not a necessary part of substantivalism about spacetime: they can be safely separated. As long as one believes that whatever is

[135] I should point out that the claim that *substantive* energy is carried by gravitational waves has been questioned by Hoefer ([1996], p. 13). His argument involves the fact that this energy must be represented by a pseudo-tensor, and is not unique (essentially because there is no canonical choice of time). To a large extent, the issues of the hole argument and, even more so, the problem of time are relevant to this problem (and it is relevant to them): how one deals with the energy content gravitational waves will depend on ones treatment of the problem of time (on which, see Chapter 7).

accorded the status of spacetime in a theory is treated as a robust entity that exists independently of matter, then this counts as a substantivalist interpretation regardless of whether the thing that is spacetime is absolute or dynamical. Clearly though, Stein is talking about something other than manifold substantivalism, for his version includes the metric field in the representation of spacetime. Hence, Newton was not a manifold substantivalist, and the old debate was founded on the fact that the (fixed) metric was an *essential* part of space and time. The importance, and distinctiveness of the metric field filters through into general relativity too, for there is no such thing as a part of space without a gravitational field—*i.e.* the metric field is nowhere vanishing.

The fact that one and the same object is being utilised to defend opposing position strongly suggests, however, that Rynasiewicz's objection is not without support. As I show in the chapter dealing with the responses to the hole argument, we find that explicit defenses of relationalism and substantivalism in fact look almost identical! This has lead some to seek a *tertium quid* between these opposing positions. For example, Mauro Dorato [2000] has outlined a position he calls "structural spacetime realism", according to which the relational nature of spacetime is accepted, but that this structure exists *independently* of physical objects and events. However, the position makes the same moves as the so-called "sophisticated substantivalists" who wish to endorse Leibniz equivalence. Moreover, it strikes me that Dorato has simply given another name to Stachel's relationalist position—see §5.3.2. We can view it as relationalist or substantivalist depending on how we view the metric field, and we are thus no further on. But I do agree with Dorato that the terms 'matter' and 'spacetime' are problematic as far as the metric field is concerned; it doesn't sit comfortably under either banner. Better, I say, is a Steinian structuralism according to which we simply be realist about the structure that is exemplified.[136] If one wants to point out the fact that the metric field is an entity that can exist independently of other fields, and so is substantival, then one will always face Rynasiewicz's problem: is it 'material-like' or is it 'space-like'? One can always find a way to defend a relationalist position or a substantivalist position from such a spot. Better to wipe the spot away. This is, of course, the underdetermination of metaphysics by physics again. Previously, I intimated that we should be structuralists about gauge theories and classical spacetime because the interpretation of the symmetries always allowed us to set up reasonable opposing views. The same point applies here too.

Let us put these complications aside until the final chapter, where I fully defend my position. For now let us grit our teeth and agree with Earman and Norton that the manifold is the best object to represent spacetime for the substantivalist, that it functions as a container for the other fields (including the metric field), and that a manifold substantivalist will be committed to the view that points of this manifold represent physical spacetime points. We still need to say something about the relation between the models of general relativity and worlds.

[136] This is more or less in keeping with the general account of structures admitting gauge-type symmetries that I have been pushing for the past two chapters. I consider Dorato's proposal, and defend a different version of structuralism in Chapter 8.

4.2. MODELS AND WORLDS

The model-theoretic formalism makes the distinctions between the various theories of spacetime particularly easy to define and comprehend. It therefore makes a very useful tool for philosophers of physics interested in spacetime. The basic idea is to split up the elements of a theory of spacetime into various classes of entity: at the most fundamental level we have the spacetime $\mathcal{M}$ represented by a differentiable manifold.[137] Then over $\mathcal{M}$ there are a number of geometric-object fields, relating the points of $\mathcal{M}$ in various ways, and assigning properties to them. The geometric-object fields come in two types: 'background' $\mathcal{B}_i$ and 'dynamical' $\mathcal{D}_j$. The former are taken to characterize the 'fixed' structure of spacetime, and the latter are taken to characterize the physical contents of spacetime. These object fields define a set of relations on $\mathcal{M}$. Hence, a complete spacetime theory is characterized by a model of the form: $\mathrm{M} = \langle \mathcal{M}, \mathcal{B}_i, \mathcal{D}_j \rangle$. In order to represent a dynamically possible spacetime of a particular theory T, M must satisfy the equations (laws) of T.[138] This latter connection highlights the view that the set of models of a theory represent the possibility space of a theory (the kinematically possible histories), and the laws select a subset of this space comprising the *physically* possible worlds (or 'T-worlds') of that theory (the dynamically possible histories).[139] Hence, a model of a spacetime theory would represent a world with a certain spatiotemporal structure as given by the $\mathcal{B}_i$-fields, with $\mathcal{D}_j$-fields distributed over the domain subject to the theory's laws that serve to relate the fields. The interesting fact about general relativity is that its models contain only $\mathcal{D}_j$-fields, so that the spatiotemporal relations over $\mathcal{M}$ are always given dynamically, by solving the equations of motion. This is what is meant by background independence: the equations of motion are of the form $\mathfrak{F}[\mathcal{D}_i] = 0$.

Earman and Norton couch their argument in terms of models, where the models are intended to represent the physically possible worlds of the theory, in the manner I suggested above. Certainly, the responses see it like this, and most physicists seem to view the models in this fashion. How is this association to be understood? It is simply an interpretation of the model: a specification of the possible worlds of the theory; the players (domain) and their roles (properties and relations), and the lawlike constraints they are subject to (equations of motion). So, for a model of general relativity $\langle \mathcal{M}, g, T \rangle$, the domain $\mathcal{M}$ is taken to represent the spacetime points (along with absolute properties and relations determined by the topological and differential structure of the manifold) that exist at a world, and g and T defined over $\mathcal{M}$ are taken to represent the (dynamical) properties and relations that the points possess. Earman and Norton see manifold substantivalism as

137 We can of course factor this structure down into finer substructures, such as a topological space and a continuum of points. However, such 'low-level' structures are too elementary to use to distinguish spacetime theories: the manifold is the highest level of structure that all spacetime theories will agree upon. This might alter, and it might become necessary to construct a theory that is independent of differential and topological structure too.

138 These equations will, in general, be of the form $\mathfrak{F}[\mathcal{D}_i, \mathcal{B}_j] = 0$.

139 Note also the connection to the definition of structure that I gave in §1.2. In this context, the structure is a domain of spacetime points, and the relations are geometrical ones. The notion of symmetry given there functions well in this context: the symmetries of a theory are those transformations that preserve geometrical relations. For such transformations, the dynamically possible histories are left invariant, and the theory is said to be invariant with respect to these transformations.

implying that the metric field (represented by g) and the stress-energy field (represented by T) are fields contained within spacetime (represented by $\mathcal{M}$), and that the points of spacetime have their identities fixed independently of these fields.

As ever, for our purposes, the crucial question concerns the nature of the relation between models and worlds. How do the models represent worlds: one-to-one, many-to-one, selectively, not at all, or in some other way? Do we take the elements in the domain of the model to denote the same thing as the elements in the domain of another model? In other words, do the domains $\mathcal{M}_i$ of distinct models contain the *same* points, or are they themselves distinct? If the former is the case then transworld identity can be represented as one and the same individual appearing in different domains in different models (*cf.* [Melia, 1999], p. 641). But in virtue of what are the points the same? The manifold substantivalist cannot make use of the g and T fields to individuate them, because the points are taken to have their identities fixed independently of these. Yet the points have no distinguishing characteristics independently of g and T, so it would seem that some non-qualitative property must be required to do the job. This line of reasoning leads to the view that manifold substantivalism is committed to haecceities or primitive thisnesses to individuate the points. Earman and Norton trade on this notion to generate the indeterminism, for if the points have their identities fixed independently of the various dynamical fields, then any redistribution of those fields will result in a distinct possibility, even though the possibilities differ purely haecceitistically.

From what I had to say in §2.4 the flaws in this reasoning should be readily apparent. Even if we assume that the manifold substantivalist's position does require haecceities, this does not imply commitment to haecceitism (and *vice versa*). One can rule out the former implication by denying transworld identities, so that haecceities serve to individuate points within a world only. The latter implication can be ruled out by using 'this worldly' counterparts to ground possibilities.[140] Hence, just as was the case in the Leibniz-shift argument, it is haecceitism that causes the problems *not* substantivalism: these theses can be safely detached from one another.

Let me express what is being suggested by Earman and Norton using the account of structure and symmetry that I presented in §1.2. A model M corresponds to a structure S; the set of spacetime points $\mathcal{M}$ corresponds to the domain D; and the tensor fields g and T on $\mathcal{M}$ correspond to the set of relations $\mathcal{R}$ defined over D. The spacetime points are the objects of the theory. The elements of the domain are indistinguishable with respect to the relations given by g. Hence, we should brace ourselves for symmetry and indifference: we can permute the points of the domain without affecting the relations. In the context of general relativity this property is called *diffeomorphism invariance*, where a diffeomorphism $\phi : \mathcal{M} \rightarrow \mathcal{N}$ simply corresponds to a smooth permutation of the points. General covariance is then a formal property of models, such that acting on a model by a diffeomorphism generates

140 Thus, for two (same worldly) points $x, x' \in \mathcal{M}$, such that $P(x)$ and $P'(x')$ are complete lists of the properties of the points, we can view x' as a counterpart of x, so that x' represents *de re* of x that it might have had the properties catalogued in P'.

another model—this is a symmetry of the theory; a purely formal requirement.[141] Since diffeomorphisms are gauge-type symmetries, they thus generate an inflated possibility space. General covariance concerns the *relative admissibility* of the elements of this space: if M is a solution of Einstein's field equations (or represents a physically possible world), and $M^* = \phi(M)$, then M^* is a solution of the field equations (or represents a physically possible world) too. The reason is symmetry: M and M^* will be indistinguishable with respect to the relations encoded in g (and T), differing merely with respect to which points play which role in the relations of the respective structures. Since the field equations (determining physical possibilities) concern g, solutions that respect this will represent physical possibilities. Since diffeomorphic models are indistinguishable with respect to g, one will automatically represent a physical possibility if the other does. Notice that neither diffeomorphism invariance nor general covariance commit one to a particular stance with regard to the representation relation between models and worlds.

However, the received view is that particular conceptions of spacetime *do* underwrite particular interpretations of the representation relation. Earman and Norton argue that since the manifold substantivalist is committed to the view that spacetime points are real and have their identities fixed independently of g and T (the relations), then any permutation of the points (or, equivalently, any redistribution of the fields over the points) will result in an ontologically distinct situation: the points will possess different properties and enter into different relations in each diffeomorphic case. This is so, even though the relations themselves are unaffected by the permutations.[142] I think that this shows that Earman and Norton are claiming that for the manifold substantivalist, the identity of the points goes beyond the roles they play in the relational structure of the world. We will see in the responses that this can be denied in a number of different ways. I will deny it too, but not to put a different notion of spacetime point in its place, nor to deny that spacetime points exist; rather, I wish to show that the theory itself does not have the conceptual resources to support any metaphysical conclusion about the ontological status of spacetime one way or the other. Focusing on the relations, and eschewing talk of the natures of objects gives us a firm response to the hole argument. Moreover, it is a response that fits the practice of physicists too (see Chapter 6). However, such a response must be supplemented with an account of symmetries and observables, for the symmetries are still present in the formalism. A firmed up conception of observables can explain how this can be so without leading to problems of the kind outlined in the hole argument.

Clearly there is something to the hole argument; but what it is no different from the similar arguments we have seen in the previous two chapters. It is a general problem concerning gauge-type symmetries. The hole argument doesn't tell us anything about the ontological nature of spacetime, but it does tell us about the conceptual limits of general relativity: the symmetries mean that metaphysical talk

[141] There are two ways to act on a model with a diffeomorphism: (1) by applying ϕ to the points of the manifold; and (2) by dragging along g by ϕ^*. The two actions are equivalent.

[142] A relationalist interpretation is seen as underwriting the view that permutations of points do not result in distinct possibilities, precisely because the relational structures of such diffeomorphs will be isomorphic. I argued in §2.4 this alignment is not a necessary part of the relationalist's position. The same point holds in the present context.

about the objects (configurations, states, possibilities, worlds, or whatever) related by symmetries will be underdetermined by the physics. I will, however, agree that the argument works to this extent: if the substantivalist is committed to the representation relation between models and worlds being one-to-one *and* is committed to there being the same objects in the different models' domains, *and* is committed to the independent existence of these objects, then, without another conception of determinism on the table, she is in trouble: many ontologically distinct futures will be compatible with an initial state and the laws. The problem is, as I mentioned, the substantivalist is committed to none of these conjuncts; so regardless of what conception of determinism we have to work with, the substantivalist is safe. Even if all conjuncts *are* accepted, there are options for taming the indeterminism, as outlined in §3.3.1. I'm getting ahead of myself here, let us now present the argument itself.

4.3. THE HOLE ARGUMENT: THE VIEW FROM GAUGE THEORY

Recall that a model of general relativity is given by a triple $\mathrm{M} = \langle \mathcal{M}, g_{\mu\nu}, T^{\mu\nu} \rangle$—where $\mathcal{M}$, $g_{\mu\nu}$, and $T^{\mu\nu}$ represent the spacetime manifold, the metric field, and the stress-energy field respectively. Such models are taken to represent the physically possible worlds of general relativity when they satisfy the field equations. The hole argument setup demands that $\langle \mathcal{M}, g_{\mu\nu} \rangle$ possesses a Cauchy surface, "so that the environment [is] as friendly as possible for determinism" (Earman [1989], p. 179).[143]

Although it is not often presented in such terms, the hole argument of Earman and Norton is based upon the initial-value formulation of general relativity. This allows one to consider general relativity as describing the evolution in (parameter) time of an initial data set comprising the canonical variables at an initial time. The initial value problem requires that spacetime $\langle \mathcal{M}, g_{\mu\nu} \rangle$ is globally hyperbolic. With this condition satisfied, we are free to foliate $\langle \mathcal{M}, g_{\mu\nu} \rangle$ by Cauchy surfaces Σ_t (parametrized by a global time function t). Viewed in this way, the Lorentzian spacetime metric $g_{\mu\nu}$ on $\langle \mathcal{M}, g_{\mu\nu} \rangle$ induces a (Riemannian) spatial 3-metric q_{ab} on each of the Σ_t, and also induces its determinant $|q|$ and its inverse q^{ab}. One obtains the spacetime 4-geometry by selecting a point on Σ_t, specified by coordinates x^i, and displacing it normally to the slice. The change in proper time $d\tau$ of the point is given by the lapse function N, so that $d\tau = N\,dt$. The spatial coordinates will generally be shifted in such a displacement, giving: $x^i(t+dt) = x^i(t) - N^i\,dt$, where N^i is the shift vector.

This formulation views spacetime as representing the history of a Riemannian metric on a hypersurface. The dynamical variable is the 3-metric on the hypersurface.[144] There are many details I have skipped over here, but we have the essential

[143] A Cauchy surface is a hypersurface that intersects every non-extendible timelike curve once only. This restriction on the models allows one to pull back a global time function $t\colon \mathcal{M} \to \mathbb{R}$, where the level surfaces of t are Cauchy surfaces, and such that t increases along the future direction of a timelike curve.

[144] There are other formulations that begin by writing down general relativity in terms of a connection, in which case the connection on a hypersurface is the dynamical variable. See [Rickles, 2005] for the hole argument translated into these other formulations.

ones required for the hole argument (a full, technical account can be found in Wald [1984], Ch. 10). General relativity should, then, be able to predict the evolution in parameter-time of the spatial metric on a hypersurface. The hole argument says that, according to the manifold substantivalist's conception of spacetime, this is not possible: general relativity cannot determine which future point will underlie a certain field value. In general, *local* field quantities cannot be predicted.

The argument is based upon a conception of determinism whereby agreement about the initial data on some initial hypersurface should suffice to uniquely pick out a single set of data on some future hypersurface. This is violated in the following way. Consider the Cauchy data (Σ, g) on an initial (Cauchy) slice Σ_t $(t = 0)$. To the future of Σ_t $(t > 0)$ define a 'hole' $\mathcal{H} \subset (\mathcal{M} - \{\Sigma_t : t \leqslant 0\})$, such that within $\mathcal{H}$ the gravitational field is non-zero and any non-gravitational fields are set to zero—that is, $\mathcal{H}$ is as empty as we can get in the context of general relativity. Then define a hole diffeomorphism $\phi_{\mathcal{H}}$ so that $\phi^*_{\mathcal{H}}$ acts as the identity on the exterior of the hole (at $x \in (\mathcal{M} - \{\Sigma_t : t \leqslant 0\})$), and smoothly differs from the identity on the hole's boundary and in the interior of the hole (at $x \in \mathcal{H}$)—*i.e.* $\phi_{\mathcal{H}}$ is a diffeomorphism with compact support. Thus, we see that $q(x) = \phi^*_{\mathcal{H}} q(x)$ $(\forall x \in (\mathcal{M} - \{\Sigma_t : t \leqslant 0\}))$ but $q(x) \neq \phi^*_{\mathcal{H}} q(x)$ $(\forall x \in \mathcal{H})$.

The general covariance of Einstein's field equations gives us the following equivalences: (1) $q' = \phi^*_{\mathcal{H}} q$ and (2) $q'(\phi_{\mathcal{H}} x) = q(x)$. The first equivalence simply means that a metric (which solves the fields equations) and its drag-along are both solutions. The second equivalence is a *local* equivalence; it says that the metrics are equivalent when the dragged-along metric is evaluated at the dragged-along point. However, the metrics at the *same* point are not equivalent; *i.e.*, $q'(x) \neq q(x)$. General covariance therefore implies that there exists a pair of solutions that agree up to an initial data slice but diverge thereafter: different evolutions of the metric into $\mathcal{H}$ are compatible with the initial data plus the field equations.[145]

The question is whether or not q and q' are to be regarded as representing physically distinct solutions (or distinct physically possible worlds). Recall that the manifold substantivalist thinks of the points of $\mathcal{M}$ as having their identity and individuality settled *independently* of any fields (representing matter or energy sources) defined with respect to them. Therefore, the solutions are irrelevant to the identity of the points, and we can speak of the *same* point entering relations with both q and q'. The relations are going to be different between these cases since, in general, $q'(x)$ and $q(x)$ represent distinct assignments of geometrical properties to the same point x. It looks as though the manifold substantivalist is going to have to say that the diffeomorphic solutions do indeed represent distinct physical possibilities.

What this implies is that a complete specification of the fields outside of the hole (given by the Cauchy data on a hypersurface) is not sufficient to uniquely determine the evolution of the fields within the hole. Hence, one cannot solve uniquely the Einstein field equations describing the fields within the hole: the evolution into the hole is underdetermined by the field equations. Instead, one has an

[145] Equivalently: we can always find a pair of models $\langle \mathcal{M}, g \rangle$ and $\langle \mathcal{M}, g' \rangle$ (such that $g' = \phi^*_{\mathcal{H}} g$) that agree up to some Σ_t (an initial temporal segment) but diverge thereafter. That $g' = \phi^*_{\mathcal{H}} g$ implies that if either of these models is admissible then so is the other.

infinite class of solutions describing diffeomorphic metric fields, all compatible with the initial data. The statement that diffeomorphically related solutions represent one and the same physical state of affairs is called *Leibniz Equivalence* by Earman and Norton; they say that it is just this equivalence that the manifold substantivalist must reject. It is this rejection that leads to the indeterminism.[146]

How does this manifest itself in a problematic way? Rovelli and Gaul give a nice example demonstrating the underdetermination of the metric field by the field equations:

> Take for example two points $P, Q \in \mathcal{M}$ and consider two metrics $g_{\mu\nu}(x)$ and $\tilde{g}_{\mu\nu}(x)$, which are both solutions of [Einstein's field equations]. Then the distance d between P and Q computed using the two metrics is different, *i.e.*, $d_g(P, Q) \neq d_{\tilde{g}}(P, Q)$. We have two *distinct* metrics on $\mathcal{M}$ which both solve Einstein's equations. ([Gaul and Rovelli, 2000], pp. 303–4)

Though we don't have a complete breakdown of determinism,[147] we do have *some* problem of predictability (concerning which points 'sit under' which bits of the fields, or are assigned which field values), and some problem of indeterminism. The problem is that we cannot *uniquely* determine the evolution of any fields into the hole if we understand the equivalence of class of metrics (under diffeomorphisms) as representing a class of distinct possibilities. This leads to the problem outlined by Rovelli and Gaul. Of course, the problem they mention fades away if we evaluate the dragged-along metric at the dragged-along (image) point. Then the computed distances would be identical. But this supposedly involves a relationalist move, for the metric field (and other fields) are taken to define the point's identities.[148] However, as we've seen, the manifold substantivalist attributes an identity to the points *over and above* the fields, and must consider the evaluation of the metrics at the same point as a possible operation. This seems to be mandated by the active diffeomorphism too, which is precisely the invariance under transformations of the fields over the same points—*i.e.* one and the same point is assigned different properties.

I think that the best way to understand what is going on in the hole argument is along the lines of the underdetermination problem of gauge theories that resulted from gauge freedom (as explained in §3.3). The standard position of physicists as regards the status of general covariance (*e.g.*, as one finds in textbooks on general relativity) is to interpret the general covariance of the field equations as expressing the gauge freedom of general relativity, which is then taken to be a gauge theory. One can find this viewpoint voiced explicitly in the following passage from Wald:

> If a theory describes nature in terms of a spacetime manifold, M, and tensor fields, $T^{(i)}$, defined on the manifold, then if $\Phi : M \rightarrow N$ is a diffeomorphism, the solutions $(M, T^{(i)})$ and $(N, \Phi^{*}T^{(i)})$ have physically identical properties. Any physically meaningful statement about $(M, T^{(i)})$ will hold with equal

146 Both the claim that the substantivalist must deny Leibniz equivalence, and that this denial leads to indeterminism can be rejected, as I show in the responses (Chapter 5).

147 It is not that we have no idea at all how the data will evolve: the initial-value problem is well posed *up to diffeomorphism*.

148 This is certainly how Rovelli sees it, and Stachel too; but Hoefer and others see this as compatible with substantivalism. I discuss their arguments in Chapter 5.

> validity for $(N, \Phi^* T^{(i)})$ ...Thus, the diffeomorphisms comprise the gauge freedom of any theory formulated in terms of tensor fields on a spacetime manifold. In particular, the diffeomorphisms comprise the gauge freedom of general relativity. ([Wald, 1984], p. 438)

Thus, even though we 'move' the tensor fields with the action of the diffeomorphisms (via the drag-along), the tensor fields 'retain' their structure, they are essentially the same. Though they may be formally very different, diffeomorphic models pick out exactly the same physically measurable properties—though the problem of explicating these remains. This is, of course, very much like the indifference situations I presented in the previous chapters, especially the kinematic shift argument from the Leibniz–Clarke correspondence: formal distinctness coupled with physical indistinguishability.

Conceived in this way, the indeterminism issuing from the hole argument is simply a natural consequence of the underdetermination resulting from the gauge freedom of the theory; as we have seen, this is something to be found in *any* gauge theory. Thus, according to this 'gauge theoretic' account, the hole argument's conclusion is simply yet another example of the underdetermination that results from direct interpretations of the representation relation. Clearly, the argument hits manifold substantivalism because that view is seen as underwriting a direct interpretation of the models of general relativity.

Bearing these points in mind, let me now present general relativity as a gauge theory, and show how the hole argument arises as a natural consequence of a direct interpretation. The following account builds on the material that I presented in Chapter 3 and the previous subsection.

First, fix a compact three manifold Σ. Consider globally hyperbolic (vacuum) solutions to the field equations for general relativity, $\langle \mathcal{M}, g \rangle$, such that the Cauchy surfaces of the solutions are diffeomorphic to Σ. Then consider an embedding of Σ in $\langle \mathcal{M}, g \rangle$, given by a diffeomorphism $\phi : \Sigma \to \mathcal{M}$—hence, $\phi(\Sigma)$ is a Cauchy surface of $\langle \mathcal{M}, g \rangle$; call it S. Then g induces a geometry on S characterized by a pair of tensors q_{ab} and K_{ab}: q is a Riemannian metric on S (aka the first fundamental form); K is the extrinsic curvature of S describing how it is embedded in $\langle \mathcal{M}, g \rangle$ (aka the second fundamental form). The map ϕ is used to pull-back q and K from S to Σ. Σ can be embedded in $\langle \mathcal{M}, g \rangle$, provided q and K satisfy the following (constraint) equations:

$$R + (K_a^a)^2 - K^{ab} K_{ab} = 0$$
$$\nabla^a K_{ab} - \nabla_b K_a^a = 0$$

These are called the Gauss and Codazzi constraints respectively—q enters only in the definition of the scalar curvature R and the covariant derivative ∇. The pair (q, K) is taken to represent a dynamical state of the gravitational field just in case it satisfies these two constraints at each point $x \in \Sigma$.

Given this setup, it is natural to choose the space of Riemannian metrics on Σ as the configuration space: $\mathcal{Q} = \text{Riem}(\Sigma)$. Hence, the extended (full) phase space is just $T^*\mathcal{Q}$. A symplectic structure is induced by taking the canonically conjugate momentum to q to be $p^{ab} =_{df} \sqrt{|q|}(K^{ab} - K^c q_c^{ab})$ (where $|q|$ is the determinant of q).

However, the extended phase space will contain points that are dynamically 'inaccessible', corresponding (at best) to models that are not solutions to the field equations. The 'true' (physical) phase space is then the constraint surface $\mathcal{C} \supset T^{*}\mathcal{Q}$ on which the following (first-class) constraints hold:

$$H_{\perp} = \sqrt{|q|}\left(p^{ab}p_{ab} - \frac{1}{2}(p_{a}^{a})^{2} - R\right) = 0 \tag{4.1}$$

$$H_{a} = \nabla_{b}p_{a}^{b} = 0 \tag{4.2}$$

These are now called the scalar (Hamiltonian) and vector (diffeomorphism) constraints respectively—there are infinitely many, since they must hold for all $x \in \Sigma$. Restricting the symplectic form to $\mathcal{C}$ and setting the Hamiltonian to zero (since it is just a sum of the constraints) yields the gauge theoretic formulation of general relativity characterized by a presymplectic geometry. This presentation contains all the information we need to extract the hole argument (and most of the details we need for the problem of frozen dynamics considered in Chapter 7).

The hole argument is generated as follows. Firstly, we note that the constraints generate Hamiltonian vector fields on the constraint surface $\mathcal{C}$, such that the vectors are null with respect to the presymplectic form on $\mathcal{C}$. Now, consider two points (p,q) and (p',q') lying in the same gauge orbit $[(p,q)]$ on $\mathcal{C}$, and such that they can be joined by an integral curve of the *vector* constraint. Then there is a diffeomorphism $\phi : \Sigma \to \Sigma$ such that $\phi^{*}p = p'$ and $\phi^{*}q = q'$, implying that (p,q) and (p',q') agree on the geometrical structure of Σ. They can be seen as disagreeing only with respect to which points of Σ play which roles; *i.e.*, as to the geometrical properties assigned to the points $x \in \Sigma$. Hence, the vector constraint generates gauge transformations that act by permuting the points of a spatial slice, rearranging their geometrical properties. This results in inflation, and if there is some commitment (for whatever reason) to the inflated possibility set then haecceitistic differences will be implicated in ones ontology.

It is obvious that Earman and Norton's manifold substantivalist will be forced into considering the different points on a gauge orbit as representing distinct states of affairs, since the points of the spatial slice have different geometrical properties and have their identities fixed independently of these properties; for example, according to (p,q) it is the point x that has the largest scalar curvature value, whereas according to (p',q') it is the point x'. If this is the case, then it is indeed true that general relativity is indeterministic for it can, at best, determine the geometrical structure of Σ, it cannot determine how this structure is distributed over the points. This conclusion follows from the premise that (p,q) and (p',q') represent distinct physically possible worlds; *i.e.* from a direct interpretation of the formalism. It is, then, no different, in kind, from the Leibniz-shift, permutation symmetry, and electromagnetism scenarios. All four are simply a manifestation symmetry, and, in the case of general relativity and electromagnetism, of gauge freedom.

Diagnosing the indeterminism of the hole argument in terms of gauge freedom leads to a number of possible resolutions of the hole argument. These options are fairly standard moves used when dealing with gauge freedom, and not surprisingly are essentially the same as those given in §3.3. However, it should be remembered that the manifold substantivalist is at liberty to uphold her views in the face of the indeterminism: it is not observable after all! (Belot [1996] has even argued that the surplus structure that results from one-one interpretations can be a positive feature when it comes to dealing with the problem of time.) In this context the methods look like this:

- One might implement Leibniz equivalence directly by 'quotienting out' the diffeomorphism symmetry and moving to the reduced space $\mathcal{Q}_0 = \text{Riem}(\Sigma)/\text{Diff}(\Sigma)$. Points of the reduced space, called "superspace", are then equivalence classes of diffeomorphic metrics on Σ. If the construction procedure given above is then repeated with $\mathcal{Q}_0$ replacing $\mathcal{Q}$ then we eradicate the need for the vector constraint: diffeomorphically related metrics are identified at the first step. This gives us a partially reduced phase space, with the scalar constraint remaining. One then reapplies the same procedure with the scalar constraint, identifying points related by gauge transformations that it generates. In this way a standard Hamiltonian system with no gauge freedom is recovered. This route faces both technical and conceptual problems when it comes to quantizing the theory (discussed in Chapter 7). There is also the interpretive question of what conception of spacetime such a move would underwrite. It is generally supposed that such a move is only available to the relationalist; however, there have been claims that a suitably modified substantivalism can also follow this option. I discuss this briefly in Chapter 6 where I argue that both parties can help themselves to this method and, indeed, that this underdetermination leads into a structuralist conception of spacetime.
- One might 'gauge fix' the theory, as we have seen, essentially amounting to choosing a particular model from an equivalence class of gauge equivalent models. There are ways of doing this in general relativity, but they involve choosing a fixed foliation of spacetime. Once more, we face the question of what conception of spacetime is underwritten by this approach. In the next section I show that certain 'modalist' substantivalist responses to the hole argument achieve something analogical to gauge fixing. Once again, relationalists too can adopt this method.
- We also have the option of a gauge invariant interpretation, according to which the observables of the theory are precisely those functions that commute with all of the constraints. According to this method, each gauge orbit is taken to represent a single physically possible state of affairs. The representation relation between the points of phase space and physically possible worlds is many-one. Clearly, the reduced phase space method would classify as gauge invariant since the observables on such a space would correspond to gauge invariant quantities on the full phase space. Such a move is generally aligned with relationalist conceptions of spacetime; however, again, it has been argued that a modified substantivalism might also be compatible with this move. I will argue that this is indeed the case.

Thus, we have a counterpart for each of our methods for dealing with gauge freedom in the context of the hole argument of general relativity. Many of the responses can be seen as implementing one or another of these methods, although their connection to gauge theory is never made explicit. But the methods are neutral with respect to the ontology of spacetime. This motivates my broadly structuralist account based upon the general account of symmetries sketched in the previous chapters. I make some headway on this position in §6.2.

CHAPTER 5

Responding to the Hole Problem

Earman and Norton interpret their hole argument as showing that a manifold substantivalist conception of spacetime is untenable because it renders general relativity indeterministic. There are many ways that the substantivalist can evade the argument. The bulk of the responses work by denying the following claim: "[S]ubstantivalists, whatever their precise flavour, will deny: *Leibniz equivalence*", namely the view that "[d]iffeomorphic models represent the same physical situation" ([1987], p. 521). There are responses that seek to show that Leibniz equivalence is a live option for substantivalists too. There are two varieties within this class: some simply characterize determinism in such a way that the diffeomorphisms do not result in physically distinct possibilities; others endorse Leibniz equivalence directly, arguing that substantivalists can take the diffeomorphs to be different modes of presentation of the same spacetime. A different class of substantivalist response accepts Earman and Norton's claim, but shows that there is more than one way to deny Leibniz equivalence: one can say that only one or some of the models represent physical possibilities. There are other responses that agree with Earman and Norton, and attempt to construct an alternative conception of spacetime: generally a relationalist or structuralist conception. I take the view that the problem is best eradicated with a gauge invariant stance, using the gauge-theoretical notion of observable such that the local field quantities appearing in the hole argument are not genuine observables at all. This is broadly in line with my indifference view of symmetries and observables, for the observables here will be relational, and the relations in question will not be able to distinguish certain (formally distinct) local field quantities. I steer clear of reduction once again, for nothing in the theory mandates it, and the surplus allows for ontological neutrality. The observables of the theory are to be understood as indifferent to which points play which roles.

Central to the interpretive options is the nature of the representation relation between models and reality (the "formal" and "physical" elements of §1.1). It seems pretty clear that a direct, one-to-one conception of the representation relation in the context of the models of general relativity leads to the kind of indeterminism (or underdetermination) as outlined by Earman and Norton. However, we saw in Chapter 3 that the indeterminism rife in such interpretations can be tamed with an appropriate measurement theory, such that all observables are gauge invariant. It is open to the manifold substantivalist to agree with Earman and Norton that a strange kind of indeterminism arises from their conception of spacetime, but that it is physically inconsequential since the *true* observables are deterministic.

As it stands, though, this is something of a cop out: we still need an account of the symmetries and the gauge freedom.

Below is a catalogue of the various ways in which we might understand general covariance and the nature of the representation relation between models and reality:[149]

- Haec: $\forall \phi^*_{\mathcal{H}}$, $\langle \mathcal{M}, g, T\rangle$ and $\langle \mathcal{M}, \phi^*_{\mathcal{H}} g, \phi^*_{\mathcal{H}} T\rangle$ represent distinct physical possibilities.
- LE: $\forall \phi^*_{\mathcal{H}}$, $\langle \mathcal{M}, g, T\rangle$ and $\langle \mathcal{M}, \phi^*_{\mathcal{H}} g, \phi^*_{\mathcal{H}} T\rangle$ represent one and the same physical possibility.
- One: $\forall \phi^*_{\mathcal{H}}$, if $\langle \mathcal{M}, g, T\rangle$ represents a physical possibility then $\langle \mathcal{M}, \phi^*_{\mathcal{H}} g, \phi^*_{\mathcal{H}} T\rangle$ does not represent a physical possibility.
- Some: If $\langle \mathcal{M}, g, T\rangle$ represents a physical possibility then $\langle \mathcal{M}, \phi^*_{\mathcal{H}} g, \phi^*_{\mathcal{H}} T\rangle$ represents a physical possibility iff $\phi^*_{\mathcal{H}}$ is an isometry for g and T.

The crucial claim of Earman and Norton's argument against manifold substantivalism is that the adherent to such a conception of spacetime is committed to Haec. The hole argument cannot function without such a set of distinct (yet indistinguishable) physical possibilities. They then claim that this commitment leads to a violation of determinism in the context of general relativity. As I mentioned above, both claims come under scrutiny in the responses, and neither claim is watertight: there's more than one way to deny LE, and LE can be endorsed by substantivalists. Note that the advocate of Some will be committed to Haec in the special case where the diffeomorphism is an isometry.[150] Note also that each of these responses fits one of the methods for dealing with gauge freedom that I introduced in the previous chapter and outlined briefly in the previous section. Haec corresponds to a direct interpretation; LE to an indirect interpretation; and One and Some to a selective interpretation.

It is important to note that our survey is not purely academic; the responses one gives to the hole argument are believed to be directly relevant to the interpretive problems of canonical quantum gravity. The general response to be discerned in the physics community is, unsurprisingly, to uphold determinism at the price of dropping substantivalism. They see this as entailing some form of relationalism (this line can be seen most clearly in Rovelli, Smolin, and many others who work on the canonical approach known as loop quantum gravity—see [Rovelli, 2004]). However, my conclusions are largely negative as regards the ontology of spacetime: the hole argument and its responses allow for both substantivalist and relationalist conceptions.

5.1. TROUBLES WITH DETERMINISM

What is Earman and Norton's minimal definition of determinism? Butterfield states it like this: "one physically possible world is singled out by the specifica-

149 These are borrowed and modified versions of those given by Butterfield [1988] and Pooley [in press]—note that Butterfield calls "Haec" "Each"; I prefer Pooley's since it makes the connection to the metaphysics of modality more explicit.
150 This fact forms the basis of one of Norton's criticisms of Maudlin's metrical essentialism, a response that denies LE by endorsing Some. I discuss this objection below.

tion of the physical state on some region of spacetime" ([1988], p. 66). That is, if we were to completely specify the physical data in some part of spacetime, on a slice say, then this specification should fit exactly one physically possible world. In the context of the hole argument, and using the language of models, we might instead couch it in these comparative terms: if a pair of models agree on physical data in some part of spacetime, then they agree everywhere else in spacetime. The hole-diffeomorphic models of general relativity violate this conception: we can make two models agree in a region of spacetime but disagree elsewhere (even if agreement is everywhere but a single point)!

Butterfield outlines two definitions of determinism, **Dm1** and **Dm2** (where **Dm** stands for 'deterministic model').[151] His claim is that while **Dm1** is violated by hole diffeomorphs, **Dm2** isn't. The definitions he gives are, as he points out (though in slightly different terms to those I use here), related to the notions of background dependence and independence, respectively. Only background independent theories—those theories that "treat metric structure on a par with matter fields ... [and] constrain metric fields and connections only by requiring them to satisfy field equations"—are subject to the hole argument. Background dependent theories cannot be subject to the hole argument, for such theories possess a metrical structure that is fixed across models, and yet hole diffeomorphs are such that they have different metrical properties and relations.

The idea is this: background dependent theories do not violate **Dm1**; because of the fixed metrical structure, a diffeomorphism that acts as the identity somewhere on the metric acts as the identity everywhere on the metric—it just *is* the identity. But background independent theories do violate **Dm1** because of the hole argument—identity somewhere does not mean identity everywhere. Background independent theories do not, however, violate **Dm2** (nor, trivially, do background dependent theories). Moreover, Butterfield claims that **Dm2** is more in keeping with physics, since "general relativity texts that discuss the initial value problem [the initial setup for the hole argument] prove a 'uniqueness of solution' result, which suggests that determinism ... holds good" ([1988], p. 67). Indeed, he claims that **Dm2** is extracted from such discussions. Hence, in addition to providing a conception that allows the substantivalist to escape the hole argument, it is also well motivated.

The definitions are based on models $\langle \mathcal{M}, O_i \rangle$ involving a manifold of points $\mathcal{M}$ and a set of geometric objects O_i (vectors, tensors, *etc.*). The setup is grounded in the notion of a "region" R (to be understood as the hole) which admits a slicing S (to be understood as the 'form' of the hole), such that "agreement on regions of a certain kind (typically sandwiches or slices [S]) implies agreement elsewhere" ([1988], p. 67). This matches up to the modal claim that worlds that agree (in a certain way) in a region agree (same way) everywhere else if they are deterministic. The sticking point, as we shall see, is in saying what is meant by "agreement".

Butterfield's interpretation of Earman and Norton's determinism goes like this:

> **Dm1** A theory with models $\langle M, O_i \rangle$ is S-deterministic, where S is a kind of region that occurs in manifolds of the kind occurring in the models, iff:

[151] Butterfield also considers two further possible definitions, **Dm** and **Dm0**, too strong and inappropriate respectively. I don't discuss these.

given any two models $\langle M, O_i \rangle$, $\langle M', O'_i \rangle$, and any diffeomorphism d from M onto M' that drags any absolute object among the O_i to the corresponding absolute object among the O'_i throughout M': and given any region R of M, of kind S:-
if $d(R)$ is of kind S, and also for all $i, d^*(O_i) = O'_i$ on $d(R)$, then:
for all $i, d^*(O_i) = O'_i$ throughout M'. ([Butterfield, 1988], p. 70)

The basic idea of the definition is this: agreement (of models) on a region of a manifold (given by a slice) with respect to the fields on the manifold implies agreement of the fields over the rest of the manifold. This definition implies that background dependent theories are deterministic since the metric and connection structure is fixed across models for such theories: background structures are absolute objects, and are, by definition, the same across all models. However, for background independent theories, **Dm1** is violated by any two hole diffeomorphs. The proof uses the identity map id_M on M. Take two models $\langle M, O_i \rangle$ and $\langle M, d^* O_i \rangle$, where d^* is a hole diffeomorphism that acts as the identity on R (R can be arbitrarily large or small). id_M is then a diffeomorphism between the models such that $\mathrm{id}^*_M(O_i) = O_i$ living on R. However, $\mathrm{id}^*_M(O_i) \neq O'_i$ within the hole, $M - R$. The fact that there are no background structures (of the relevant kind: metrics and connections) in the theories under consideration, of which general relativity is an example, means that **Dm1** is violated for id_M.

Butterfield goes on to provide a weaker definition that is not violated by *any* pair of hole diffeomorphs:

Dm2 A theory with models $\langle M, O_i \rangle$ is S-deterministic, where S is a kind of region that occurs in manifolds of the kind occurring in the models, if:
given any two models $\langle M, O_i \rangle$, $\langle M', O'_i \rangle$ containing regions R, R' of kind S respectively, and any diffeomorphism d whose domain of definition includes R and which maps R onto R':-
if $d^*(O_i) = O'_i$ on $d(R)$, then:
there is an isomorphism f from M onto M' that sends R to R',
i.e., $f^*(O_i) = O'_i$ throughout M' and $f(R) = R'$. ([Butterfield, 1988], p. 71)

Butterfield claims that "**Dm2** is the general definition of determinism implicit in modern presentations of general relativity's initial value problem" ([1988], *ibid.*; see also, Butterfield [1989]). The crucial differences between this and Earman and Norton's definition concern the generality of the diffeomorphisms allowed (global and local), and the removal of background structures from the definition. I agree that Butterfield's **Dm2** is indeed a more appropriate definition for general relativity (and background independent theories in general). Taken on its own, however, it does not resolve the hole problem for manifold substantivalism, as I shall now explain.

Butterfield's claim is that **Dm2** is well motivated because it is in line with the initial-value problem of general relativity. So it is. However, in such a context the gauge freedom is often explicitly violated, and a gauge choice must be made (*cf.* Wald [1984], §10.2). Different gauge choices yield different solutions related by a gauge transformation. Hence, the "uniqueness" of which Butterfield speaks is uniqueness 'up to a gauge transformation' or uniqueness 'given a gauge condition'. But how does a definition of determinism that matches this lead us out of the problem of the hole argument? The various gauge choices would represent different physically possible worlds, and according to Earman and Norton, the substantivalist is committed to their existence. Hence, unless Butterfield has some well-grounded gauge condition, the choice will be arbitrary and there will be a set of worlds in limbo. This arbitrariness does not bother physicists for they choose one in the knowledge that they all correspond to the same observable state of affairs; convenience is usually the sole criterion.

As it stands, the only way that I can see Butterfield's proposal working is if it is coupled to a Bergmann-style conception of observables (*i.e.*, one restricted to diffeomorphism-invariant quantities—*cf.* [Bergmann, 1961]) that would function as the initial data set. This gives another method for tackling the initial-value problem: one takes as observables quantities that do not 'see' the differences brought about by the action of the diffeomorphisms, though they remain in the formalism as surplus structure. These quantities would be insensitive to the kinds of differences brought about by the hole diffeomorphisms. Only then is **Dm2** able to eradicate the indeterminism, but if this latter view is accepted then **Dm2** is idle; the diffeomorphism-invariant quantities can avoid the hole problem independently of it.

Moreover, if either path is followed (gauge fixing or Bergmann style measurement theory), then it is incumbent on the substantivalist to show that their conception sits well with these. This isn't obviously the case, and many physicists assume that the latter (Bergmann conception) underwrites a relationalist conception of spacetime. The gauge fixing proposal quite clearly clashes with Earman and Norton's assumption that the substantivalist is committed to each model as representing a physical possibility. (In fact both are compatible with both relationalism and substantivalism. I return to this below in Chapter 6.) In §5.2.2 I discuss Butterfield's use of counterpart theory to resolve the hole problem. This matches the gauge choice option above. However, although it does indeed give a philosophical model for Butterfield's **Dm2**, there is still no well motivated reason for a particular gauge choice—indeed, I argue that most 'reasonable' gauge choices require *material* objects such as dust variables; for non-material gauge fixings there is a Gribov effect in general relativity (*cf.* §3.3). This, I argue, is a very serious stumbling block for Butterfield's approach (and there are many other problems too, the explicit breaking of general covariance for example).

Carolyn Brighouse offers a similar resolution to Butterfield, but the details differ in important ways. I think that Brighouse's proposal is in almost decisive against the hole argument—though incomplete, as I explain below. In a nutshell, Brighouse argues that the kinds of differences brought about by the application of diffeomorphisms should be seen as irrelevant in the analysis of physical determinism. Recall that, if they represented worlds, the hole diffeomorphs would represent worlds that differed non-qualitatively. Brighouse writes: "the haecceitistic features of the qualitatively isomorphic histories simply don't feature in the way the physical states of the worlds evolve. Haecceitistic differences between objects make no difference to the way objects behave" ([1997], p. 478). This proposal is in line with Butterfield's **Dm2**, for it is qualitative similarity that counts (*i.e.* isomorphisms between the object fields). But one still needs to give an account of the sense in which the theory is deterministic, and what this means for the ontology. Brighouse doesn't flesh out her proposal with answers to these questions, and so it has to be seen as incomplete. There are all the options from the previous chapter available: we can literally identify the worlds that differ haecceitistically (*e.g.* by taking equivalence classes (gauge orbits) as representing physical possibilities: working in $\mathrm{Riem}(\Sigma)/\mathrm{Diff}(\Sigma)$), we can treat the models as standing many-to-one with worlds, or we can even take the models as standing one-to-one with possible worlds but define the observables to be diffeomorphism-invariant quantities. Each one will be in line with Brighouse's proposal and Butterfield's **Dm2**, but according to the prevalent opinion they support quite distinct conceptions of spacetime.[152]

Thus, Earman and Norton clearly think that futures differing only with respect to which objects play which role are relevant in deciding whether or not a theory is deterministic (*cf.* Belot [1995] for a similar, though milder stance made explicit). But both Butterfield [1989] and Brighouse [1997] argue that Earman and Norton's brand of determinism countenances too many possibilities, ones that should be irrelevant in the analysis of determinism: only qualitative features should be relevant in questions of physical determinism.[153]

Joseph Melia argues that both parties are wrong: "distinct yet qualitatively similar futures *can* threaten determinism ... [b]ut ... not *all* futures differing only over which objects play which roles threaten determinism" ([1999], p. 640). He draws on examples given by Belot in his defense of this conclusion, and argues that there is a conception of determinism that treats these cases as indeterministic, but treats the hole argument and those like it as deterministic; thus respecting our intuitions in such matters. Hence, as regards the hole argument, like Butterfield and Brighouse, he believes that "the fault ... lies not with substantivalism, but with the conception of determinism Earman & Norton assume the substantivalist must accept" (*ibid.*, p. 641). (In fact, this is but one fault among many, and as I mentioned above.)

Melia labels "D-haecceitistic" the view "that a theory may be indeterministic, even if all the different possible futures open to any world which makes the theory true are qualitatively identical" (*ibid.*, p. 640). Thus, D-haecceitism says that there can be theories that are classed as indeterministic purely as a result of hacceitistic differences between possible futures. There is however, as Melia points out, no direct implication between believing that there can be purely haecceitistic differences between distinct worlds and accepting D-haecceitism: one can "accept haecceitism whilst rejecting D-haecceitism" (*ibid.*).[154] Clearly, for Earman and Nor-

[152] One of the main points I wish to get across in this book, as should be evident by now, is that these interpretive moves do not underwrite particular conceptions of spacetime: they are compatible with both substantivalism and relationalism. Note that the moves *do* appear to support various conceptions of the physical possibility space.

[153] Lewis [1983b] argues that only qualitative differences between possibilities should count in the analysis of determinism, for only these represent distinct possible worlds. Butterfield and Brighouse both draw their views from Lewis' account. They argue that the differences brought about by the application of a diffeomorphism are too mundane to be brought to bear on the issue of deciding whether or not a certain theory counts as deterministic or not: non-qualitative differences should not be counted as relevant to determinism. (Maidens [1993], Hoefer [1996], and Hoefer & Cartwright [1993] argue similarly). Hence, we have two options so far: either *all* non-qualitative differences are relevant to determinism or *none* are.

[154] He cites Lewis [1983b] as someone who believes that there *can* be worlds with differences simply amounting to a 'property-swap' across some objects, but who doesn't allow such property-swapped worlds to count as distinct possibilities

ton, general relativity with its spacetime points substantivally conceived implies D-haecceitism in the theory.

Melia first outlines Lewis' idea of what it is for a theory to be deterministic. He notes that a branching worlds conception is not available, for given Lewis' counterpart theory there cannot be overlapping worlds. The idea of branching clearly mandates that a pair of worlds share a common history (initial temporal segment) up to the point where branching takes place. But Lewis can model branching using his notion of duplicates (defined, roughly, by maximum qualitative similarity between two things). The idea is spelled out by Melia as follows:

> take a pair of worlds, w and w^*, which both satisfy the theory T; take any two histories h (from w) and h^* (from w^*), each segment representing the complete history of each world up to some times t and t^*; and ask the following two questions: (1) are h and h^* duplicates of each other? (In other words, are exactly the same qualitative properties and relations instantiated in exactly that same way in both h and h^*?); (2) are w and w^* duplicates of each other? T is indeterministic if and only if there are some T-worlds and some histories for which (1) is answered 'yes' and (2) is answered 'no'. ([Melia, 1999], p. 649)

The notion of duplicates implies that only qualitative differences enter into questions of deciding whether or not some theory is indeterministic. The basic idea of determinism is this: if a pair of initial temporal segments (both satisfying the laws of some theory) are duplicates, then the worlds of which they are a part are duplicates. If not, then the theory is indeterministic. As Melia points out, this conception of determinism rejects D-haecceitism, and just this feature has been utilised by Butterfield and Brighouse as a response to the hole argument.[155] The reason is clear: the diffeomorphic models of general relativity differ only in a non-qualitative (haecceitistic) way, but such differences are ignored on this account: we only look at *global* features concerning whole histories and worlds.

Melia gives an example of a world and a theory where this outlawing of non-qualitative differences apparently conflicts with intuition.[156] We are asked to consider a world w containing a "single symmetrical homogeneous cylindrical tower standing on a single symmetrical homogeneous planet for one second before toppling over and coming to rest upon the ground" (*ibid.*, p. 649). The similarity to a Buridan's Ass-type scenario is obvious: the initial conditions exhibit perfect symmetry, so the evolution to any subsequent state is going to be indeterministic: the system might have done something else. But each of the possible outcomes for the tower will be qualitatively identical, because of the symmetry of the situation. A God wanting to make a PSR informed decision on the basis of where the tower

in considerations of determinism. Lewis' point is that the same possible world suffices to represent both possibilities. In the case of a maximal property swap of my properties and my doctoral supervisor's, it is my 'this worldly' supervisor who represents this possibility for me.

155 I should point out that Brighouse's position differs significantly from Butterfield's in that she claims that *all* diffeomorphs represent a genuine physical possibility (the same), but it is given by one world. This is contrasted with Butterfield's claim that only one represents a genuine possibility. Thus, Brighouse endorses **LE** while Butterfield endorses **One**. I return to this difference when I consider the counterpart theoretic response below.

156 Melia credits the example to Belot [1995], who himself credits it to Mark Wilson ([1993], pp. 215–6).

fell (such as where to place a system of matter in another world) would be unable to do so, for example. But surely we want to say that each possible outcome is a *distinct* possibility don't we? On one future it lands *here* in another it lands *there*. Melia thinks we should and so does Belot. To bolster intuition, imagine that a circle of philosophers are sitting around the tower (preserving the symmetry), so that one of them will get squished by the falling tower. According to one possible future it hits *this* philosopher, to another *that* philosopher: the 'theory' describing this system cannot determine which philosopher is hit. If we do want to say this is a case of indeterminism, then we are forced to admit that the theory is D-haecceitistic. But on the account of determinism given by Butterfield, Brighouse and Lewis, this is ruled out: only qualitative differences matter, and the futures for the tower scenario are indistinguishable with respect to such properties for exactly the same qualitative properties and relations occur in the same way in these worlds. Hence, their account seems to conflict with our intuitions in this case. It seems that there is a real sense in which there are different ways the tower could fall, even though these ways are "qualitatively isomorphic". It seems that we have a case of indeterminism concerning which objects play which roles: the example is thus conducive to D-haecceitism. But the account given by Lewis *et al.* says that it is deterministic. As such, although their account does indeed escape the hole argument, it is too strong, it does too much.

Melia outlines a conception that he thinks will respect our intuitions in the above examples, but will not fall prey to the hole argument; one that gives the 'right' answers to both cases. His idea is to use a conception of determinism that involves the Lewisian notion of duplicates at a *local* level (the level of inhabitants of worlds) as well as at a *global* level (the worlds and histories themselves). This essentially forms the basis of his rejection of the Lewisian account: the focus there is on complete histories of worlds, and worlds themselves, but not on the individuals in the histories and at the worlds. Melia's suggestion is as follows:

> [I]t's not enough to insist just that any two worlds which match qualitatively before t must match qualitatively throughout time—any two such worlds must meet a further condition: each of the parts of these worlds which match qualitatively before t must match qualitatively throughout time. ([Melia, 1999], p. 654)

Hence, with reference to the above tower world, two worlds that differ in which philosopher gets hit on the head by the tower are global duplicates, but they are not local duplicates; there are philosophers from the worlds that are local duplicates up to the time of collapse, but are not local duplicates thereafter: one philosopher gets hit, the other doesn't. This world is indeterministic according to this conception. Worlds represented by hole diffeomorphs are deterministic according to this conception because histories and objects within them that agree qualitatively up to some time *will* agree qualitatively thereafter: there is a global isomorphism between such models that matches up object fields and objects. This is supposed to accord with intuition: it offers a conception of determinism that gives the required split between haecceitistic differences that matter and those that don't.

I find Melia's conception of determinism rather puzzling: certainly our intuitions do not match on this example. My concern has to do with the idea that one must be able to refer to the parts of worlds, even though those parts are qualitatively identical. Surely, by construction, the parts are referentially indeterminate, for were they not there would be some asymmetry present in the world to ground the reference. Clearly any descriptivist account will not do, for many objects will satisfy one and the same description. Given this, one cannot individually refer to any of them. Perhaps Melia has in mind an ostensive account? For example, he claims that "[f]ocusing attention on some part of the ground in the tower world, it is not determined whether that part will be hit by the tower or whether it will remain unscathed" ([1999], p. 654). However, it is certainly not clear to me that this is a possible procedure. Do we have access to a Verneoscope[157] that will allow us to look into such worlds that is sensitive enough to distinguish between qualitatively identical parts? I am reminded at this point of an old paper of N.L. Wilson's [1959], where he poses the question "What would the world be like if Julius Ceasar had all the properties of Mark Anthony and Mark Anthony had all the properties of Julius Ceasar?" (p. 522). Wilson's answer matches the Lewisian anti-haecceitist line: "our attempt to describe a distinct possible world has produced just the same old world all over again" (*ibid.*, p. 523). This applies to the tower world, for the putatively distinct scenarios simply concern which individuals get which properties, which philosopher gets his head bashed in. But at least in the Wilson world we have two qualitatively distinct individuals to start off with; in this case we do not have such a luxury. Allowing the philosophers to refer to themselves—thus giving an indexical account along the lines of Adams' modified Blackian 2-sphere world [1979]—will not do either: though they will determinately refer to themselves uniquely and the symmetry will be preserved, Melia's 'test' for indeterminism requires that we be able to focus in on the parts and check their evolutions for future qualitative differences. However, I say, the setup strictly forbids this procedure. Thus, I do not think the tower world counts as an example of indeterminism any more than the hole argument constitutes a case of indeterminism according to Melia's account: in each case, were it possible to refer to the homogeneous parts there would be clear grounds for claiming that some asymmetry caused the final result and in the absence of determinate reference there is no saying that some particular part gets some qualitative property not possessed by the others.[158]

Even if my objection can be dealt with, and Melia's proposal can be shown to work, I do not think it is necessary. Brighouse [1997] has said enough: such haecceitistic differences between futures are simply not relevant in the analysis of determinism *in the context of physics*. In fact, this is not quite right. It isn't that such differences are not relevant, it is simply that the physics is indifferent to such possibilities. Physics cannot be brought to bear on such matters. Hence, though the examples Belot gives can be said to be indeterministic, the sense in which they

157 As Lewis describes it "[t]he Verneoscope is an impossible device, invented by David Kaplan, for inspecting other possible worlds" ([1983a], p. 383).

158 Both of Melia's proposed resolutions to what he takes to be a case of indeterminism make the same mistake in supposing that we can simply refer uniquely and determinately to the indiscernible parts of symmetric worlds like those characterized by diffeomorphic models of the hole argument and the tower world.

are so is not one that could matter in the context of physical theories, at least not in general relativity.

5.2. THE MODALIST TURN

The responses we consider in this section work by denying that each element from the full set of hole diffeomorphic models represents a distinct physically possible world. The idea common to these schemes is one we met in each of the previous chapters: namely, that the mathematical structure (models and solutions) is not in one-to-one correspondence with physical structure (worlds). The claim is, then, that substantivalists are not committed to a direct interpretation of the formalism, *i.e.*, to Haec. The first response, Maudlin's metrical essentialist scheme, argues that the diffeomorphic models cannot represent physically possible worlds because that would involve points of spacetime losing essential metrical properties. The second, Butterfield's counterpart theoretic scheme, argues that if one model out of a pair of diffeomorphic models represents a physically possible world then the other cannot because it would contain the chosen model's points; this is impossible because the points are worldbound. The third, Hoefer's rejection of primitive thisness differs radically from the first two because it endorses LE whereas the first two deny it. For Hoefer, the diffeomorphic models represent one and the same possible world because the identities of the points 'follow' the metric field. Thus, the first two are selective and the third is indirect.

This might at first sight appear to grant Earman and Norton their conclusion, and signal a shift from substantivalism to an alternative conception of spacetime. For surely relationalism just is the denial of the Leibniz–Clarke counterfactuals.[159] This is not the case, for there are two (realist) ways of denying the model-world correspondence: either one can say that the models represent one and the same world or one can say that not all models represent worlds. The first two responses I consider here take the latter route. Relationalist options tend to take the former route, but as I mentioned, Hoefer defends a substantivalist position that takes this route too. Recall too that I detached the claim that relationalists are necessarily committed to an endorsing of LE in Chapter 2.

We saw earlier that Leibniz advocated LE for the type of case we are considering here; namely, those for which a set of mathematical elements would represent physically indistinguishable situations. His reasons had to do with his two great principles, PII and PSR. A form of relationalism followed from his commitment to these, but the converse is not true: one can be a relationalist and not be committed to either (compare this claim with Saunders [2003b], p. 16).

In Maudlin's case the representing set is restricted to those models for which the set of spacetime points (*i.e.*, elements of $\mathcal{M}$) share a common set of essential (geometric) properties—hence, he advocates Some (for there may be isometric

[159] Belot is someone who strongly defends this thesis [1996; 2000]. I can certainly understand his insistence, for he wants a clear cut debate between substantivalists and relationalists and thinks that this is the best way to achieve it. However, it goes squarely against my minimal definitions of these positions as I outlined in Chapter 2, which I think function as well, and get to the heart of the ontological disagreement better. I have already argued why I think Belot's alignment doesn't work in that chapter, so I won't go over it again here. I will, however, return to Belot's claim again in the subsequent chapters, where I attempt to further erode it.

models that preserve the essential properties—in which case Maudlin is an haecceitist). Butterfield restricts the representing set down to just one model—hence, he advocates **One**. Hoefer views all of the diffeomorphic models as representing one and the same world—hence, he endorses LE. So, we have four separate substantivalist interpretations, distinguished by the nature of the representation relation they specify: (1) all models represent distinct possibilities; (2) one model represents a genuine possibility; (3) some models represent (distinct) genuine possibilities; (4) all models represent the same possibility.

5.2.1 Maudlin's metrical essentialism

At the heart of Maudlin's scheme [1988; 1990] is the idea that only *one* model from an equivalence class of hole diffeomorphs represents a physically possible world.[160] Butterfield too takes this line, though he grounds it in a different way, using counterpart theory. For Maudlin the hole diffeomorphism (or *any* non-trivial diffeomorphism) does not allow one to generate physically admissible models from models. And the reason for this is that a diffeomorphic copy of a model does not, in general, represent a physically possible state of affairs for the objects represented in its domain—though it does represent a distinct state. For Maudlin, the metrical properties of actual spacetime points are essential; and, in general, the essential properties of points are altered by a diffeomorphism. For example, Maudlin must admit that the value of the curvature scalar $R(x)$ at a point $x \in M_{@}$ (representing an actual point) is one of its essential properties, for it is built from the metric. Suppose that we act on x with a non-trivial diffeomorphism ϕ, so that $\phi(x) \neq x$, where $\phi(x) = y \in M_{\phi}$. Then it follows, given reasonable assumptions, that $R(x) \neq R(y)$, and the point x loses an essential property, so that M_{ϕ} does not represent a metaphysically possible state of affairs. The spacetime points are still viewed along broadly manifold substantivalist lines, they are robust existents, and only an aspect of their modal semantics is revised. In particular, the names of the points function as rigid designators, referring to the same points in all models in which they occur.

Andreas Bartels ([1996], pp. 35–6) claims that LE would be the more appropriate interpretation of the representation relation for Maudlin, since if ϕ maps $x \in M_{@}$ onto $y \in M_2$, then x and y have the same metrical properties, and so represent the same spacetime point. However, Maudlin wants to apply his approach to the case of the Leibniz-shift argument, where he argues that substantivalists should side with Newton (or Clarke) in saying that the shifted worlds represent genuine physically possible worlds. Maudlin clearly concurs with Earman and Norton with regard to their 'acid test' as far as the Leibniz–Clarke controversy goes: the substantivalist is committed to the multiple indistinguishable possible worlds represented by translations of matter. LE is ruled out in this case, but the thesis of metrical essentialism supports it, for the points keep their essential properties across models, since the metric is a background structure and does not vary

[160] In fact, as I mentioned above, in isometric cases, where the points retain their essential properties, there can be multiple models that represent physically possible worlds. This amounts to an endorsement of **Some**. However, both **Some** and **One** are examples of *selective* interpretations.

the way it does in general relativity. In this case there is no indeterminism problem, for the metrical structure is absolute. Clearly, however, in the case of general relativity we can pick a hole diffeomorphism for ϕ, and this will give x and y different metrical properties. This leads us into the indeterminism, and Maudlin's essentialism nips it in the bud by disallowing ϕ to generate physical possibilities for x, unless ϕ is an isometry in which case the essential properties are preserved. Since the metric *is* dynamical in general relativity, one might think that this will result in Maudlin's metrical essentialism taking on 'hyper-essentialist' proportions: the points are rendered worldbound and cannot appear in *any* other world, so it looks like all of a point's properties are essential ones. I don't think this is true. One can imagine cases where the points retain whatever properties are deemed essential in the actual world, but the fields representing matter differ in some way that is non-destructive to the essential properties. The cases where spacetime possesses exact symmetries are enough to show that Maudlin's approach is not necessarily hyper-essentialist.

Norton [1988] puts Maudlin in the camp of 'manifold-plus-further-structure' [MPFS] substantivalism. The idea is that the manifold structure itself is not sufficient to individuate the points of spacetime, but if some further structure is placed atop of this manifold, then individuation is possible. Maudlin suggests the metric, and argues that properties of points inherited from the metric are essential. Norton's objection to this is that such a position would only work if the spacetime admitted no symmetries. For in the presence of symmetries we get the problems of the hole argument once more: there would be points with the same set of essential properties, the diffeomorphisms could hit these points giving worlds that Maudlin would have to accept were distinct (hence Some). Maudlin would indeed accept them as distinct genuine possibilities, as can be discerned from his insistence that the substantivalist is committed to accepting Earman and Norton's acid test. However, the indeterminism threatens once we admit symmetries, and since Maudlin does not accept LE he does not have the option of maintaining that the models represent one and the same state of affairs. The determined MPSF-substantivalist might simply dig her heels in and argue that given the non-trivial nature of our world, such symmetries are highly unlikely to occur. This weakens the position somewhat, since the question of determinism and the number of general relativistic possible worlds containing the same spacetime points becomes dependent upon the existence or not of symmetries. However, it seems decidedly odd that two worlds differing only in the presence of a single symmetry could be so ontologically different.[161] More likely, Maudlin will accept that there is indeterminism here, and that there can be worlds that differ haecceitistically, as he does in the case of the Leibniz-shift scenarios.

161 Compare with Adams' [1979] counterexample to the PII, wherein a world with two qualitatively distinguishable spheres are gradually—by successive passages to other possible worlds—made to approach qualitative similarity, with the aim of reaching qualitative identity. Eventually, one gets to a stage (a world) in which the spheres differ with respect to, say, just one particle: this is surely a genuine possibility? (Though Leibniz would not accept this if the sphere's were composed out of identical particles, save the one differing particle.) The question is, can we proceed and remove the particle so that the spheres are qualitatively identical: is this a genuine possibility? If not, then the question is how could the removal of a single particle have such devastating ontological consequences (perhaps taking a world out of existence!)?

Maudlin argues for his metrical essentialism from the following oft-quoted passage from Newton's *De Gravitatione*:[162]

> [T]he parts of space derive their character from their positions, so that if any two could change their position, they would change their character at the same time and each would be converted numerically into the other. The parts of duration and space are understood to be the same as they really are because of their mutual order and position; not do they have any hint of individuality apart from that order and position which consequently cannot be altered. ([Newton, 1962], p. 136)

This passage seems to fly in the face of Each, and might appear to undermine Earman and Norton's claim that the substantivalist is committed to this position. However, Newton is making crucial use of the metric, which, in a background dependent theory like his, is the same across all of its models. Without the metric, the points of a manifold do not have any individuality, and do not represent points of space and time. It is precisely the fact that the metrical relations are deemed necessary to distinguish the (otherwise indistinguishable) points from one another that leads Maudlin to his position. The parts of space and time are taken to "bear their metrical relations essentially" ([Maudlin, 1988], p. 86), it is these relations that individuate them. What does seem clear from the above passage, however, is that Newton would deny Each along with Leibniz were it not for the inclusion of matter in the models; for without matter, any permutation of the points of the manifold would permute the identities: identity follows character in the matter free case. It is precisely this latter aspect that Stachel draws his position from, though he sees the metric of general relativity more along the lines of matter fields.

As Brighouse [1994] notes, the essentialism of Maudlin's stance renders his approach ill-equipped to deal with simple counterfactuals that we would like to accept (see also Earman [1989], p. 201). She gives as an example: "if the sun had had extra mass then the curvature around the sun would have been different". I would be inclined to say that such a counterfactual expresses a truth, but as Brighouse points out, Maudlin's metrical essentialism forces him to claim that it is false, as our example with the scalar curvature at a point demonstrated. Pooley too follows a similar line to Brighouse, though he argues that we can agree with Maudlin that *some* metrical relations are essential for points of space and time, but it does not follow that a *particular* set of metrical properties are essential ([Pooley, in press], pp. 95–6). That is, accepting that metrical properties are indeed essential properties of spacetime points does not imply that a certain point could not have existed with a different set of metrical properties from the ones it actually has, just that it needs some to exist. Take as an analogy the property 'having mass-energy'. We might take this to be an essential property of material objects; my laptop is a material object, and it weighs about 5.6 lbs. But it does not follow that my laptop is essentially a 5.6 lbs thing: a new harddrive might have been invented

[162] It is interesting to note that Howard Stein reads off a structuralism from the very same passage. I side with Stein, though I think the reasons can be best expressed using Stachel's notion of an "individuating field"—on which, see below §5.3.2. Stachel uses similar reasoning to argue for relationalism. So we see that one and the same passage has been used to defend all the major flavours of interpretation!

that reduced the overall weight. Hence, Maudlin's stance is unable to deal with our modal intuitions, it goes too far.

Maudlin does have a response to the problem of the above counterfactuals. Let me first state his response, and then outline the further objection. Maudlin claims that when it comes to assessing such counterfactuals as 'the curvature at this point might have been different than its actual value'—something Maudlin has to rule false—we should view the worlds with different values of the curvature (represented by non-isometric models) as representing genuine possibilities, but not for the actual world. Rather, the distinct possibilities "are just different possible space–times, not different possible states of *this* space–time" ([Maudlin, 1988], p. 90). He then claims that the counterfactuals can be explained using counterpart theory, so that the different spacetimes can represent *de re* of the actual spacetime that it might have had other properties.

Brighouse notes that this counterpart-theoretic analysis does not sit easily with Maudlin's metrical essentialism; there is something "rather schizophrenic" about it ([Brighouse, 1994], pp. 119–20). The counterpart theorist will understand the essentialism in the following terms: for any essential property *P* of some object, any counterpart of that object, under any counterpart relation, will also have the property *P*. Hence, if I have my humanity essentially then any counterpart of mine will also have that property, and if a spacetime point has its metrical properties essentially, then any counterpart of this point will also have those properties. But this doesn't seem to give Maudlin what he wants, namely an account that respects our modal intuitions concerning the problematic counterfactuals. The problem is that the worlds represented by diffeomorphic models will not contain kosher counterparts for actual points (*cf.* Pooley [in press], p. 97). The counterfactuals will still come out false since any point that does not possess the metrical properties possessed by the actual point will not be a counterpart of that point; and yet just such a point is what is needed to make the counterfactuals come out true. The only counterparts that would be available would be points possessing the exact same metrical properties, but they cannot deal with the counterfactuals. In fact, I think Brighouse and Pooley slightly overstate the case against Maudlin. Perhaps Maudlin is entitled to the claim that we can use some other points to function as counterparts in a Pickwickian sense, not as representing *de re* of the actual point that it might have had metrical properties other than it actually has, but that there is a way that *some* point similar to this actual point that has those properties.

Even if my treatment of Maudlin's counterpart theoretic analysis of the counterfactuals is wrong, there is also the objection that mixing counterpart and metrical essentialism to evade the hole argument is to commit overkill, since the counterpart theorist can do the job without essentialism (as we see in the next subsection). Hence, we also have good reasons to reject Maudlin's account on methodological grounds.

5.2.2 Butterfield's **One**

The second 'modalist' option we consider is that of Jeremy Butterfield [1988], who utilizes David Lewis' counterpart theoretic response to the problem of transworld identity (as applied to spacetime points) to avoid the hole argument. His 'complete' response has two aspects: (1) to sharpen up Earman and Norton's imprecise definition of determinism in such a way so that it is not violated by hole diffeomorphs; (2) to philosophically ground this sharpened conception of determinism. I have already dealt with (1) above, where I argued that although the definition Butterfield offers implies that general relativity is deterministic, it does so in a way that might be incompatible with the substantivalist position he was seeking to defend, it also rules against cases that we would intuitively like to say *are* indeterministic. Fortunately, the two aspects can stand (or fall) on their own, regardless of the cogency of the other. Let us now consider the second aspect.

Recall that the problems for the substantivalist arose from the supposed requirement that she must deny Leibniz equivalence. This requirement is tantamount to the pair of claims (1) that hole diffeomorphs represent worlds, but (2) do not represent the same world. Rather, each diffeomorph represents a distinct world. These worlds are such that they differ haecceitistically, in a non-qualitative way concerning which object plays which role. The commitment to this view is supposed to flow naturally from the substantivalist's conception of spacetime points as robust individuals. This is the position that Leibniz had Newton occupying—and Newton may have been happy with his—that the 'non-individualists' had the 'individualists' occupying in the context of quantum mechanics; and that the 'gauge invariantists' had the 'literalists' occupying in Maxwell's theory. The latter stances were seen to entail a commitment to inflation, whereas the former stances were seen to be deflationary. In each case I argued against these entailments, but that is by the by for the moment.

Butterfield offers an ingenious method of preventing inflation in the case of substantivalism about the spacetime points of general relativity, though his method is part of his general denial that transworld identities are strict identities. He distinguishes two ways in which one can deny Leibniz equivalence; he calls these ways "**Each**" and "**One**": as above, **Each** says that each model (from an equivalence class of such) represents a different physically possible world of the theory, **One** says that at most one of the models represents a physically possible world (*ibid.*, p. 7). Note that Butterfield's **Each** and **One** correspond more or less with my notions of direct and selective interpretations of gauge theories (§3.3). The hole argument works by supposing that the substantivalist is committed to the direct, one-to-one interpretation of the models suggested by **Each**.

Butterfield's argument rests upon showing that **One** is available to the substantivalist, provided one works with an alternative modal semantics for spacetime points. Counterpart theory is the semantics for the job. For according to this approach there simply isn't transworld identity, and therefore we are not at liberty to speak of the *same* point possessing different properties in different worlds. Thus, for any pair of hole diffeomorphs, "if one ... model represents a world, then the other cannot since it contains the first model's points" ([1989], p. 22).

In the context of counterpart theoretic semantics, instead of allowing one and the same thing to exist in multiple worlds, it is only allowed to exist in one: no thing is in two or more worlds. Strict transworld identities are thus outlawed and replaced with similarity. As Lewis himself puts it:

> Where some would say that you are in several different worlds, in which you have somewhat different properties and somewhat different things happen to you, I prefer to say that you are in the actual world and no other, but you have counterparts in several other worlds. ([1986a], p. 316)

Lewis achieves this feature by substituting a counterpart relation for the identity relation that one finds in standard quantified modal logic (as in the systems of Carnap, Kripke, and Montague). This move alleviates at a stroke many of the problems to do with 'transworld identity,' and how best to cash this notion out when identity is taken in a strict sense (*i.e.*, as an equivalence relation). The counterpart theoretic solution to these problems is simply to say that if a thing exists in some world W_x then it does not exist in W_y $(\forall y,\ y \neq x)$.

Applying counterpart theory to spacetime points is to "deny that a spacetime point can be an inhabitant of two possible worlds" ([Butterfield, 1988], p. 66). The points' being confined to one world, or being "worldbound individuals" (as Plantinga expresses it: [1973], p. 195), implies that all but one of the diffeomorphs do not represent genuine metaphysical possibilities. Instead of a pair of diffeomorphs containing the same set of spacetime points (*i.e.*, the same manifold), they consist of counterparts.

Though I agree that this does circumvent the hole argument, there are a number of reasons why I think this approach is inadequate. The first two objections I consider are due Norton ([1988], p. 62). Norton is worried that Butterfield's suggestion strays too far from the practice of science. It is standard, in general relativistic physics, to employ the 'gauge theorem' (or something like it): if M is a model, then any diffeomorphic relation M′ of M is a model too, and models represent worlds. Without it, general covariance is hard to implement or understand. Indeed, as Norton points out, "[t]he ... problem is that "One" directly contradicts the active general covariance of ... local spacetime theories" (*ibid.*). And without active general covariance we have surely strayed too far from general relativity. This objection is interesting given that Butterfield claimed that his **Dm2** was motivated by scientific practice—namely, the standard setting up of the initial value problem.

Just how cogent is this first objection? I think we have to agree with Norton that Butterfield's **One** implies that active general covariance is put under some pressure.[163] For Butterfield is claiming that not all models represent. Saying this is to concede something to the anti-realist, for it implies that not all of the mathematical structure represents physical possibilities. Crucially, the part that does represent (a single model from an equivalence class of models under the application of a hole diffeomorphism) cannot, it seems, be distinguished by any means from the part that does not (the remaining models). Hence, whatever it is that grounds the fact that only one model represents a physical possibility, it is not a qualitative property, and as such is completely opaque to physics. This objection leads on to the next.

The second objection is an *epistemological* one: how do we determine which, out of a set of models, represents the actual world? It cannot be an empirically

[163] I consider below the question of whether or not we can we retain the content of the principle, while rejecting its physical significance, for the same criticism can be applied to all of the modalist options that uphold either **One** or **Some**.

manifest property as mentioned, for by construction (of the hole argument) there are none. Norton puts the point as follows:

> Given two diffeomorphic structures, at most one is a model of the theory according to "One". Take the case in which one of them is a model. How are we to distinguish the real model from the imposter? There must be some property which distinguishes them and the property must be physically significant as it tells us which structure represents a physically possible world. Since the real model and the imposter are diffeomorphic, this property cannot have observational consequences. ([Norton, 1988], p. 63)

Norton thinks that this empirical inaccessibility puts Butterfield's counterpart theoretic substantivalist in much the same position as the manifold substantivalist as regards the hole argument. The reason for this is that the "field equations will be unable to distinguish between the development into The Hole of the real model or of one of its infinitely many diffeomorphic copy-imposters" (*ibid.*). Both positions, according to Norton, use physically significant properties that are both empirically inaccessible and opaque to the laws of the spacetime theories in question. We should surely be leery of such properties, especially in the context of physical theories.

On this, I think Norton misreads Butterfield's position. Butterfield's proposal apparently amounts to the following claim: choose one model out of a pair of diffeomorphs to represent the actual world; then the other model does not represent a possible world. We don't need to distinguish the 'real' one from the 'imposter', for once we have chosen a representative model, the other model represents an impossibility. Hence, what are the imposters is conditional on what is taken to be the representing model.

Butterfield responds to Norton's first objection by wheeling out his distinction between models and worlds. His claim is that the endorser of One is free to make use of the "plethora of models," so long as the distinction between models and worlds is kept in mind: use the models, but remember that only one represents a physically possible world. Hence, Butterfield advocates what Rynasiewicz calls "Isomorphism Closure", but denies "Model Literalism" ([1994], p. 409): isomorphic models are admissible models but not all models represent worlds. This is part and parcel of the denial of Leibniz equivalence says Butterfield. Choose a representative model and rule the others out as world representers. Thus, model literalism is avoided, but not at the expense of LE; rather, the one-to-one connection between models and worlds is severed by having one of any pair represent.

My main objection to Butterfield's position is that it makes little sense to speak of only one out of a set of observationally indistinguishable models as representing a physical possibility, and the others as being physically impossible. If a model M represents a physically possible world, and M^* is physically indistinguishable from M, then surely M^* too represents a physically possible world: one and the same world (LE: indirect), or one indistinguishable from it (Each: direct). If M represents a physically possible world, and M is isomorphic to M^*, then M^* should represent a physically possible world too. What this shows, I think, is that an anti-haecceitism is required in order to make Butterfield's position work. The problem is that anti-haecceitism is not a consequence of counterpart theory.

I mentioned that I think that Butterfield's response (and, though to a lesser extent, Maudlin's) can be best understood as corresponding to the *gauge fixing* option for dealing with gauge freedom. Recall that such a move involved taking a gauge slice through the set of orbits so that exactly one point from each orbit represents a physically possible state. This way, a one-to-one correspondence can be set up between elements of the gauge slice and physically possible worlds. Hence, in order to fully formalise his response he would have to find a suitable set of *physically motivated* gauge conditions—this should involve some reason for privileging one such choice of gauge (generally it is simplicity that carries the weight, but that won't do here).[164] Until Butterfield has accomplished this, I suggest that his proposal is merely tentative.

Perhaps more philosophically damming is Brighouse's [1994] objection that, unlike identity, the counterpart relation itself is not unique. There are many ways to define a counterpart relation. Butterfield's, like Lewis', choice of counterpart relation is that which agrees on all qualitative properties. Hence, it is possible that we could find a relation such that two points with different geometrical properties come out as counterparts, or even that a point has multiple counterparts, thus leading us back into the hole problem.[165]

5.2.3 Denying primitive identity

Recall that the manifold substantivalist was supposed to get into bother with indeterminism because of the alleged commitment to the existence of worlds that are qualitatively identical but differ with respect to how the geometrical properties are spread over the points of space. The general covariance of general relativity implies that the equations of the theory cannot uniquely determine this spreading of the geometrical properties over the points. If the manifold substantivalist is so committed then the indeterminism surely follows. We have just seen two ways in which the substantivalist can escape this commitment, one manifold substantivalist (Butterfield's) and one involving metrical properties (Maudlin's). Both, however, deny LE. In this subsection I outline a proposal that is both substantivalist and endorses LE. It does this simply by denying primitive identities for spacetime points.[166]

To the best of my knowledge, the approach is question began with Maidens [1993], and variants have since been defended by Stachel [1993], Brighouse [1994],

[164] Recall that in some cases, due to the Gribov obstruction, the gauge conditions will intersect some orbits more than once. Such a scenario would correspond to Maudlin's response.

[165] For Lewis' expression of the problem see ([1975], p. 39).

[166] Primitive identities are non-qualitative individuating properties, such as $a = a$—the basic details were given back in §2.4. An immediate problem with the denial of primitive identities is, then, that it is unclear how one is able to support set theory on which general relativity's models are based (I owe this point to Steven French). There are ways of accommodating the denial of primitive identities through the use of 'quasi-set theory' in which the identity relation is not a well-formed formula for indistinguishable objects (see French & Krause [1999] and Krause [1992]). However, how this is supposed to be implemented in the context of general relativity is beyond me. There are perhaps two ways to accommodate this problem though: (1) claim that one's usage of standard set theory, with identity, is merely heuristic; (2) claim that standard set theory *can* be applied just fine given that the points have been individuated by, *e.g.*, the metric field or some matter field.

Hoefer [1996], Saunders [2003a], and Pooley [in press]—Earman [1989] mentions it, only to reject it.[167] Since the points of space are absolutely indistinguishable it must be a non-qualitative property of this kind that individuates them. According to this approach, it is not manifold substantivalism *per se* that causes the indeterminism of the hole argument; the real culprit is an extra unnecessary component regarding the identity conditions of spacetime points. In particular, the hole argument assumes that the points can be identified, distinguished, and labeled from world to world and within a world. That is, it assumes a notion of transworld identity for spacetime points allowing us to speak of *this* point having certain properties in this world and having different properties in another world, and of *this* point being different from *that* point. Therefore, if we jettison this component from manifold substantivalism, the hole argument is avoided and substantivalism is rescued.

Anna Maidens suggests that whether or not a notion of transworld identity holds depends upon physical theory. She compares the role of spacetime points in the hole argument with the case of indistinguishable particles in quantum statistics. In the classical scenario we can speak of reversing the results of outcomes, since the physics is based upon Maxwell–Boltzmann statistics. Hence, it makes sense to speak of the world in which *this* coin was heads and *that* coin was tails. This cannot be the case with indistinguishable particles like electrons. For if we say of a pair of electrons that their states could have been swapped, we would have to count this in the statistics, but this would simply yield Maxwell–Boltzmann statistics instead of Fermi-Dirac statistics. Redhead and Teller [1992] argue that the case of quantum statistics highlights a crucial difference between the identity claims that can be made in classical and quantum mechanics. In the former we can assign labels to the particles that pick out the same particle in many possible worlds, they designate their referents rigidly. They call this property "transcendental label identity" (p. 203), this is, roughly, another way of saying 'primitive identity'. This labeling is, they say, not workable in the quantum scenario.[168] Maidens suggests that the points of spacetime are analogous to quantum particles in terms of their transworld properties, we can't think of them as having names that function as Kripkean rigid designators. Crucial to this way of thinking is that there is no denial that spacetime points exist. They do exist, just as much as quantum particles exist; only, they don't possess that kind of identity that make transworld identifications (or, indeed, independent intra-world distinctions) possible. In the case of quantum particles, it was the puzzling quantum statistics that pointed to this conclusion, in the case of spacetime points it is the hole argument that points to this conclusion.[169]

167 Note, however, that Stachel and Saunders associate their view with *relationalism* rather than substantivalism as the others do.

168 In fact it is, and Redhead and Teller admit that it is possible to think of the particles as individuals in a robust classical sense, provided we impose certain conditions on the form of their wave-functions (as argued in French & Redhead [1988]). (*Cf.* French and Rickles [2003], for an overview of this and other philosophical aspects of quantum statistics.) Note that Redhead and Teller, though acknowledging the underdetermination, defend the view that particles should be viewed as 'non-individuals' (field quanta) on "methodological grounds" (*ibid.*). I'm not convinced by their transition from methodology to ontology, but I cannot go in to this here.

169 There are many points of contact between the issues surrounding the hole argument and those of quantum statistics. A natural alignment would seem to be between an individualistic and a substantivalist package on the one hand and a non-

Let us get clearer on what the ascription of primitive identity to objects amounts too. For this, I switch to Hoefer's discussion [1996]. He writes that the commitment to primitive identity is well illustrated by the following example:

> Suppose I have two dice and name them A and B by pointing to them. I now ask: Does it make sense to ask whether die A could have been located where die B is with all B's actual properties? ([Hoefer, 1996], p. 14)

The example matches, more or less, the Leibniz-shift scenario. Clearly, if this 'property swap' scenario were a possibility it would be one qualitatively identical to the actual one. Hoefer thinks that the question is senseless, that the "question rests on the presupposition that the names 'A' and 'B' can be used to talk about particulars that have a primitive identity wholly independent of the properties these particulars actually possess" (*ibid.*, pp. 15–6). Consequently, if it can be shown that "primitive identity is a mistake, then names cannot in fact be used in this way" (*ibid.*, p. 15). Hoefer calls upon Lewis definition of haecceitism to further illustrate the notion of primitive identity. Recall that Lewis' definition amounted to the claim that there are cases of haecceitistic differences between worlds, where haecceitistic differences are cashed out in terms of worlds differing in what they represent *de re* of some individual, but do not differ in any qualitative way—anti-haecceitism is simply the denial of this claim (*cf.* [Lewis, 1986a], p. 221). Hoefer claims that primitive identities are a necessary condition for haecceitism, though the converse is not true—hence, reject primitive identities and you rule out haecceitism. Now, the crucial claim is that the diffeomorphic models of the hole argument can only represent different physically possible worlds if spacetime points exist and have primitive identities. Why? Because the models do not represent any qualitative difference, and the difference amounts to a swapping of geometrical properties between actual spacetime points.[170] Hence, Hoefer sees primitive identities as responsible for the haecceitistic differences that make up the multiple futures responsible for the indeterminism. If this layer of metaphysics can be stripped away from substantivalism, then we have a solution; for without primitive identities we have no worlds that differ by haecceity, and without those there is a unique world, as required by determinism. In other worlds, the way is opened up for the substantivalist to endorse LE.

The argument that Hoefer gives to detach primitive identity from substantivalism is rather brief: The fact that two objects (of the same type) have contingent properties does not license the conclusion that a property swapped situation is a possible situation distinct from the actual one. If by "possible situation" Hoefer means "possible world" then I think we can agree. Lewis, for example, will want

individuals and relationalist package on the other (see, for instance, Stachel [2002]). However, on Maidens' account we see that the substantivalism and non-individualistic packages are aligned. This is an aspect of the transformability of positions by revising conceptions of identity and modality that I mentioned earlier: interpretive positions can help themselves to features generally assumed to be unique to their opposites by tweaking certain parts concerning modality and identity.

[170] Recall that Maudlin avoided the problem by arguing that the swapping operation is not metaphysically permissible because the metrical properties of spacetime points are essential ones. The swapping only gets a purchase on contingent properties. Thus, only the model representing the actual world represents a physically possible world. Butterfield avoided the problem by arguing that swapping operation should not be construed as generating a physical possibility for the actual points, because the actual points cannot figure in the domain of any other world. Thus, if we fix a model to represent the actual world, then no other model represents a physically possible world. Both options deny LE like good Earman and Norton defined substantivalists.

to say that the actual world contains the possibility that Hoefer outlines: the actual object B represents *de re* of the actual object A that it might have had all of Bs properties. If, however, he means "possibility", then I think that he is wrong. We are asked to suppose that the properties of A and B are contingent, and that there exists A and B in the actual world, and we wish to consider them having their properties switched. Surely this is a distinct possibility, a different way that A and B could have been, given contingency? I think we should side with Lewis in saying that it is a possibility, though not a possible world: B is As counterpart and represents of A that it might have had all of the properties B in fact has. There is no proliferation of possible worlds, but there is a proliferation of possibilities, many to each world. Let us assume that Hoefer intends the former reading, then we have severed primitive identity from substances, for A and B can be substances capable of possessing properties and being named, but we aren't forced into the property swapped situations representing distinct possible worlds. This latter feature is taken by Hoefer to go hand in hand with the notion of primitive identity. There is a second aspect to Hoefer's argument, based upon another use of primitive identity, namely to ground the identities of distinct qualitatively identical objects, such as spacetime points. Hoefer rejects the need to use primitive identity in this context because he doesn't believe that one needs a principle of individuation to individuate such objects. He cites Black's counterexample to Leibniz's PII involving two qualitatively identical iron spheres to make his case. The reason there are two spheres rather than one is "because, as we stipulated, there are *two* of them and not *one*" (*ibid.*, p. 19). Clearly, PII is being rejected here because that requires that distinct objects must differ in some qualitative (intrinsic or extrinsic) respect, yet the spheres, nor spacetime points (in the vacuum case) do not. PII prevents us from speaking about qualitatively identical objects that differ numerically. Hoefer rejects it within worlds, for we can stipulate such worlds. There is, I think, a better way of grounding the distinctness between the spheres, and that is simply that there is an observable relation between them, being two miles apart, for example.[171] Saunders [2003b] has recently described a new way to understand (Quine's) PII that does not admit cases like Black's spheres as counterexamples. The idea is that the two spheres are distinguishable because, although the two objects share their monadic properties, and a symmetry holds between them preventing us from using relations they bear to other objects to distinguish them, they do satisfy an irreflexive relation: they are two miles from each other, but not two miles from themselves. Such objects Saunders calls "weakly discernible". But since Hoefer explicitly rejects PII such moves are not necessary.

Hoefer is satisfied that he has stripped primitive identity from spacetime points, and substances in general. But the view that results has been associated with a structural role theory of the identity of spacetime points. Hoefer admits that this theory is problematic for reasons outlined by Earman ([1989], pp. 198–9), but seeks to distance the denial of primitive identity from such a theory. What exactly does this theory say? Hoefer claims it is the view that

[171] This relation will not be part of the ontology of a Hacking style redescription in terms of a single object in a highly curved spacetime. See French [1994] for a critique of Hacking redescriptions.

> the identities of points are determined by some subset of the properties and relations ascribed to the points by a model of the space–time theory. So, to be point A in a world described by model $\langle M, g, T \rangle$ is just to have the metrical (or metrical plus material) properties and relations to other points that A has in the model. A's structural role in the model constitutes *what it is to be A* ... "Identity follows isomorphism," in this case diffeomorphism: two models related by the right kind of isomorphism have the same points, and the isomorphism shows how to identify the points in one model with points in the other. ([Hoefer, 1996], pp. 20–1)

A problem of 'multiple isomorphisms' faces any such account of the points' identities. Earman describes the problem as follows:

> If $\psi_1 : W \rightarrow W'$ and $\psi_2 : W \rightarrow W'$ are relevant isomorphisms, total or partial as the view of identity requires, and if i is an individual of W, it follows that i is identical with $\psi_1(i)$ and with $\psi_2(i)$. And so by transitivity of identity, $\psi_1(i) = \psi_2(i)$, which gives a contradiction if ψ_1 and ψ_2 are distinct. ([Earman, 1989], p. 199)

Hence, if identity does indeed follow isomorphism, then domain points will be identical with its image points for any isomorphism. If there are multiple isomorphisms, as in general relativity, then it will be identical with multiple image points. But when the range is the same for the multiple isomorphisms then we get the absurd conclusion that distinct image points are identical. To escape this problem Hoefer refers back to Lewis' anti-haecceitism with its qualitative notion of transworld identification: "no two worlds can represent *de re* different things about the very same objects, unless the worlds differ qualitatively in the properties and relations ascribed to the objects" (*ibid.*, p. 21). This is the same in all essentials to Brighouse's response to a similar problem. Speaking in terms of counterpart theory, she says that the substantivalist will say that the isomorphic models will represent one and the same world because "the counterpart of any point in any of the qualitatively indiscernible worlds will have all the same qualitative properties as that point has" ([1994], p. 122). However, I think Saunders ([2003a], p. 25), defending a similar position (see below) says the right thing here (if we eschew counterpart theory): the isomorphisms if distinct will map the point into nominally distinct worlds that are related by isomorphism, not the same world in which two objects are related by isomorphism.

This leads me to a problem with Hoefer's approach. He wants to help himself to tools from Lewis' workshop, but he nowhere claims that he advocates counterpart theory. He believes that denying primitive identity gives him what he wants: anti-haecceitism. However, if Hoefer is a closet counterpart theorist, then he needs to come out and say so. For denying primitive identities isn't enough to escape the possibility of haecceitism; I showed as much in my discussion of the Leibniz-shift argument. French ([2001], p. 21) also draws attention to this problem, noting that "[a]ccording to Lewis, a belief in haecceities is neither necessary nor sufficient for haecceitism [in Hoefer's sense] ... one might assert haecceitism but deny primitive identity". Presumably French has in mind Lewis' notion of "haecceitism on the cheap", according to which the case where A and B swap their qualitative properties is a possible world, but is the same world as the actual world; *i.e.*, B *de re* represents of A that it might have had all of Bs properties. French suggests that it is not so much primitive identity that needs to be rejected, but haecceitism *simpliciter* (*ibid.*, p. 22). But I showed earlier that this can be done just as easily.

A crucial question with this proposal is whether or not this haecceitism (assuming that this is really what Hoefer wants rid of) can be jettisoned from substantivalism without thereby destroying the latter. Oliver Pooley has recently argued at great length that "haecceitism is not an obvious concomitant of viewing space as a genuine, substantival entity" ([in press], p. 191). Pooley calls the anti-haecceitist substantivalist position that results "sophisticated substantivalism" (rescuing the term from Earman and Norton's derogatory usage and meaning anti-haecceitistic substantivalism). For the most part I agree with Pooley and Hoefer that haecceitism is not a necessary part of the substantivalist's position, and that such a modification offers a way out of the hole argument. I argued in my discussion of the Leibniz-shift argument that the substantivalist's position can be disentangled from all manner of metaphysical baggage, and that the relationalist can be saddled with it. However, I believe that there are independent reasons—primarily to do with the nature of observables in general relativity—to reject both substantivalism and relationalism in the case of general relativity and its quantization. I make a start on these reasons in the next section.

I agree with Hoefer's general conclusion that primitive identities are not a necessary part of the substantivalist's position. I agree also with the claim that what results might nonetheless be construed realistically (*i.e.* robustly).[172] We have, then, a variety of substantivalism that endorses LE. However, as we saw earlier, Hoefer does not believe that the manifold of points, primitive identities or not, possesses sufficient structure to represent spacetime. Instead, he defends the view that "[w]hen it comes to representing spacetime, literally and to the best of our current abilities, the metric field of GTR is the only game in town" ([1998], p. 464). It is only the metric that has all "the usual spatiotemporal features" ([1996], p. 24). I agree for the reasons Hoefer gives, and the reasons I presented earlier. Given this, he characterises substantivalism about general relativity as follows[173]:

> The metric field as presented in GTR literally describes a substantial, quasi-absolute entity that interacts with ordinary matter. It is 'quasi-absolute' because, unlike earlier absolute spaces, its structure is partially determined by the coarse-grained material contents, in the way specified by Einstein's field equations. It merits the term 'absolute', however, because according to GTR it can exist without any material contents, and with a variety of structures; it merits the term 'substantial' for this reason and because our causal-explanatory understanding of (gravitational) mechanics involves spacetime both acting on matter and being acted on in turn by matter. ([Hoefer, 1998], p. 464)

[172] I differ from Hoefer in that I think that both substantivalists and relationalists can be said to be realists about spacetime, merely differing in *how* they think that spacetime exists, *qua* independent substantial entity or relational entity.

[173] In the paper that this passage comes from, Hoefer is at pains to set up a decent formulation of the substantivalism/relationalism debate in a bid to overcome Rynasiewicz's objection that it has become outmoded.

Relationalism is then characterised by the contrary claim that "[t]he basic constituents of our universe do not include an independent, substantial space or spacetime" (*ibid.*, p. 465). I think that there are some terminological problems here: for "absolute" I would say "dynamical"; absolute implies fixed or model independent. This aside, I think that the way Hoefer sets it up loads the dice heavily in favour of substantivalism, for in identifying the metric field with a substantival spacetime, he is forcing the relationalist out of a position he would naturally like to occupy, *viz.* treating the metric field along the lines of a matter field. The relationalist must adopt the absurd view that the universe does not contain a 'real' metric field, for he doesn't believe in a substantivalist spacetime, and in this case that just is the metric field. Substantivalism is sure to win! But Stachel advocates a very similar view to Hoefer's metric field substantivalism and calls it relationalist. Hoefer admits too that Teller outlines a 'relationalist' position that "arrives at essentially the same place" ([Hoefer, 1996], p. 25). What gives? We seem to be right back in Rynasiewicz's problem.

Now, I agree that the best chance that substantivalism has in general relativity is by treating the metric field as representing spacetime. Only it has all of the properties that we expect a spacetime to have. Recall that Newton was not a manifold substantivalist; he too was a substantivalist along similar lines: that is, his attributions of independent reality concerned $\mathbb{E}^3 \times \mathbb{E}$ *not* $\mathbb{R}^3 \times \mathbb{R}$. The reasons are clear: it is the metric that provides the structures that we associate with spacetime, *viz.*, past, future, distances between events, and so on. Of course, his spacetime (space and time) was not dynamical; the metric was a background structure; it does not possess energy in any kind of way. Hence, the problems we face in deciding what represents spacetime were not applicable in that context (remember Earman's witty remark about having the luxury of knowing what they were talking about). The relationalist was not in a position to help himself to the metric as an object, because it didn't have any properties associated with material fields. But the spacetime relations encoded in the metric had to be reduced to relations between objects *not including the metric*. Hoefer is right to think that the metric is where the debate should be focused, but the formulation is unfair: it is lopsided, and the poor relationalist is almost certain to lose. A question is surely being begged when Hoefer claims that the "metric field of GTR ... [does] not seem to be eliminable in favour of some set of primitive relations holding among material things (whether fields or particles)" ([Hoefer, 1998], p. 463). Surely the relationalist's relationalism will involve the metric field as the "material thing" that determines the "set of primitive relations". For example, Stachel writes that

> [s]everal philosophers of science have argued that the general theory of relativity actually supports spacetime substantivalism ... since it allows solutions consisting of nothing but a differentiable manifold with a metric tensor field and no other fields present (empty spacetimes). This claim, however, ignores the second role of the metric tensor field; if it is there chronogeometrically, it inescapably generates all the gravitational field structures. Perhaps the culprit here is the words "empty spacetime". An empty spacetime could also be called a pure gravitational field, and it seems to me that the gravitational field is just as real a physical field as any other. To ignore

its reality in the philosophy of spacetime is just as perilous as to ignore it in everyday life. ([Stachel, 1993], p. 144)

No wonder Hoefer suggests that the "debate ... may well be settled"! But Stachel is no better, for he claims the metric field all to his relationalist self by treating it as a physical field like any other: clearly it isn't quite like any other, for we can set any other fields to zero (*i.e.* $T = 0$), but not the metric field. A fairer way would be to focus attention on the metric field, as Hoefer suggests, but to allow the relationalist access to it *as a physical field*. But not in the way that Hoefer implies, for what he seems to be calling for is a reinstating of the doctrines of "material metric field" versus "physical metric field". According to the former the metric of general relativity spacetime is *reducible* to the behaviour of material objects, such as clocks and rods and so on; according to the latter view the metric field is an *irreducible* physical field (*cf.* Weingard [1975], pp. 426–7).[174] Hoefer takes the physical metric thesis as corresponding to substantivalism. But no relationalist in their right mind will accept this: the material metric thesis is simply false, because there are solutions even when the stress-energy field is set to zero. I suggest formulating the debate precisely in terms of whether or not the metric field is 'space-like' or 'matter-like', whether the chronogeometrical properties outweigh the gravitational properties. Then many of Hoefer's arguments concerning the difference of the metric to the other fields might prove themselves successful.

Hoefer's claim, then, is that the metric field is the best candidate for a substantivalist's spacetime.[175] He suggests that such a view can be attributed to Einstein (*loc. cit.*: 6). This interpretive option does not face the hole argument because it is natural on this account to say that the members of an equivalence class of diffeomorphic models represent one and the same state of affairs: all invariants are preserved and the identities of the points are carried along with the metric transformations—remember there are no primitive identities. The view that it is the metric field that represents spacetime is well motivated too by the fact that there is no such thing as spacetime until a metric has been defined—he draws this point from Einstein—and it is the metric that does all the explanatory work in general relativity. But once again, the relationalist can agree with both of these and still use them to support their own view, as indeed Stachel does, drawing on the same passages from Einstein that Hoefer uses.

As I have intimated, depending on one's views about the ontological status of the metric field, Hoefer's position might come out as a rather uncomfortable looking relationalism, more than a substantivalist position—indeed, Belot ([2000], p. 576, fn. 36) characterizes him as a "crypto-relationalist". Belot and Earman [2001] see (LE-endorsing) sophisticated substantivalism of this sort as being aligned with relationalism. This is certainly how Stachel sees his position, and it is virtually identical to Hoefer's aside from terminological niceties. More importantly than this word mongering is what I take to be a crucial aspect that Hoefer does not properly discuss: the transition to quantum theory. Hoefer believes that the time is not yet right; but as I argued in the introduction to this

174 This is largely the debate as conceived of by Grünbaum [1973] who himself advocated the material metric field thesis.

175 Note that Hoefer himself is a Machian relationalist.

book, we can apply our interpretive skills just fine in a number of proposals. Especially important, I think, is the assessment of how well a position does, and what kind of approach results when the interpretation is applied in quantum general relativity—canonical quantum gravity being the best candidate. On the front runner from that approach, the metric is a derived object and it is the connection that forms the basic ontological category. On the canonical approach in which the metric does appear, it is quantized, and this will surely have a bearing on whether or not it is substantival or relational. In fact, I don't think it definitively supports either, but it should still be a consideration for any conception these days. I will have more to say on this in later chapters.

Before I leave this approach, I should point out that there are other LE endorsing substantivalist options related to this proposal. One such is the view that the manifold points represent spacetime points, but which spacetime point a particular manifold point represents is dependent upon how the geometric fields are distributed over the manifold (see Brighouse [1994]). Each diffeomorph will then represent the same possibility because the properties and relations of spacetime points are carried over with the diffeomorphisms. Hence, we are not necessarily dropping primitive identities, but we are dropping haecceitism, for it is qualitative similarity that counts. This is grounded in a counterpart relation. We see that both are also very closely related to Stachel's idea that the points of spacetime are individuated dynamically by solving for the metric field, so that the points of the manifold cannot be considered to be named rigidly *à la* Kripke. (I turn to Stachel's view in the next subsection.) Belot has argued against this form of substantivalism[176] on the grounds that it is a mere "variation on relationalism" ([1996], p. 184). As he goes on to say:

> Lockeanism is not the saviour of substantivalism: rather it is a pallid imitation of relationalism which should be of interest only to those substantivalists who are too cowardly to wager that quantum gravity will carry full-blown non-Lockean substantivalism to a decisive victory over relationalism. ([Belot, 1996], p. 184)

The resolution to the hole argument comes about because the formally distinct spaces that result from diffeomorphisms are seen as numerically identical. Models differing only in terms of which points of space possess which geometric properties are identified. Our substantivalist claims are to be applied to this *intrinsic* space, rather than the individual diffeomorphs. Hence, the substantivalist can accept Leibniz equivalence, as long as haecceitism is ruled out.

5.3. VARIETIES OF RELATIONALISM

The final class of response we consider comprises those that see the lesson of the hole argument as implying an endorsement of LE, and see that as implying relationalism. We have already seen that there are forms of substantivalism that fall

[176] Though he uses the terminology "Lockean" and "non-Lockean" to denote the anti-haecceitistic and haecceitistic brands of substantivalism.

under this mantle too. Hence, it is possible for both relationalists and substantivalists to endorse LE. I begin with a quick look at Einstein's *point-coincidence* argument. This resolution involves an understanding of points in terms of intersections of particles: quantities defined at these points are deterministic. Stachel's proposal is superficially similar: his central idea is that points of spacetime are dynamically individuated. This can carried out using material objects and fields as Einstein suggested, or one can use the metric itself. There are similarities to Hoefer's response too, but where Hoefer drew a substantivalist conclusion (by denying primitive identities for points, but allowing the metric to individuate them), Einstein and Stachel draw a relationalist conclusion (claiming that the relations are ontologically prior to, or at least on all fours with points). Saunders' response is relational, but not reductive or eliminative in its treatment of spacetime points: the idea is that points can be distinguished by their relational properties.

5.3.1 Einstein's point-coincidence argument

It is now well known that the hole argument was pivotal in Einstein's eventual completion and understanding of general relativity.[177] The argument led him to at first reject generally covariant field equations for gravity (on account of the fact that different fields were compatible with the extension of data into an arbitrary region of spacetime), and then led him to reinstate the principle, by understanding the general covariance in a deeper way. The basic idea that led to his change of mind revolves around an idealised conception of measurement according to which any possible observation reduces to the intersection points of systems (observers, apparatuses, and the objects to be measured—though no absolute distinction exists between these categories). This theory of measurement is expressed in what has been called the 'point-coincidence argument'. Einstein states it as follows:

> That the requirement of general covariance, *which takes away from space and time the last remnant of physical objectivity*, is a natural one, will be seen from the following reflexion. All our space–time verifications invariably amount to a determination of space–time coincidences. If, for example, events consisted merely in the motion of material points, then ultimately nothing would be observable but the meetings of two or more of these points. Moreover, the results of our measurings are nothing but verifications of such meetings of the material points of our measuring instruments with other material points, coincidences between the hands of a clock and points on the clock dial, and observed point-events happening at the same place at the same time. The introduction of a system of reference serves no other purpose than to facilitate the description of the totality of such coincidences. ([Einstein, 1916], p. 117)

Einstein's suggestion superficially smacks of positivism and relationalism: positivism since evidence is reduced to observations, and relationalism since material points are considered prior to spacetime. There is no mention made of vacuum

[177] See Toretti [1987] for a nice, highly readable account of this history.

spacetimes; the implication of Einstein's view is, however, that a vacuum spacetime is meaningless! There is a passage of Einstein's that, on the surface at least, looks like a more explicit espousal of relationalism, and that contains an answer to the vacuum spacetime scenario:

> If we imagine the gravitational field, *i.e.*, the functions g_{ik}, to be removed, there does not remain a space of the type (1) [Minkowski space–time], but absolutely *nothing*, and also no "topological space." For the functions g_{ik} describe not only the field, but at the same time also the topological and metrical structural properties of the manifold... There is no such thing as an empty space, *i.e.*, a space without field. Space–time does not claim existence on its own, but only as a structural quality of the field. [1961: 155]

This quotation superficially looks like relationalism again. Indeed, it contains a passage that explicitly endorses the material dependency thesis: "There is no such thing as empty space". However, substantivalists have an option to redefine what they mean by 'space', and 'empty space', and so uphold the material dependency thesis. They might say that it is true that a spacetime cannot exist without a metric field, but that the metric field itself is a necessary part of space; we have been here already with metric-field substantivalism, of course: for Hoefer, the metric field just *is* spacetime (this is how many general relativists speak too). Recall that in the original debate between Leibniz and Clarke the space came ready equipped with a fixed metric. When we said there could be empty spaces according to substantivalism, we meant that there could be empty spaces that had the form of a metric manifold, not a 'bare' manifold. Indeed, it was the symmetries of this metric that allowed for the possibility of the Leibniz-shift scenarios in the first place. It is true that the metric is no longer a background structure in general relativity, but it might nonetheless be seen as an essential component of spacetime; it determines those crucial features that make spacetime what it is. Clearly, on this account we are no longer taking spacetime simply to be a container for matter to occupy; rather, it is something dynamic that interacts with matter. As we have seen, this might be construed as a plus point for the substantivalist; for if spacetime is capable of dynamical action, then surely it is all the more substantial for it. We might also read off a structuralist position from this passage in an obvious way: spacetime is structurally constituted by the metric field. Hence, like the famous passage of Newton's, concerning the individuality of points as depending upon their pattern in a relational structure, we can read off all the major positions from this passage.

Stachel suggests that these passages cannot be considered as independent of the issues taken up in the hole argument. When this is taken into account, Stachel thinks that the view expressed in the above passages points not to an operationalism, but to the thesis that

> only a physical process can individuate the events that make up space–time ... a manifold only becomes a space–time with a gravitational field after the specification of the metric tensor field, and that, prior to such a specification, there is no physical distinction between the elements of the manifold ... his comments were not meant ... to indicate that space and time have *no*

physical reality, but that they no longer have any *independent* reality, apart from their significance as the spatial and temporal aspects of the metrical field. ([Stachel, 1989], pp. 87–8)

Hence, we see Stachel explicitly connecting Einstein's view to a denial of the independent reality of spacetime: relationalism. I consider Stachel's development of this line of thought through his notion of an "individuating field" in the next subsection.

The earlier point-coincidence argument, I think, sits better, not with operationalism, but with a revamped notion of 'observability' and what kinds of observables will be possible in general relativity: general relativity cannot deal with *local* field quantities (defined at points of spacetime). In fact, although Stachel argues that the second argument is more sophisticated, I think that it is the first argument that most quantum gravity physicists follow. Compare it with this passage from Rovelli:

> in a general relativistic theory ... [o]nly quantities that *do not* depend on the coordinates may correspond to concretely physically observable quantities. Localization with respect to a background spacetime, or with respect to a fixed external reference system, has no meaning. What has physical meaning is only the *relative* localization of the dynamical objects of the theory (the gravitational field among them) with respect to one another. The physical picture of the world provided by general relativity is not that of physical objects and fields over a spatiotemporal stage. Rather, it is a more subtle picture of interacting entities (fields and particles) for which spatiotemporal coincidences only, and not spacetime localization, have physical significance. ([Rovelli, 2000], p. 3779)

I consider the relevance of the hole argument to the problem of the observables of general relativity in more detail, including the bearing of this issue on spacetime ontology in the next chapter. For now, it will suffice to know where the view expressed by Einstein and Rovelli springs from. For Einstein it was simply general covariance; but, as Kretschmann demonstrated, this isn't quite sufficient to get the desired conclusion.[178] Rovelli suggests that it is the active diffeomorphism invariance underlying the general covariance of general relativity that leads to the physical insignificance of the points of spacetime. In other words, it is background independence that leads to this view. This seems to be the 'majority view' of physicists who consider the hole argument and the meaning of general covariance and diffeomorphism invariance.

Note that what is being suggested here is not that spacetime itself is relational, but only that *localization* is relational (see §6.2). Hence, we cannot consider quantities at manifold points to be observable, but we can consider quantities at materially (possibly including the metric itself) defined points to be observable precisely because they are diffeomorphism invariant. This is a significant departure from previous theories. For example, the observables of, *e.g.*, QED are explicitly local,

[178] Recall that Kretschmann's objection is that *any* (local) spacetime theory can be given a generally covariant formulation. It can, but active diffeomorphism invariance cannot be implemented except in a highly trivial way (the derivatives for metrics and connections must be made to vanish).

they are dependent on a fixed metric on spacetime, as are the observables of most other classical and quantum field theories. The suggestion here amounts to the claim that the observables of general relativity (and its quantization) will not possess such spacetime dependence. Einstein, Stachel, and Rovelli see this as pointing towards relationalism. I say that it doesn't actually have a bearing on the ontology of spacetime *vis-à-vis* relationalism and substantivalism: both can be made compatible with it—substantivalists can simply adopt the view that the metric field represents spacetime and so, since localization relative to *it* is possible, relational localization does not rule out substantivalism.[179]

Since structuralism is fairly neutral (relationalists and substantivalists will agree about the physical relational *structure*), I believe that it is the best candidate for the ontology if we are trying to read our ontology off the physics. However, my argument is not bound to this ontology; and this stance has to be seen as not strictly flowing directly from the interpretation of the theory. I consider this further in the next chapter, in the next subsection I examine Stachel's take on the hole argument.

5.3.2 Stachel's relationalism and individuating fields

The problem of the hole argument is a result of the general covariance of the field equations for gravity. This property of the equations allows us to derive both g and $\phi^* g$ as solutions. But $g(x) \neq \phi^* g(x)$, and so the manifold substantivalist is supposedly forced to admit that these solutions represent distinct possibilities. However, the curious feature of the problem is that if we consider $x' = \phi(x)$, then we find that $g(x) = \phi^* g(x')$. We want to say in this case that g and $\phi^* g$ represent one and the same gravitational field. It is only when we view the points of the manifold as possessing their identity independently of the solutions that we get the result that they represent distinct possibilities. Stachel proposes that we should instead view the points not as independently individuated, but as individuated once a metric tensor has been specified, once the equations have been solved. Thus, he claims that diffeomorphism invariance has the effect of:

> precluding the existence of any pre-assigned (kinematical) spatio-temporal properties of the points of the manifold (even locally) that are independent of the choice of a solution to the field equations (no kinematics before dynamics). The physical points of space–time thus play a secondary, derivative role in the theory, and cannot be used in the formulation of physical questions within the theory (they are part of the answer, not part of the question). ([Stachel, 2003], p. 16)

Hence, we should forget about $g(x) \neq \phi^* g(x)$, because this implies that we can continue to refer to the *same* point of spacetime independently of the metric, and we should deal only with $g(x) = \phi^* g(x')$. Stachel suggests that we should look

[179] Even in the quantum theory this move will be possible. For example, in loop quantum gravity certain objects known as 's-knots' take the place of the classical metric as the best representation of quantum space—see [Rovelli, 2004] for details. Like the metric-field substantivalist, a determined substantivalist could avail herself of the s-knots using the same kind of reasoning. I describe this move in detail in [Rickles, 2005].

upon this latter relation as showing that the identity of the point x 'tracks' the metric, so that it brings its identity over with it.

This all sounds very similar to Newton's position on the identities of space points. Recall that he claimed that if two points were to swap places they would swap their identities. This claim flowed from the fact that he was using the metric to individuate the points, and of course the metric in Newton's theory was a background structure, it was non-dynamical, so it could perform such a function. Presumably Stachel would agree with this, for all such models will be isometric. The points can be individuated independently of solving the equations of motion for the dynamical objects. But in general relativity the metric is a dynamical object, we need to solve its equations of motion, so we cannot individuate the points independently of a solution if we follow Newton's idea—and, in general, distinct models will not be isometric.

The key difference for Stachel lies in what he calls an "individuating field"—a set of properties that can serve to uniquely distinguish the points of a manifold. In the case of Newton's theory, as in general relativity, the metric allows for an individuation of the points of the manifold. However, in the former case the metric is non-dynamical, so it can individuate independently of solving any equations. In this case the metric is what Stachel calls a "non-dynamical individuating field". Not so in general relativity, for the metric is dynamical. One cannot use the metric to individuate until one has solved for it. In this case we have a "dynamical individuating field". It doesn't have to be the metric that performs the function; a physical matter field might do the job just as well. But if we deal with vacuum general relativity then such an option is clearly unavailable and the metric must functions as the unique individuator of points. This goes back to my point in §1.1 that it is background independence that distinguishes general relativity from pre-GR theories of spacetime.

Stachel's response is directed as much at Einstein's hole argument as it is at Earman and Norton's. According to Stachel, manifold substantivalism is tantamount to the possibility of being able to provide a "kinematical coordinatization" of spacetime. The lesson of the hole argument for Stachel is that such a possibility leads to an ill-posed Cauchy problem for general relativity. Since it was manifold substantivalism that allowed for the coordinatization, it should be rejected for the reasons advanced by Earman and Norton. Stachel puts the point as follows:

> If the points of the manifold were physically distinguished *kinematically* (*i.e.*, independently of the solutions to the field equations), we should have to regard these solutions [hole diffeomorphs] as physically distinct. ([Stachel, 2003], p. 24)

Of course, it is just this excess that leads to the indeterminism. But by using the metric field to individuate the points, the indeterminism is eradicated, for the points cannot be labeled before a solution, they don't 'sit' under the metric waiting for a redistribution, they 'come about' in virtue of a certain distribution, and diffeomorphic distributions carry the points identities over. In particular, it makes no sense to speak of the same point existing in different models and worlds. The Kripkean rigid naming of points cannot operate in this context because the referents of the names get their individuality from the structure of the *dynamic* metric field.

Now, Stachel calls this relationalism, for he sees spacetime points as coming into existence relationally. This is clearly a further step from supposing that they are preexisting, but somehow indeterminate (more or less Hoefer's suggestion for the substantivalist). But it is important to note that they are not brought into existence by relations between material objects. They are brought about by relations between parts of the metric field; as Stachel puts it (echoing Einstein):

> until a metric tensor field is specified there is no spacetime ... even the topology of the differentiable manifold associated with a model of the theory cannot be specified a priori, but must be chosen so that it is compatible with the metric of that model. ([Stachel, 1993], p. 143)

Stachel isn't explicit as to exactly *how* the points are individuated by the metric. He mentions that one might use the values of the four invariant scalar fields built from contractions of the Weyl tensor, but leaves out the exact details.[180] We saw earlier that there are substantivalists who would like to help themselves to the metric field as forming part of substantival spacetime and applying a similar method to Stachel for understanding the points of spacetime—Hoefer's proposal can be seen along these lines. Stachel is adamant that such a position is at odds with substantivalism; he admits the metric field provides the chronogeometric structure of spacetime, but he argues that they ignore "the second role of the metric tensor field [that it] generates all the gravitational field structures ... and the gravitational field is just as real a physical field as any other" ([1993], p. 144). He views the approach as relational precisely because the spacetime relations and points are derivative from the "metric-cum-gravitational field". This is verging on word mongering: the same object is being used in more or the less the same way to defend opposite positions simply by viewing at as 'space-like' or 'matter-like'. Both parties agree that the metric field is a *real* physical field: Stachel views this aspect as debarring it from having the status of substantival spacetime and Hoefer views this aspect as all the more in favour of viewing it as a substantival spacetime! I think that a position that eschews or collapses the opposition is better suited, for there really is no opposition here: both parties agree it is real and (ontologically) free standing! Notice that the question is no longer one concerning the reality of spacetime points, but whether or not the metric is spacetime or whether the metric is material and determines spacetime (relations and points). Hoefer opts for the former, and takes spacetime points to be individuated by the metric; Stachel opts for the latter and takes spacetime points to come into existence with the metric. We can see that there is a difference in that Hoefer views the points as existing *independently* of the metric (but without primitive identity) whereas Stachel does not.[181] But this difference is doing no work, and it is not drawn from what general relativity tells us.

[180] Dorato and Pauri [2006], have recently filled in these details. They argue, as I do, though for different reasons, that this way of individuating points suits a structuralist conception of spacetime, rather than relationalism as Stachel claims. I consider their proposal in the final chapter. See also the subsection on Saunders' position below, §5.3.3.

[181] In fact, it isn't entirely clear that Hoefer endorses *this* form of substantivalism; presumably he would see the fact that there is *always* a metric field present as making this independence claim incoherent. If this is the case then the division between his and Stachel's position is ever more blurred.

general relativity can tell us nothing about the natures of the points: all we have to work with is the metric and the relations and properties it has. The question of which points play which role is inconsequential.

5.3.3 Saunders' non-reductive relationalism

The hole argument can be viewed as showing that the particle coordinates at a given time are underdetermined by the field equations; they are arbitrary functions of time—or, in other words: gauge. Likewise for the values of fields. Shifting from a coordinate dependent approach, we can couch the argument in terms of the points of the manifold so that the values of local fields or particle position are underdetermined. According to the hole argument, general relativity cannot predict such quantities. As Saunders ([2003a], p. 152) rightly points out, the hole argument—like its ancestor the Leibniz-shift argument—targets absolute quantities. The manifold is a background structure in general relativity, and position relative to it is an absolute quantity. The natural solution is to shift focus away from absolute quantities, which are not invariant under the transformations of general relativity, to those that are invariant. These happen to be relational quantities, as they were in the resolution to the Leibniz-shift argument. That is, the invariant quantities of general relativity are those that are not defined relative to the manifold, but with respect to physical fields or objects.

Now, Saunders argues that the sort of relationalism underwritten by such a response to these symmetry arguments "has nothing to do with a *reductionist* doctrine of space or spacetime" (*ibid.*), *i.e.* with what I have been calling relationalism and what Saunders calls "eliminative relationalism". He claims further that the response can be applied to "any exact symmetry in physics". I fully concur with Saunders, and indeed much of what I have been trying to get at in my discussion of symmetries is expressed by Saunders. However, I associate the view with structuralism, whereas Saunders advocates a position that resembles a hybrid between Stachel's relationalism and sophisticated substantivalism. More contentious is Saunders' claim that this position is a "natural expression of Leibniz's ... principle of identity of indiscernibles". This uses a 'modernized' version of PII informed by modern logic: from this version there follows Saunders' position, that he calls "non-reductive relationalism". This move forces Saunders to take a stance concerning the ontological status of those objects related by exact symmetries. I have argued that since incompatible positions can be held with respect to such objects, we should choose a neutral position. Such a position flows from my idea that physics operates with a version of PSR, such that the natures of the objects are irrelevant to the structures that possess the symmetries.

Saunders' method is quite ingenious. He argues that Leibniz was led to his eliminative relationalism because of the logic of his time, based as it was on the notion that propositions were of subject-predicate form. When relations are considered, the proposition is still taken to be of subject-predicate form, and applies to a single subject. The relations had to be reduced to monadic properties of their relata. This view of relations naturally underwrites what Saunders calls "Leibniz's independence thesis", the claim that a description of a thing should be intrinsic, containing no reference to other things or relations (*ibid.*, p. 13). Now Saunders points out that when we consider Frege's logic there is no such privileging of predicates, or "1-place concepts" in the terminology of Frege's *Begriffschrift*, with its distinction between 'object' and 'concept'. Relations are free standing and propositions aren't restricted to subject-predicate form. Saunders then examines how this shift in logic affects PII. Firstly, he notes that if one deals solely in 1-place predicates then PII says that objects with exactly the same properties are (numerically) identical. Adding higher-order predicates into one's language weakens the principle since then PII says that objects with exactly the same properties and relations are identical—there is another level of 'similarity' the objects have to satisfy. This gives us the strong and weak forms of PII respectively; clearly, Leibniz's logic forced him to endorse the strong form, and it is this overly stringent form that lends itself to easy counterexamples. Working with relations and adding identity to our language, Saunders presents an axiom schema formalising the indiscernibility of identicals as follows (*ibid.*, p. 18):

$$x = x; \quad x = y \rightarrow (Fx \rightarrow Fy) \tag{5.1}$$

This schema implies that terms with the same reference can be substituted *salva veritae*. Now Saunders (*ibid.*, p. 19) proceeds to give a definition of identity[182] using only terms 'x' and 'y', and unary predicates A (*i.e.* properties), binary predicates B (*i.e.* relations), up to n-ary predicates P (*i.e.* higher order relations), such that $x = y$ iff:

$$\begin{array}{ccc} & A(x) \longleftrightarrow A(y) & \\ B(x,u_1) \leftrightarrow B(y,u_1), & & B(u_1,x) \leftrightarrow B(u_1,y) \\ & \vdots \quad \vdots & \\ P(x,u_1,\ldots,u_{n-1}) \leftrightarrow P(y,u_1,\ldots,u_{n-1}), & & \text{and permutations.} \end{array}$$

The definition simply says that two things are identical if they match up on properties and relations. The relation conditions are defined so that, whatever relations x stands in (for some free variable: u_1 in the binary case u_1 to u_{n-1} in the n-ary case) y stands in too. I mentioned that for languages with only 1-place predicates one gets a strong principle of identity, such that: $[\exists F, (Fx \wedge \neg Fy) \rightarrow (x \neq y)]$. For more general languages, admitting higher order predicates, Saunders distinguishes three ways to get $x \neq y$, *i.e.* non-identity (*ibid.*, pp. 19–20). Firstly, he says that two objects are "absolutely discernible" if there is a formula (*e.g.* $P(z, u_1, \ldots, u_{n-1})$) with some free variable u_i that applies to one, x say, but not the other, y. In which case $x \neq y$. Secondly, two objects are "relatively discernible" if there is a formula in two free variables (*e.g.* $P(z, u_1, \ldots, u_2)$) that applies to x and y only in one order.[183] Thirdly, two objects are "weakly discernible" if $B(x,y)$ is true, B is a symmetric predicate (*i.e.* $B(x,y)$ iff $B(y,x)$), and B is irreflexive, so that $B(x,x)$ is always false—

[182] He credits the definition to Hilbert and Bernays, and notes that it has been defended by Quine.

[183] An obvious example is the 'taller than' relation: Joseph is taller than Dean, but Dean isn't taller than Joseph, hence, Joseph and Dean bear a different relation to one another. Clearly, asymmetry is at the root of this case of non-identity.

this counts as non-identity according to the definition because it implies that there is a u_1 such that $B(u_1, x)$ is true and $B(u_1, y)$ is false, namely for $u_1 = y$.[184]

This is Saunders' modernised version of PII: "objects are numerically distinct only if absolutely, relatively, or weakly distinct" (*ibid.*, p. 20). With this definition of identity, and with his ways of getting non-identity, Saunders is able to show that the standard counterexamples to PII are in fact examples of weak discernibles, and so do not violate his PII. Black's two qualitatively identical iron spheres in empty space are weakly discernible according to Saunders' account because there exists an irreflexive (distance) relation. Symmetry and qualitative identity are not sufficient to secure indistinguishability, though they are clearly necessary. He notes that there is a counterexample to it in the form of two or more bosons in exactly the same state; so PII is neither necessary nor contingent, it still faces difficulties in quantum mechanics. Though it can accommodate fermions, for even in the most symmetrical scenario—"where the spatial part has exact spherical symmetry, and the spin state is spherically symmetric too" (*ibid.*)—the fermions will satisfy the relation of having opposite component of spin to one another but not to themselves. This is clearly irreflexive and so any two fermions will be weakly discernible. Saunders draws metaphysical conclusions from the violation of his PII by bosons. He advocates a non-individualistic view according to which bosons are modes or excitations of a gauge field (with the exception of the Higgs boson). Note that given his PII, it is not possible to advocate the 'state restriction' view whereby bosons are individuals whose wave-functions are subject to symmetrization as an initial condition. Now, I am quite sympathetic to this view for it makes a principled distinction between 'matter' and 'force', a difference that seems to occur in nature, but which is conflated on most other conceptions of quantum particles (*cf.* Saunders [2003b], pp. 294–5). However, Saunders continues to refer to them as "objects", even though he claims that "one cannot *refer* to any one of them singly", and suggests that they be called "*referentially indeterminate*" (*ibid.*). I think as far as bosons go, he'd do better to drop all talk of objects at a fundamental level entirely, possibly in favour of pure structure. The latter might seem vaguer, but at least one can refer to it singly and determinately! As pointed out in French & Rickles ([2003], p. 228), the non-reductive nature of this form of relationalism sits well with the structuralist notion of individuation of relata by relations, according to which the relata do not have ontological primacy over relations but are understood in terms of "intersections of relations".[185]

Saunders puts his modernised version of PII to work in the context of spacetime theory. The position that results is, as he notes, similar to Hoefer's position; I think it is similar to Stachel's too. His claim is that although spatial points in a homogeneous space cannot be absolutely discerned by that subset of properties that apply to them (for they will share these properties), they can be discerned by relations to material objects and events (*ibid.*, p. 23). This is where he differs

184 An obvious example here is to choose B as a distance relation between two objects: Steve is 10 meters away from Dean, and Dean is 10 meters away from Steve (so they are not relatively discernible), but Dean is not 10 meters away from himself, and neither is Steve.

185 I take this up again in the final chapter, §9, where I argue for a position that I call 'minimal structuralism', based upon the Eddingtonian idea that neither relations nor relata should be seen as ontologically fundamental.

from Hoefer: Saunders points out that his notion of referential indeterminacy corresponds to Hoefer's idea that primitive identity fails for spatial points. It follows that there is no hole-type indeterminism or shift-type underdetermination regarding which points are occupied by a system of matter for "there is no fact of the matter as to which point is occupied" (*ibid.*, p. 25; see also Saunders [2003b], p. 304). However, Saunders claims that the points referred to by a matter distribution *can* be uniquely picked out, *contra* Hoefer. The reason: Hoefer denies PII and embraces Leibniz's independence thesis. Hence, Hoefer calls his position 'substantivalist' whereas Saunders calls his 'relationalist'. Hence, Saunders is able to avoid the hole argument by endorsing LE; he believes in spacetime points but in a way similar to Hoefer (they are not absolutely discernible, and do not possess primitive identity, but they are weakly discernible so that empty spaces are possible); appealing to his PII, however, relations to matter can serve to individuate points uniquely. The hole and shift arguments are avoided, but the symmetries are retained. Hence, the relationalism refers to the relations between spacetime points and matter, not to the reductive doctrine that deals only in relations between material objects. In this way, Saunders can be seen as providing a response to Grünbaum's problem of individuating points in homogeneous empty spaces; he writes that:

> If there are no *extra*-geochronometric physical entities to specify (individuate) the *homogeneous* elements of space–time ... then whence do these elements of otherwise equivalent punctal constitution derive their individual identities? ... I see no answer to this question as to the principle of individuation here within the framework of the ontology of Leibnizian identity of indiscernibles. Nor do I know of any other ontology which provides an intelligible answer to this *particular* problem of individuating avowedly *homogeneous* individuals. ([Grünbaum, 1970], p. 587)

Hoefer, of course, simply denied the PII and denied primitive identity, in effect agreeing with Grünbaum as regards the bare manifold. But Hoefer doesn't see that as representing spacetime at all. Saunders, however, has a clear response even for the case of a bare manifold: the points are weakly discernible and so distinct individuals. Clearly, we can simply disagree that Grünbaum is correct in speaking of the points of a manifold as representing spacetime; we can say that the metric field either is spacetime, is part of spacetime, or can serve to individuate the points of the manifold. We have seen that Hoefer opts for the former, Maudlin for the middle option, and Stachel for the latter option. Saunders' position is more subtle, as we have seen.

Saunders considers the Bergmann–Komar idea of using the four non-vanishing invariants of the Riemann tensor to individuate points, as did Stachel (see Bergmann & Komar [1972]). For non-symmetric cases, this procedure will give four real numbers that (hopefully!) differ at distinct points of the manifold. Macroscopic objects, fields and so on, can then be referred to their values, so that for the four invariants $(\xi_1, \xi_2, \xi_3, \xi_4)$, we can speak of the field ρ having the value λ at the point $(\xi_1, \xi_2, \xi_3, \xi_4)$: $\rho(\xi_1, \xi_2, \xi_3, \xi_4) = \lambda$. Saunders implicitly implies that this works for the case of general relativity, but he rejects it on the grounds that it fails in the Newtonian theory of gravity, for the invariants will be constant along the integral curves of the Killing vector fields. The values of the invariants will then be the same for all points of an orbit of the symmetry group. One must refer to matter again to individuate the points (*cf.* Stachel [1993], p. 143). Recall that this is slightly similar to the problem of symmetries of the metric faced by Maudlin.

This brings me to a serious problem with Saunders' approach, and with those approaches that use the Bergmann–Komar procedure in general. Classically, the method works, I admit that: relations either to matter or the invariants of the metric can serve to individuate points. However, when we turn to quantum theories, of matter and gravity, we face a problem: the matter and metric fields will be quantized. If we want to individuate the points by the value of some fields, then these fields had better have well defined values. But, generically, they won't in a quantum theory. We could only get a well defined value for the relevant observable if we measured the field. This would require a position whereby the points of space are individuated by making whatever one wishes to use to individuate the points *interact* with another thing. This is relationalism one level up from Saunders'. I don't say this isn't workable, but it means that if we take the points to be individuated in general by matter or metric, then in non-interacting situations, the points will still be indeterminate. Saunders will still be able to make use of his PII to ground the identities of the points of space, but this will be dependent upon an interaction. That strikes me as a strange position to hold. Again, I say, better to do away with points as fundamental existents *à la* Saunders, and if one thinks that they are indispensable, to view them as nodes in a network of relations. There will still be quantumness, be we don't view this as applying to independent points: material and points are part of one and the same network.[186] I should point out that Saunders does appear to switch to something like this view ([2003b], p. 305); but insists on clinging to a "thin" notion of object. Neither objects nor relations are given full ontological weight over the other; rather, they are on all fours—hence the *sobriquet* "non-reductive" to describe this form of relationalism. We need to ask just what work is being done by this thin notion of object; indeed, it is so thin as to be almost worthless. Perhaps Saunders is merely concerned to have a framework that accommodates many *types* of object, and to give them a satisfactory treatment: from spacetime points and quantum particles to cabbages and kings. Fair enough, but if this work can be done without them—using the relational structure and deriving 'objects' from it—then why claim that the objects are to be included in our ontology as well?

[186] I suppose there is the possibility of shifting to an interpretation of quantum theory that views position as determinate, such as the Bohm interpretation. Or shifting to an alternative conception of logic, such as quaset theory (this option was suggested to me by Dalla Chiara, private communication). These will have to be shown to be compatible with quantum general relativity: the Bohmian proposal faces problems here since the metric will be quantized, and a classical metric determines certain crucial features of the formalism of that approach. Knowing that Saunders' preferred interpretation is the relative state theory, I would be interested to find out how this would apply in this context. However, this would take me too far afield.

CHAPTER 6

What Is an Observable in General Relativity?

It is a curious fact about the hole argument that the indeterminism is not an observable feature; it is not an *empirical* kind of indeterminism that we could in any way detect. Even if it was true that general relativity was indeterministic in the sense that Earman and Norton say the manifold substantivalist is committed to, it is of such a strange kind that we could never tell one way or the other whether we lived in a world that ran according to such a scheme. For example, Hoefer writes:

> Note that this indeterminism is just about *what individual points will underlie what material processes* (matter and metric fields). It does not entail a failure of determinism in terms of the nonindividualistically-expressed happenings in space and time, that is, nothing observable is made indeterminate by the hole argument. ([Hoefer, 1996], p. 9)

Why is this so? It clearly depends on what we take 'observable' to mean. Recall the two fields, g and $\phi^{*}g$, that resulted from an active diffeomorphism.[187] The fields are mathematically distinct: they are spread differently over the manifold, such that $g(x) \neq \phi^{*}g(x)$. But this is the *only* way they are distinct, only by their localization relative to the manifold—*i.e.* with respect to their *absolute* location. The crucial question is: do they represent the same physical situation, the same possible world? As Norton explains, "It would be very odd if they did not. Both systems of fields agree completely in all invariants; they are just spread differently on the manifold. Since observables are given by invariants, they agree in everything observable" ([2003], p. 114). But what things are observable? According to Einstein, following Kretschmann's bashing, only spacetime coincidences of fields and particles. This rather narrows down the space of observables, perhaps far too much, but it avoids the indeterminism and the unmeasurability of the quantities defined with respect to the manifold's points (by simply doing away with such quanti-

[187] By active diffeomorphism here I mean a mapping of the manifold to itself that preserves the topological and differential structure of the manifold. We made use of the carrying along (by the diffeomorphism) of geometric object fields on the manifold, such that if our diffeomorphism sends the point x to the point y then field values at y (post-diffeomorphism) look the same as they did at x (pre-diffeomorphism). Or, as Rovelli puts it, "[a] field theory is formulated in manner invariant under passive diffs (or change of co-ordinates), if we can change the coordinates of the manifold, re-express all the geometric quantities (dynamical *and non-dynamical*) in the new co-ordinates, and the form of the equations of motion does not change. A theory is invariant under active diffs, when a smooth displacement of the dynamical fields (*the dynamical fields alone*) over the manifold, sends solutions of the equations of motion into solutions of the equations of motion" ([Rovelli, 2001], p. 122).

ties). But the modern answer is pretty close: not necessarily *coincidences* between particles and fields, but *correlations* between field-values (one of which will be the gravitational field)—the latter presumably include the former.

This, or something analogous, was the case with the Leibniz-shift argument, permutations of indistinguishable quantum particles, and the indeterminism of Maxwell's theory. All agree that the theories *are* indeed deterministic or unproblematic at the level of empirically observable ontology—putting aside quantum indeterminism of course—and that we can make well confirmed predictions within each theoretical framework. The same goes for general relativity too; it has yet to be disconfirmed in any of its predictions about the behaviour of empirically observable objects. The equations of motion of the theory of general relativity are sufficient to propagate all empirically observable components of that theory. The problem concerns the empirically *unobservable* (unmeasurable[188]) ontology (if, indeed, we take there to be such): which individuals play which roles in the structure? If we chose as our motto 'what we cannot see cannot hurt us', there would be no problem with any of the cases I have examined so far. But surely we have progressed beyond such straight-jacketed empiricism? Perhaps, but any interpretation that takes a stand with respect to the ontological status of these "individual" elements—such as spacetime points and absolute location with respect to them—*on the basis of physics* will have the spectre of the Quine–Duhem problem to deal with: there will be multiple incompatible interpretations compatible with the theory and the evidence.

Observables are generally understood to be those quantities described by physical theories that are measured in physical interactions between systems; but they go beyond the empirically observable, for is no necessity that *we* can observe them. They encode information about the state of a system, and their values should be able to be predicted by the theory. Clearly, some of the candidates for observable ontology cannot be predicted, for one has at best a range of possible values connected by symmetry; one cannot determine the unique value from within this range. What I have been pushing for is an indifference or insensitivity of the laws of physics (specifically concerning observables) to *certain kinds* of unobservable ontology (qualitatively indistinguishable individuals). Specifically, those involving elements (objects) connected by gauge-type symmetries, such that if those elements were to appear permuted in different scenarios then the scenarios would be indistinguishable. In other words, the observables should not register haecceitistic differences—but this does not imply that there are none, simply that the physics is, or ought to be, a qualitative enterprise. This is borne out by gauge invariance principles, where this condition is built-in: in gauge theories, the observables give the same value on such elements, so that for some observable $\mathcal{O}$ and elements x and y related by a gauge transformation, *i.e.* $x \sim y$, we have $\mathcal{O}(x) = \mathcal{O}(y)$. General relativity is a gauge theory, with the gauge freedom given by the dif-

188 Recall that the unmeasurability stems from the fact that the local values of certain fields are underdetermined by the theory and the world's qualitative properties and relations; since there will be many assignments of values to the fields that are compatible with the equations of the theory and the empirical structure (*e.g.* the electromagnetic field in the case of Maxwell's theory). Thus, one will not know whether one has measured A_μ or $A_\mu + \partial f$ in the electromagnetic case, so one cannot view measurement as possible in this case: what is the *result* of the measurement? No measurement (observable) could ever distinguish between these cases.

feomorphisms of spacetime; a diffeomorphism's action on the fields amounts to a gauge transformation. However, in order to make proper sense of this proposal, we need to know what the observables are. I have dealt with the case of electromagnetism already, and showed that there were a number of candidates; if we wanted to escape the indeterminism we had to make them gauge invariant (invariant under U(1)-transformations of the fields). In this section I discuss the much more complicated problem of observables in general relativity.[189] In this case the observables will have to be diffeomorphism-invariant if the theory is to avoid the indeterminism. Since the diffeomorphisms comprise the gauge part of the theory, such observables will be thereby classed as gauge invariant as well. Any fields related by diffeomorphism will thus be understood as the same as far as the observables of the theory are concerned (*i.e.* as far as the physics goes). Since the observables will be the same in such cases, and since just such cases comprise the problem cases of the hole argument (namely metrics painted differently, yet diffeomorphically, on to the points of $\mathcal{M}$), the hole argument is avoided. Let us discuss this resolution further, before considering its philosophical consequences.

One of the first things we notice about when we consider general relativity as a gauge theory is that the metric variables with respect to which the hole argument is defined do not class as observables, and so any local (*i.e.* defined with respect to manifold points[190]) quantity constructed from the metric cannot be observable either.[191] Why not? Because the observables must be indifferent to permutations of the points, for such permutations (invertible C^∞ ones) comprise the gauge freedom in the theory. The diffeomorphisms underlie the general covariance of general relativity, and this tells us that the models that are thus related will be indistinguishable as far as quantities not connected to the pointwise location of the fields are concerned. Therefore, the hole argument fails to get a grip: general relativity isn't meant to predict values for these quantities, nor for any quantities localized to points of the (bare) manifold regardless of how well we can specify them in connected regions of spacetime. The kinds of quantity that do class as observables in general relativity are relational (or highly non-local), much as Einstein argued, though not quite so narrowly defined. (I agree with this classification, but below I tame the view of many physicists that this implies relationalism about spacetime.) Let us say a little more about these observables, and give some examples that connect with measurements.

[189] I only discuss the problem briefly here, focusing specifically on how the problem relates to the hole argument. In the next chapter I discuss the relations between the observables, the hole argument and the problem of time. This section can be seen as a bridge connecting the hole argument with the problems of time since those problems result from the application of the resolution of the hole argument presented in this section. In the final chapter I home in on ontological issues.

[190] There is a proof (for the case of closed vacuum solutions of general relativity) that there can be no *local* observables at all [Torre, 1993]—'local' here means that the observable is constructed as a spatial integral of local functions of the initial data and their derivatives.

[191] Though, you will recall, the metric does contain invariant components. However, these are best used to define local events and coordinates, rather than functioning as self-contained observables. One localizes observables with respect to these invariants.

6.1. DEFINING OBSERVABLES

It is observables that give us our connection to the world in the context of physics; they are the things we measure and whose values we predict. They form the qualitative character of a world in the sense that two worlds that are duplicates in terms of the observables they contain, and in terms of their assigned values, are qualitatively indistinguishable. I think that the resolution of the hole argument can only come about once we have a proper grasp on what the theory of general relativity is about and what it is supposed to predict. This might sound rather obvious, but this involves the problem of defining the observables of general relativity—a notorious problem that I discuss more fully in the next chapter. In this section I simply skirt over many 'deep' issues, and highlight the relevance of the definition of observables to the hole problem (including a natural resolution) and, in a preliminary way, to the problem of time.

In general relativity, the physically relevant quantities are taken to be the diffeomorphism invariant quantities, or Bergmann observables. These are any quantities that spit out the same values under 'draggings' of the dynamical fields (the building blocks of the observables) by diffeomorphisms. If we accept the view that general relativity is a gauge theory, with the gauge part given by the diffeomorphisms, then the observables will be gauge invariant quantities, or Dirac observables. This is usually taken to mean that any quantities that are related by a gauge-transformation describe the same physical state; one and the same physical possibility is being multiply represented, and the excess representational machinery is deemed to be mere surplus structure. But this is not necessary. As far as the physics goes, I think that such differences should not make a difference; the observables grasp on to the qualitative character of the world, and any quantities that fail to distinguish between qualitative duplicates should be ruled out on the grounds that they would not be measurable or predictable in the context of the theory. This, I submit, is what is desired from gauge invariance; when physicists say *physical* equivalence, they simply mean to imply *qualitative*, or *empirical* equivalence. Gauge invariance of the observables then implies that those differences that are non-qualitative do not get counted in, say, statistical algorithms or in the initial-value problem of the theory (or it does so only 'up to a gauge transformation'). But this does not imply that there are not such differences, or that the ontological elements that are allowing for the qualitatively indistinguishable worlds (*e.g.* the manifold and its symmetries) do not exist. This cannot be read off the physics, but the observables don't care about such things.[192] These quantities, as we have seen, avoid the hole argument, for that argument only targeted those quantities that were changed by a diffeomorphism, quantities localized at points of the spacetime manifold (*e.g.* the scalar curvature at the point $x^i \in \mathcal{M}$). In the case

[192] This might sound like nothing more than empiricism, or at best constructive empiricism. However, the conception of observables I am talking about here goes way beyond what an empiricist, of any stripe, would be willing to countenance. The reason is that 'qualitative' is bound up with the technical sense of 'observable', and this includes magnitudes that would not be measurable by us, and certainly not with the direct, unaided senses. The understanding is, I say, *realist*: the observables give us directly the structure of the world. Metaphysical arguments then have to be used to convince us whether or not there is anything else besides this structure, physics alone is not up to that job. I discuss this matter in detail in §9.4.

of Earman and Norton's version of the argument, it was the absolute positions of the dynamical fields relative to absolute spacetime that was targeted.

What is an example of an observable of general relativity? Rovelli [2000] gives the following examples: 'the distance between Earth and Venus during the last solar eclipse'; 'the number of pulses of a pulsar in a binary system that reach the Earth during one revolution of the system'; and, 'the energy deposited on a gravitational antenna by a gravitational wave'. Clearly such quantities are independent of individual points of the manifold: they don't mention coordinates or spacetime at all. Rather, they highlight a dependency on dynamical objects, including the gravitational field. Now, one will doubtless compute each of these quantities *in practice* by using a coordinate chart, by using a model containing labeled spacetime points. One can cite the events at certain points in the chart and compute distances using a metric associated to the chart. But the quantities themselves are independent of this choice, they don't care which of a diffeomorphic set of charts we use; if they weren't thus independent, they would not be measurable.

Of course, we will need to say what these observables actually are *ontologically* speaking. There are two natural candidates for observables: non-local (global) quantities[193] and relational (correlational) quantities (as mentioned above), or what Earman, following Einstein, calls "coincidence quantities" ([2002], p. 15). As Earman notes, there is something "rather strange" about an ontology composed of such correlations since "[w]e are used to thinking of an event as the taking on (or losing) of a property by a subject, whether that subject is a concrete object or an immaterial spacetime point or region" (*ibid.*), whereas the correlation events are "subjectless'—Stachel would, of course, speak of the subject, the point, itself being determined (individuated) by the correlation. I defend something very similar to Earman's idea in the final chapter, where I line it up with structuralism.

This puzzling quality highlighted by Earman is shared by the holonomy and Wilson loop observables that we discussed in Chapter 3. In fact, the Wilson loops are the most natural choice for the fundamental variables in the best canonical approach to general relativity and for its quantization, so a non-localized ontology of the kind envisaged in the correlations fits this rather well—the loops are naturally gauge invariant and can be made diffeomorphism-invariant. I will argue in the final chapter that this feature sits well with a structuralist position but rubs against both relationalist and substantivalist positions: the latter positions require a notion of individual subjects that speaks against the preferred loop variables. The point is that we should not think of the correlates as being *composed of* independent individuals since, in isolation, these 'components' are not gauge invariant (*cf.* Earman, *ibid.*, p. 16): they have no reality outside of the correlation in which they figure.[194] I put this feature (combined with structuralism) to work in the next chapter where

[193] An example would be the 4-volume of the universe: $V(\mathcal{M}) = \int_{\mathcal{M}} d^4x\sqrt{|\det g|}$. Clearly such a quantity is not changed by the kind of diffeomorphisms appearing in the hole argument; the volume of an entire spacetime is not the sort of thing that depends on which point plays what role.

[194] We find something like this view appearing in Einstein's work shortly after his completion of general relativity: "the gravitational field at a *certain location* represents nothing 'physically real,' but the gravitational field together with other data does" ([Einstein, 1918], p. 71). Hence, the gravitational field, and any other fields, are 'ontologically entangled' in the correlations.

I argue that it can be used to evade a pair of serious objections to the correlation view in the context of the problem of time.

Let us briefly return to the more formal definition of observables to make it absolutely plain how the gauge invariant conception evades the hole argument, and how it relates to the methods of the previous chapters. Following this, I consider the relevance of these observables, along with the gauge invariant interpretation, to the issue of spacetime ontology.

According to the theory of constrained Hamiltonian systems the classical observables are just those dynamical variables that have weakly vanishing (*i.e.* on the constraint surface) Poisson brackets with all of the first class constraints. Thus, for any dynamical variables, $\mathcal{O}_i$, on the phase space of our constrained Hamiltonian system, with first class constraints ϕ_j, $\mathcal{O}$ is an observable iff $\{\phi_j, \mathcal{O}\} \approx 0$. We have seen this definition at work in the context of electrodynamics, and something like it in quantum statistical mechanics. The core idea is that any states related by the symmetries generated by the constraints (=gauge transformations) are physically indistinguishable, therefore the observables of a theory should not distinguish between such states. These states themselves are not observables of the theory of course, for if they were we could not measure them: there is no way to distinguish, in a qualitative way, the individual elements of a gauge orbit. This motivates the idea of gauge invariance and the idea that the observables of a theory should be precisely those quantities that embody gauge invariance.

Recall that the constraints define a hypersurface $\mathcal{C}$ within the full phase space Γ of general relativity, picking out the dynamically possible histories. The gauge motions, corresponding to the action of the diffeomorphisms, are defined everywhere on Γ; however, they leave $\mathcal{C}$ invariant. Since the Hamiltonian of general relativity is a sum of first class constraints, the dynamics takes place on $\mathcal{C}$. The orbit of any point $x \in \mathcal{C}$ under diffeomorphisms (of the kind considered in the hole argument) will be a curve contained within $\mathcal{C}$. In particular, the diffeomorphisms evolve data along the gauge orbits—this is the root of the problems of time to be discussed in the next chapter: time evolution is just one kind of diffeomorphism. The hole argument is then simply a result of the diffeomorphism (vector) constraint on the initial data.[195] The change generated by this constraint corresponds to a spatial diffeomorphism on a hypersurface Σ of just the kind envisaged by Earman and Norton—*i.e.* it corresponds to an alteration in the properties of the points of the manifold. The manifold substantivalist was supposedly committed to points whose properties and individualities were fixed independently of such alterations of properties of the hypersurface, and so the change brought about by the action of the constraint was seen to lead to distinct physical possibilities for the points.[196]

[195] I restrict attention to the diffeomorphism constraint purely for convenience, in order to distinguish the hole problem from the problem of time. Strictly speaking, the hole argument is generated by all of the constraints. However, we can set the Hamiltonian constraint to zero and yet still generate a hole-type problem.

[196] Recall that the intrinsic properties of the points are exhausted by the topological and differential structure of the space, and as such are themselves diffeomorphism invariant. The properties utilized by the hole argument are relational, or extrinsic, properties of the points. For example, in two solutions of the field equations that differ by the action of the diffeomorphism constraint, the point x might have scalar curvature R according to one solution, but a distinct curvature R^* in the other. The response of the sophisticated substantivalists is to question the grounds for saying that the point x is the *same* point in the pair of solutions. If identities are grounded qualitatively (including relational properties) then the point x

However, the natural take on the constraints is precisely that the theory is invariant under the changes they generate (*i.e.* under Diff(Σ) motions). The physically (*i.e.* as far as the observables go) relevant content of the theory of (vacuum) general relativity is given by the intrinsic and extrinsic *geometry* of the hypersurface, not the individually localized metric and extrinsic curvature. Wheeler explains the concept of a 3-geometry by using the concept of a fender (a 2-geometry):

> In whatever way our coordinates are painted on its surface, in whatever way the points of that surface are named or renamed, the fender keeps the same 2-geometry. Similarly for a 3-geometry. In mathematical terms, a ${}^{(3)}\mathcal{G}$ is not a positive definite 3×3 matrix; instead, it is an equivalence class of such metrics that are transformable, one into another by diffeomorphisms. ([Wheeler, 1968], p. 246)

This is tantamount to the gauge-invariance view. This structure is what is invariant under diffeomorphisms (indeed, each geometry corresponds to a whole gauge orbit of the symmetries generated by the constraints).

In other words, the observables should commute with the constraints. Once this way of defining the observables is taken on board, the indeterminism that threatened in the hole argument is washed away. The idea is that it is the 'proper' observables, as opposed to the gauge variant variables (*e.g.* the value of the metric field at a point), that are indifferent to symmetries: metric fields at points are just not diffeomorphism invariant, and therefore are not gauge invariant. Thus, the observables in quantum mechanics were permutation invariant, they were insensitive to matters of particle role; this meant that any physically measurable stuff (expectation values, transition amplitudes, and so on) was independent of such permutations. The observables of Newton's theory were invariant under the group $\mathcal{G}_N$, they were insensitive to matters of absolute position; this meant that the physically measurable things (relative distances between particles, the force on a particle, and so on) were independent of the operations corresponding to elements of $\mathcal{G}_N$. The observables of Maxwell's theory were invariant under U(1) gauge transformations, they were insensitive to which of a smoothly related set of vector potentials was used; this meant that the physically measurable things (magnetic field strength, holonomies, and so on) were independent of such alterations in the vector potential. Likewise, general relativity's observables will be insensitive to diffeomorphisms: the specific roles played by the individual points are not physically relevant; this means that the physically measurable things are independent of Diff($\mathcal{M}$)-transformations—these will be quantities such as those Rovelli mentioned on p. 131.

This state of affairs has led, in each case, to a battle over the nature of the individuals and whether they exist or not. The exact empirical equivalence of those interpretations that treat them as existing and those that do not, and as having well defined identities and those that do not, means that both can claim compatibility with theory. All types of substantivalist and relationalist, and indeed

will be just that point that has scalar curvature R (Saunders showed that this strategy does not require that each point has a distinct qualitative character to be counted as a numerically distinct individual—so long as relations are included, the points can be individuated without being identifiable).

structuralist, can adopt the gauge invariant position on observables and thus avoid the hole argument with ease—just as various incompatible ontological positions could each adopt the gauge invariant view of the observables of the other theories discussed. This is the resolution I favour, and the fact that all of the distinct ontological positions can be accommodated by this framework suggests that some kind of structuralist position should be adopted (I outline this position *passim* in what follows, and in detail in §9). Next I consider what physicists have made of the gauge interpretation of general relativity's observables *vis-à-vis* spacetime ontology.

6.2. WHAT IS THE SIGNIFICANCE OF RELATIONAL LOCALIZATION?

The standard view amongst physicists is that a gauge-theoretical understanding of diffeomorphism invariance implies that the localization of fields is relational (*i.e.* grounded by relations between fields rather than manifold points), and that this in turn implies spacetime relationalism, or at least anti-substantivalism. Rovelli sketches the supposed implication as follows:

> [Diffeomorphism invariance] implies that spacetime localization is relational, for the following reason. If (ψ, X_n) is a solution of the equations of motion, then so is $(\phi(\psi), \phi(X_n))$ [where ϕ is a diffeomorphism]. But ϕ might be the identity for all coordinate times t before a given t_0 and differ from the identity for some $t > t_0$. the value of a field at a given point in $\mathcal{M}$, or the position of a particle in $\mathcal{M}$, change under the active diffeomorphism ϕ. If they were observable, determinism would be lost, because equal initial data could evolve in physically distinguishable ways respecting the equations of motion. Therefore classical determinism forces us to interpret the invariance under $\text{Diff}_{\mathcal{M}}$ as a gauge invariance: we must assume that diffeomorphic configurations are physically indistinguishable. ([Rovelli, 2000], p. 3779)

This is then taken to imply relationalism since diffeomorphic configurations

> are only distinguished by their localization on the manifold. They are different in the sense that they ascribe different properties to the manifold points. However, if we demand that localization is defined only with respect to the fields and particles themselves, then there is nothing that distinguishes the two solutions physically ... It follows that localization on the manifold has no physical meaning ... In GR, general covariance is compatible with determinism only assuming that individual spacetime points have no physical meaning by themselves ... Reality is not made up of particles and fields on a spacetime: it is made up of particles and fields (including the gravitational fields), that can only be localized with respect to one another. No more fields on spacetime: just fields on fields. ([Rovelli, 2004], pp. 70–1)

Hence, the 'physical' aspects of a system are not given by specifying a single field configuration, but instead by the "equivalence class of field configurations

... related by diffeomorphisms" (*ibid.*), a geometry. The observables of such a system are then given by diffeomorphism invariant quantities. Such specifications of states and observables are clearly independent of any background spacetime: only gauge invariant quantities are to enter into such specification, and any reference to a background metric (via, for example, fixed coordinates or functions on $\mathcal{M}$) yields non-gauge invariant quantities. Thus, diffeomorphisms change the localization of dynamical fields on $\mathcal{M}$; this is represented in the Hamiltonian scenario by the action of the constraints. However, the localization is a gauge freedom, so any states or quantities involving localization to points will not be measurable. Rovelli, like so many others, sees a direct connection between taking the equivalence class of metrics and relationalism about spacetime since because the equivalence class appears to imply that the metric field is entangled with spacetime points. (But this entanglement can work for both relationalists and substantivalists alike, as we saw earlier.)

Rovelli is not alone in interpreting the hole argument as an argument for relationalism. As I said, it seems to be the majority view amongst physicists. For example, Lee Smolin writes that "the basic postulate that makes GR a relational theory is" that "*[a] physical spacetime is defined to correspond, not to a single* (M, g_{ab}, f), *but to an equivalence class of manifolds, metrics, and fields under the action of* Diff(M)" ([2006], p. 206). The idea here is that removing the symmetries (by 'modding out' by the diffeomorphisms) is taken to correspond to relationalism. In other words, relationalism is being aligned with reduction: we saw in Chapter 2 that this is a *non sequitur*; we shall return to the matter again in Chapter 8.

What status are we to attribute to the manifold once remove dependency upon its coordinates? Rovelli suggests that it is an "auxiliary mathematical device for describing spatiotemporal relations between dynamical objects" ([Rovelli, 2000], p. 3780). It is not without usefulness. Of course, spacetime coordinates enter into many areas of physics, especially mechanics and field theories, *i.e.* as positions of objects (particles, string excitations, *etc.*) or as the argument of a local field operator. Many physicists believe that general relativity rules out just such absolute local quantities—I agree. This is, again, seen to follow from, or imply, the practice of taking an equivalence class of manifolds and metrics under diffeomorphisms as the correct description of a world of general relativity. Smolin claims that a consequence of this view is that

> there are no points in a physical spacetime ... [since] a point is not a diffeomorphism invariant entity, for diffeomorphisms move the points around. There are hence no observables of the form of the value of some field at a given point of a manifold, x. ([Smolin, 2000], p. 5)

The latter point, that there are no local (*i.e.* localized to a *particular* spacetime point) observables in general relativity, is, as I have said, perfectly true of course. I think this is the real 'lesson' of the hole argument. Unfortunately, Smolin, like Rovelli, explicitly draws relationalist conclusions from the fact that the observables of general relativity are relational. Firstly, it does not follow that there are no points: that the observables are indifferent to matters of spacetime point role does not imply there are no spacetime points. Secondly, the fact that points of the manifold are problematic does not mean that there is no other notion of a point on the table. As Robert Dicke remarks, describing the view of J.L. Synge:

> general relativity describes an absolute space ... certain things are measurable about this space in an absolute way. There exist curvature invariants that characterize this space, and one can, in principle, measure these invariants. Bergmann has pointed out that the mapping of these invariants throughout space is, in a sense, labeling of the points of this space with invariant labels (independent of coordinate system). These are concepts of an absolute space, and we have here a return to the old notions of an absolute space. ([Dicke, 1964], pp. 124–5)

Here, as we mentioned in the previous chapter, the idea is to get a set of coordinate conditions that allow one to define a set of intrinsic coordinates. One constructs the complete set of scalars from the metric and its first and second derivatives, which for the matter-free case leaves four non-zero scalars that take different values at different points of the manifold. Hence, one achieves a complete labeling of the manifold in an intrinsic gauge invariant way—this follows from the fact that we are dealing with *scalars* which do not change their values under diffeomorphisms. These points can then be used to localize quantities which become gauge invariant as a result of the gauge invariance of the scalars. For Synge, the only difference between this space and Newton's is that the geometric properties of the Einsteinian space are "influenced by the matter contained therein"—that is, the latter is background independent. Of course, since we are dealing with invariants of the metric here, it is open to the relationalist to call this a 'material' field. So continues the interminable interpretive tug-of-war! The point is, it is always open for the substantivalist to construct an alternative interpretation within the framework that is supposed to be fit only for relationalists.

Connecting relational localization explicitly to the observables, Rovelli writes that since

> the only physically meaningful definition of location within GR is relational ... GR describes the world as a set of interacting fields including $g_{\mu\nu}(x)$, and possibly other objects, and motion can be defined only by positions and displacements of these dynamical objects relative to each other. [...] All this is coded in the active diffeomorphism invariance ... of GR. Because active diff invariance is gauge, the physical content of GR is expressed only by those quantities, derived from the basic dynamical variables, which are fully independent from the points of the manifold. [.] [Diff invariance] gets rid of the manifold. ([Rovelli, 2001], p. 108)

This is a fairly common view too, as I intimated above, but it is also a *non sequitur*: substantivalism is perfectly compatible with the view that observables of general relativity are relational, with a gauge invariant conception of the observables of the theory, the sophisticated substantivalists have demonstrated this—I gave two strategies above. There is no necessity to "get rid of the manifold" to help oneself to gauge invariance: Saunders' PII and his preservation of PSR could be wielded by the determined substantivalist here. Rovelli assumes that the view that the observables are relative beasts describing the relative location and evolution of dynamical objects implies relationalism about spacetime, but all this shows is that the observables are indifferent to matters concerning spacetime points. The *gauge variant* is by no means restricted to physicists. Weinstein too makes the same mistake: "formulating observables in [general relativity] as diffeomorphism-invariant objects eliminates any reference to the underlying spacetime manifold (arguably making the theory non-absolutist)" ([Weinstein, 2001], p. 68)—I assume that by "non-absolutist" Weinstein means "non-substantivalist". Once again, we have an interpretative underdetermination: both substantivalists and relationalists can lay claim to this feature of gauge invariance.[197]

However, aside from the fact that Rovelli wrongly aligns his position to relationalism, I think that there is much to recommend it. However, I prefer a conception based on neither substantivalism nor relationalism, nor on a distinction between matter and spacetime. The best way to overcome the problems raised by the symmetry arguments considered so far, and still to come, is to 'go structural' and eschew the 'subject-based' foundations upon which the traditional interpretations are predicated: the relational gauge invariant (or, what I shall refer to from now on as the *correlation*) view of observables is formally correct but the interpretation is wrong (based as it is on a position, relationalism, that is well underdetermined by this view of the observables). Thus, a kosher observable will be a correlation between field values (*e.g.* the value of the curvature of the gravitational field where and when the electromagnetic field strength takes on the value *F*[198]). These contain no dependence on the manifold, and determine locations and times via relations between quantities. The relational aspect suggests that the correlation can be broken down into components that have some ontological weight independently of their role in the correlation.[199] This is what a structuralist position denies. I elaborate this view further in the next chapter where I use it to defend Rovelli's evolving constants of motion and partial observables strategies for dealing with the problem of time from objections that focus upon the relational interpretation. I then defend the position in more detail, and compare it with other views, in the final chapters.

197 Of course, the gauge invariance here takes on a special form since it is mixed up with background independence. Strictly speaking, it is the way these two interact that lies underneath the various claims that relationalism is uniquely selected. Because there are no background fields we get the diffeomorphism symmetry from the manifold; hence, the states and observables of the theory must be constructed entirely from the dynamical degrees of freedom.

198 There are obvious symmetry problems here: if there are parts of the field with the same value then the approach fails. This is a problem with the Bergmann–Komar approach too: it rules out symmetries. However, one can form the observable with more than a pair of fields; one can bring in other degrees of freedom too.

199 Indeed, we see that this forms a class of objection to these positions in the context of the problem of time (*cf.* §7.4.2).

CHAPTER 7

Time, Change, and Gauge

The interpretive problems of gauge theory take on their most pathological form in the context of the problem of time.[200] I will argue that this problem is essentially the hole argument focused upon the Hamiltonian, rather than the diffeomorphism constraint. I therefore respond to the problem in much the same way as I responded to the hole argument: a non-reductive gauge invariant conception of observables coupled with structuralism. My two aims in this chapter are (1) to disentangle the debate between substantivalists and relationalists from the problem of time[201] and (2) to defend a structuralist resolution of the problem of time (and, therefore, to defend structuralism itself, in virtue of its 'good deeds'). The final chapter will then defend the structuralist account further.

7.1. HOLES AND GAUGE: A BRIEF RECAP

In their recent survey of the problem of time in quantum gravity, Belot and Earman note that there is a "sentiment—which is widespread among physicists working on canonical quantum gravity—that there is a tight connection between the interpretive problems of general relativity and the technical and conceptual problems of quantum gravity" ([2001], p. 214). Belot and Earman share this sentiment, and go even further in claiming that certain proposals for understanding the general covariance of general relativity *underwrite* specific proposals for quantizing gravity. These proposals are then linked to "interpretive views concerning the ontological status of spacetime" (*ibid.*). I agree with their former claim, but disagree with the latter: such proposals cannot be seen as linked with stances concerning the ontological status of spacetime *vis à vis* relationalism vs substantivalism.

The crucial claim for this chapter is that the gauge invariance reading of the general covariance of general relativity "seems to force us to accept that change is not a fundamental reality in classical and quantum gravity" (*ibid.*). I agree with Belot and Earman that, like the hole argument, the problem of time is an aspect of the more general problem of interpreting gauge theories. I also agree with Ear-

[200] The best places to learn about this problem are still Isham [1993] and Kuchař [1992]. Belot and Earman [1999; 2001] give two excellent philosophical examinations of the problem; the latter is more comprehensive and technically demanding than the former. I am indebted to these four articles, from which I learned much.

[201] Though the problems have an important bearing on the *structure* of time, by analogy with previous chapters, they cannot be decisive in questions concerning the *nature* of the existence and identity of space and time.

man's claim that the problems do not only have teeth in the quantum context, but bite in the classical context too (see Earman [2002], p. 6).[202]

The problem posed to the substantivalist by the hole argument was diagnosed in terms of a direct interpretation of the (unconstrained) phase space of general relativity (conceived as a gauge system: constrained Hamiltonian system with first class constraints). Recall that a direct interpretation takes each and every point of phase space as representing a distinct physical possibility so that, when gauge-type symmetries are present, there are indistinguishable physically possible worlds (dynamically possible histories) connected by the gauge symmetries. The implication is that for an initial data set (q_0, K_0), on a hypersurface Σ, Einstein's equations fail to determine a unique future dynamical state: there are many diffeomorphisms ϕ, to the 'future' of Σ, such that $\phi(q_t, K_t)$ is a solution too. This is simply the basic content of the general covariance (active diffeomorphism invariance) of the theory. This brings in to sharp focus the connection with the interpretive problems of gauge theory, for the indeterminism of electromagnetism is understood in precisely the same terms: given an initial state (A, E), Maxwell's equations cannot uniquely determine the future dynamical state, for there are many smooth functions f such that $(A + \nabla \times f, E)$ is a solution. Applied to the hole argument, the gauge framework suggested that the diffeomorphism constraint be viewed, as Belot and Earman put it, "as shuffling the geometrical roles played by the points on Σ" ([2001], p. 226). The equations of motion, and the observables are indifferent as to which point plays which role, so long as the intrinsic geometry of Σ is preserved.

There are strategies for dealing with this problem of indeterminism (better: underdetermination) in gauge theory. In the previous chapter I adopted a gauge invariant account (on the unconstrained phase space) according to which the observables (the *true* physical quantities) are indifferent as to which gauge related element performs a particular function—or, as to which object contributing to the generation of gauge related states performs a particular function. The observable quantities were taken to be relational, formed by correlations between two *unobservable* quantities, gauge variant dynamical variables. This stance was neutral with respect to the nature of the gauge orbits: the orbits might contain elements that represent one-to-one, or the whole orbit might represent a single possibility. The standard (indirect) gauge invariant accounts, however, take physical states as standing one-to-one with whole gauge orbits. This strategy is taken to the extreme in reduced phase space methods, according to which the gauge orbits become single points in a new phase space. This space can then be given a direct interpretation without the worry of indeterminism and surplus structure. Indeed, as Belot and Earman point out, this direct interpretation of the reduced phase space coincides

[202] The reason that the problem is supposed to be a problem only for the quantum theory of gravity is that it is thought that one requires a unique, global time variable for the definition of the quantum dynamics, to construct the inner product, and so on. The belief is that the dynamics has to be unfrozen to allow for the quantum dynamics to be defined at all. I think this is wrong in three ways: (1) a background time variable is not *necessary* for quantization and for quantum theories—true, the quantum theories we have so far all have this feature, but generalized quantum theories are possible, whereby inner products are selected without a time parameter (see Ashtekar & Tate [1996]). (2) The problem is an artifact of the gauge interpretation of general relativity which is, of course, a classical feature too. (3) There are approaches to quantum gravity and to the problem of time that do not unfreeze the dynamics, but that proceed without a fundamental time parameter (as we see in §7.4.2).

with an (indirect) gauge invariant interpretation of the original extended phase space. However, my conclusion in the previous chapter was that, as far as the gauge symmetries of the hole problem go, general relativity offers good support to neither relationalism nor substantivalism. The fact that we can get the same results using both the reduced and extended phase spaces, coupled with the fact that the substantivalist and relationalist can make use of either formulation, means that the hole argument is inert when it comes to the question of the nature of spacetime *qua* 'substance versus relational entity'. The structuralism I offered up resolved the hole problem, since it was associated with the gauge invariant observables, and was neutral with respect to the elements connected by the symmetries and, therefore, I think it offers the best hope for an ontological foundation. I apply similar reasoning in what follows and show, in addition, that this interpretation can evade serious problems facing a popular response to the problem of time.

7.2. WHAT IS THE PROBLEM OF TIME?

There are two ways of understanding the problem of time: (1) in terms of states; (2) in terms of observables. These lead to quite distinct conceptual problems: the former leads to a problem of time and the latter leads to a problem of change.[203] The first problem concerns the fact that distinct Cauchy surfaces of the same model will be connected by the Hamiltonian constraint, and therefore will be gauge related. The gauge invariant view demands that we view them as representing one and the same state of affairs. The second problem concerns the observables: no gauge invariant quantity will distinguish between Cauchy surfaces of the above kind. Together, these problems constitute the *frozen formalism* problem of classical general relativity. Each of these classical problems transforms into a quantum version.

Let us fix some formalism so we can see how these two problems arise. We are working in the Hamiltonian formulation so we start by splitting spacetime into a space part and a time part. Thus, the spacetime manifold $\mathcal{M}$ is a background structure with the topological structure $\mathcal{M} = \mathbb{R} \times \Sigma$, with Σ a spatially compact 3-manifold. We begin with a phase space Γ, which we shall take to be the cotangent bundle defined over the space of Riemannian metrics on Σ.[204] Points in phase space are then given by pairs (q_{ab}, p^{ab}), with q_{ab} a 3-metric on Σ and p^{ab}

[203] If one believes that change is a necessary condition for time then the second problem will naturally pose a problem of time too, and *vice versa*. The necessity of time for change is fairly obvious, but the (Aristotelian) converse, that time requires change, has been rightly questioned in the philosophical literature (*e.g.* Shoemaker [1969]).

[204] I follow 'standard procedure' of couching my discussion in terms of the metric variables. However, I should point out that the canonical approach based on these variables (*i.e.* geometrodynamics) has been superseded by the connection (Ashtekar) variables and loop representations—a nice introduction is Ashtekar & Rovelli [1992]; for more detail see [Ashtekar and Tate, 1996]. These result in simpler expressions for the constraints and solutions for the Hamiltonian constraint (none were known for the metric variables!). The justification for sticking with the metric variables is simply that the problem of time afflicts any canonical approach and takes on much the same form regardless of which variables one coordinatizes the phase space with. Generally, one can simply imagine replacing any expression involving functionals of the metric with functionals of these other variables. I should qualify this by saying that the preceding remarks are valid only for the case of spatially compact spacetimes—see Lusanna [2005] for the state of the art in non-spatially compact spacetimes (but note that there is asymptotic background structure lurking in the boundary conditions of these models). My thanks to Lusanna for drawing this qualification to my attention.

a symmetric tensor on Σ. The physical (instantaneous) states of the gravitational field are given by points $x \in \tilde{\Gamma} \subset \Gamma$, where $\tilde{\Gamma}$ is the constraint surface consisting of points that satisfy the diffeomorphism (vector) and Hamiltonian (scalar) constraints: $\mathcal{H}_a = \mathcal{H}_\perp = 0$. These two constraints allow data to be evolved by taking the Poisson bracket of the latter with the former; thus $\{\mathrm{O}, \mathcal{H}_a\}$ changes $\mathrm{O} \in C^\infty \Gamma$ by a Lie derivative tangent to Σ and is generated by a spatial diffeomorphism, while $\{\mathrm{O}, \mathcal{H}_\perp\}$ changes O in the direction normal to Σ. The Hamiltonian for the theory is given by $\mathrm{H} = \int_\Sigma d^3x N^a \mathcal{H}_a + N\mathcal{H}_\perp$, where N^a and N are Lagrange multipliers called the shift vector and lapse function respectively. The dynamics is thus entirely generated by (first class) constraints.[205] The implication is that the evolution of states (*i.e.* motion) is pure gauge!

What I have described above is general relativity as a constrained Hamiltonian system. The observables $\mathcal{O}_i$ for such $\mathrm{H} = 0$ systems are defined as follows:

$$\mathrm{O} \in \mathcal{O}_i \quad \text{iff} \quad \{\mathrm{O}, \mathrm{H}\} \approx 0 \tag{7.1}$$

This condition states that observables must have weakly vanishing Poisson brackets with all of the constraints; *i.e.* they must vanish on the constraint surface. From this vantage point, the observables argument is ineluctable. I mentioned above that the dynamics is generated by constraints; or, in other words, the dynamics takes place on the constraint surface, and evolution is along the Hamiltonian vector fields X_H generated by the constraints on this surface (*i.e.* the gauge orbits). Therefore, the observables are constants of the motion: $\frac{d\mathrm{O}}{dt}(q(t), p(t)) = 0$ (where t is associated to some foliation given by a choice of lapse and shift). This much gives us our two problems in the classical context. As Earman sums it up, "the Hamiltonian constraints generate the motion, motion is pure gauge, and the observables of the theory are constants of the motion in the sense that they are constant along the gauge orbits" ([2003b], p. 152). Now to the quantum problems.

Depending upon one's interpretive strategy with regard to the constraints at the classical level, there will be distinct quantization methods for the classical theory, and these correspond to different strategies for tackling the problem of time.[206] Quantization along such lines splits into two types: one can either quantize on the extended phase space or on the reduced phase space. The former method, "constrained quantization", is due to Dirac [1964]: classical constraints are imposed as operator constraints on the physical states of the quantum theory. The latter method reduces the number of degrees of freedom of the extended phase space by factoring out the action of the symmetries generated by the constraints. Hence, the reduced space is the space of orbits of the extended space; it is a (quotient) manifold and inherits a symplectic structure (see Marsden and Weinstein [1974]): gauge invariance is automatic on the reduced phase space. The extended and reduced phase spaces are equivalent on a classical level, but gener-

205 Dirac's 'conjecture' for such constraints is that they generate gauge transformations: "transformations ... corresponding to no change in the physical state, are transformations for which the generating function is a first class constraint" ([1964], p. 23).

206 Since they associate methods of dealing with the constraints with particular interpretational stances on spacetime ontology, it is in just this way that Belot and Earman claim that quantization methods are linked to the substantivalism/relationalism debate.

ally they will be inequivalent on a quantum level (*cf.* Gotay [1984]), so the choice is non-trivial.

In brief, the constrained (extended phase space) quantization method runs as follows:

- Choose quantum states (representation space $\mathcal{F}$)[207]:

$$\psi[q] \in \mathcal{L}^2(\mathrm{Riem}(\Sigma,\mu)) \tag{7.2}$$

- Represent the canonical variables q_{ab}, p^{ab} on $\mathcal{F}$ as:

$$\hat{q}_{ab}(x)\psi[q] = q_{ab}\psi[q] \tag{7.3}$$

$$\hat{p}^{ab}(x)\psi[q] = i\left(\frac{\partial}{\partial q_{ab}}\right)\psi[q] \tag{7.4}$$

- Impose the diffeomorphism and Hamiltonian constraints:

$$\hat{\mathcal{H}}_a\psi[q] = {}^3\hat{\nabla}_b\hat{p}^b_a\psi[q] = 0 \tag{7.5}$$

$$\hat{\mathcal{H}}_\perp\psi[q] = \mathcal{G}_{abcd}\frac{\partial^2}{\partial q_{ac}\partial q_{bd}}\psi[q] - {}^3R(q)\psi[q] = 0^{208} \tag{7.6}$$

- Find a representation of a subset of classical variables on the physical state space, such that the operators commute with all of the quantum constraints. (One must also find an inner product making these self-adjoint.)

The classical observables argument filters through into this quantum setup since, by analogy with the classical observables, the quantum observables $\hat{\mathcal{O}}_i$ are defined as follows:

$$\hat{\mathrm{O}} \in \mathcal{O}_i \quad \text{iff} \quad [\hat{\mathrm{O}}, \hat{\mathrm{H}}] \approx 0 \tag{7.7}$$

Note that the weak equality '$\approx$' is now defined on the solution space of the quantum constraints; *i.e.* $\mathcal{F}_0 = \{\Psi : \hat{\mathrm{H}}\Psi = 0\}$. Clearly, if Eq. (7.7) did not hold, then there could be possible observables whose measurement would knock a state Ψ out of $\mathcal{F}_0$. The state version of the problem then follows simply from the fact that the quantum Hamiltonian annihilates physical states: $\hat{\mathrm{H}}\Psi = 0$. What motivates this view is the idea common to gauge theories that if a pair of classical configurations q and q' are gauge related then $\mathrm{O}(q) = \mathrm{O}(q')$, so we should impose gauge invariance at the level of quantum states too: $\psi(q) = \psi(q')$. The diffeomorphism constraint, Eq. (7.5), is particularly easy to comprehend along such lines; it simply says that for any diffeomorphism $d : \Sigma \to \Sigma$, and state $\Psi[q]$, $\Psi[q] = \Psi[d^*q]$. Were this not the case, one could use the quantum theory to distinguish between classically indistinguishable states. The Hamiltonian constraint is more problematic, for it generates changes in data 'flowing off' Σ, and is seen as generating evolution. If

[207] Where $\mathcal{L}^2(\mathrm{Riem}(\Sigma,\mu))$ are the square integrable functions on the space of Riemannian metrics on a 3-surface, with suitable measure μ.

[208] $\mathcal{G}_{abcd}$ is the DeWitt supermetric defined by $|\det q|^{1/2}[(q_{ab}q_{cd} - \frac{1}{2}q_{ac}q_{bd})]$, and ${}^3R(q)$ is the scalar curvature of q. Eq. (7.6) is known as the 'Wheeler–DeWitt equation'.

we forbid quantum states to distinguish between states related by the Hamiltonian constraint, then there is no evolution.

According to the alternative method—reduced phase space quantization—the constraints are solved for *prior* to quantization (*i.e.* at the classical level). To solve the constraints, one divides $\tilde{\Gamma}$ by its gauge orbits $[x]_i$. This yields a space $\tilde{\Gamma}_{\text{red}}$ equipped with a symplectic form $\tilde{\omega}$. The resulting symplectic geometry $(\tilde{\Gamma}_{\text{red}}, \tilde{\omega})$ is the reduced phase space, and in the case of general relativity corresponds to the space of non-isometric (vacuum) spacetimes. Thus, the symmetries generated by the constraints are factored out and one is left with an *intrinsic* geometrical structure of standard Hamiltonian form. In this form the canonical quantization is carried out as usual, and the observables are automatically gauge invariant when considered as functions on the extended (unreduced) space. However, since one of the constraints (the Hamiltonian constraint) was associated with time evolution, in factoring its action out the dynamics is eliminated, since time evolution unfolded along a gauge orbit (*i.e.* instants of time correspond to points in a gauge orbit). Thus, on this approach, states of general relativity are given by points in the reduced phase space, as opposed to the extended phase space used in constrained quantization approach.[209]

Of course, one can *completely* remove the ambiguity associated with gauge freedom by imposing gauge conditions, thus allowing for an unproblematic direct interpretation. However, in the case of general relativity (and other non-Abelian gauge theories) the geometrical structure of the constraint surface and the gauge orbits can prohibit the implementation of gauge conditions, so that some gauge slices will intersect some gauge orbits more than once, or not at all. One frequently finds that the reduced phase space method is mixed with gauge fixation methods, so that one has a partially reduced space, with the remaining gauge freedom frozen by imposing gauge conditions. Such an approach is used by a number of *internal time* responses to the problem of time. The idea is that one first solves the diffeomorphism constraint and then imposes gauge conditions on the gauge freedom generated by the Hamiltonian constraint. This is essentially the position of Kuchař (see below), and *constant mean curvature* approaches (see Carlip [1998] for a clear and thorough review).

7.3. A SNAPSHOT OF THE PHILOSOPHICAL DEBATE

The philosophical debate on the problem of time (what little there is of it!) has, I think, tended to misunderstand the kind of problem it is; often taking it to be nothing more than a result of eradicating indeterminism by applying the quotienting procedure for dealing with gauge freedom. This point of view can be seen quite clearly in action in a recent 'mini-debate' between John Earman [2002]

[209] Little is known about the structure of the space of 3-geometries; the (Wilson) loop variables offer the best hope of carrying out the proposed reduction, or, rather, coordinatizing the reduced space. The diffeomorphism constraint is solved by stipulating that the quantum states be knot invariants. The Gauss constraint that is picked up in the loop representation is easily solved since the Wilson loops are gauge invariant. However, the Hamiltonian constraint is still problematic, though at least *some* solutions can be found. See Brügmann [1994] for more details on these points.

and Tim Maudlin [2002], where both authors see the restoration of determinism via hole argument type considerations as playing a central role. Thus, Earman writes that "[i]n a constrained Hamiltonian system the intrinsic dynamics ... is obtained by passing to the reduced phase space by quotienting out the gauge orbits. When this is done for a theory in which motion is pure gauge, there is an "elimination of time" in that the dynamics on the reduced phase space is frozen" (*ibid.*, p. 14).[210] Before I outline some of the 'standard' responses, and my own response, it will prove instructive to examine Maudlin's views and his criticism of Earman's account. I will argue that Maudlin seriously misunderstands the nature of the problem of time.[211] Let us begin with Earman's account of the problem, and highlight its relation to other conceptions of time and change.

7.3.1 Time series from A to D

Before we consider Maudlin's assessment of the problem of time, and of Earman's account of it, we had better have a grip on what is at stake, on what exactly the problem is saying about time and change (at least, according to Earman). To do this it will be useful to compare and contrast the various ways in which time and change have been understood, to see what the problem rules out. We introduce Earman's preferred account, based on his notion of a "D-series", and show how it matches up to the A-, B-, and C-series accounts in the philosophical literature on the philosophy of time.

According to McTaggart ([1927], §305–6), "positions in time ... as time appears to us *prima facie*, are distinguished in two ways": firstly, "each position is Earlier than some and Later than some of the others"; secondly, "each position is either Past, Present or Future". The distinctions encoded in the first category are permanent, while the latter category are not: "If *M* is ever earlier than *N*, it is always earlier. But an event, which is now present, was future, and will be past". The "movement" or 'flow' of time is understood as 'later and later terms [passing] into the present", or, equivalently, "as presentness [passing] into later and later terms". The first way of understanding temporal flow corresponds to sliding the B-series 'backwards' over a *fixed* A-series (a fixed present); the second way corresponds to the opposite, the sliding of the A-series 'forwards' over a fixed B-series. McTaggart then famously argues from this basis to the unreality of time. Firstly, he argues—from the premise that time involves change—that the B-series depends upon the A-series, so that the only way that events can change is with respect to their A-determinations, not their B-relations. The event 'death of Queen Anne' does not change *per se*; it changes by becoming ever more past, having been future. A-series (tensed) propositions, such as 'the Battle of Waterloo is past', are

[210] However, it is not entirely clear from the text whether Earman endorses the view that it is *only* when reduction is carried out that there is a problem of time. In any case, this is wrong since the problem remains whether or not one reduces the phase space; the problem concerns the gauge-equivalence of states that are supposed to represent different instants of time: how can there be time and change if time evolution is along a gauge orbit, if it is a gauge transformation?

[211] As I just mentioned, Earman too appears to agree with the claim that it is quotienting in a bid to restore determinism that leads to the eradication of time evolution. This is false, as I have said, and as I shall argue in more detail below; however, I think the resolution Earman gives is along the right lines (as I explain in §7.4.2). I should point out that both Earman and Maudlin do, however, give the correct presentation of the observables argument as a problem of *change*; indeed, as I shall explain below, Earman and Maudlin appear to converge at this point, though they claim to fundamentally differ.

true at some times (those *after* the battle), but not at others (those *before* the battle). Replacing this with a B-series version differs in this respect; the proposition 'the Battle of Waterloo is earlier than this judgment' is either always true or always false, it does not change its value as a result of the permanency of the B-series. On these grounds, McTaggart concludes that time requires the A-series: if time requires change, and events change only in terms of their A-series determinations, then time requires the A-series. But then McTaggart argues that the A-series, and therefore time, is inherently contradictory, for the A-determinations are mutually exclusive, and yet any and "every event has them all" (*ibid.*, §329). A single event is present, will be past, and has been future. However, there is, as McTaggart realizes, no contradiction here: no event has these simultaneously. But these tensed verbs, *is* present, *will be* past, and *has been* future, need cashing out. McTaggart (*cf.*, *ibid.* §330) suggests that 'X has been Y' is tantamount to 'X is Y at $t < t_0$'; 'X will be Y' is tantamount to 'X is Y at $t > t_0$'; and 'X (temporally) is Y' is tantamount to 'X is Y at $t = t_0$'. In other words, the analysis requires there to be 'moments' of past, present, and future time. But, McTaggart asks, what are these moments? The A-determinations cannot fix them once and for all, for the same reasons as with events. If we attempt to say that the moments do not have their A-determinations simultaneously, then the analysis must be reapplied: a moment M is future and will be present and then past, which we then rewrite as above, thus courting the same problem, producing 'higher-order' moments, *ad infinitum*.

So, without the A-series there is no change; the B-series alone is not sufficient for time, because time involves change. Moreover, the B-series depends on the A-series, since the former is essentially temporal ([McTaggart, 1908], p. 461): the distinctions it marks out are temporal, and yet without the A-series there is no time; therefore, there is no B-series! So much for the A- and B-series'; what is left to put in their place? McTaggart suggests that an ordering remains, the *C-series*, but it cannot be temporal, for it does not involve change. The C-series consists of an ordering of events themselves.[212] That we have a string of events, X, Y, Z, implies that there is any change no more than the ordering of the letters of the Alphabet change. When the A-series is superimposed on the C-series, however, then the C-series becomes a B-series.

McTaggart's analysis relies on the notion that change applies to events; the argument is grounded in times and events. What of objects? It seems that objects change their properties. Indeed, this is what most people mean by change. Change in events, if it is of the kind suggested by McTaggart is somewhat spurious; it does not *exhaust* what is meant by change. But, in any case, McTaggart believes that *any* kind of change, including the changes that objects undergo, require an A-series. Modern philosophers of time are divided on this point, and the schools of thought can be split according to whether they agree with McTaggart or not about the necessity of the A-series for change. The naysayers are grouped into the category of B-theorists or detensers, and the yaysayers are grouped into the category of A-theorists or tensers. The A-theorists will say that the B-theorists cannot properly accommodate the notion of the passage of time—that they claim,

[212] For a relationalist about time this is simply what time *really* is. There is little sense in saying that the relationalist is an anti-realist about time; he simply *reduces* it, or at least re*defines* time in terms of material happenings.

following McTaggart, is essential—and can, at best, allow that it is an illusion. The B-theorist denies that passage is necessary for time and change, and is happy to see it done away with. Both sides claim support from physics: B-theorists generally wield spacetime theories (special relativity) and the A-theorists wield mechanical theories such as quantum mechanics.

The A-theorists and B-theorists theories are often said to underwrite a 'dynamic' and 'static' conception of time respectively. The static conception represents the moments of time as an 'eternally' existing line, such that each individual moment is equally as real as any other. No fundamental ontological distinction is to be made regarding any 'elements' or 'sections' of the line. This is not the case with the dynamic conception according to which the different times are assigned different ontological status: Broad's 'growing block' theory, for example, views the future as unreal, and the past and present as real; the presentist denies reality to any times other than the present.

There are many and varied ways of responding to McTaggart (token-reflexive or indexical analyses of tenses[213]; presentism, non-property-based becoming, *etc.*); however, the most important for my purposes is the class of responses that attempt to ground a notion of real time and change within the B-series alone (see, for example, Mellor [1998]). Both Earman and I agree that the B-series is sufficient for change in the sense that different properties and relations are instantiated at different times, such that if those times were equal we would have a contradiction—in Wheeler's words "time is what stops everything happening at once"! But strip the dynamical A-determinations from the world and one is left with a *static* block of events ordered earlier to later; indeed, it isn't clear that 'earlier' and 'later' can be anything other than arbitrary directions, for it is the dynamical *flow* that gives direction to the B-series ordering a direction, and this belongs to the A-series.

Earman introduces a character called "Modern McTaggart", who attempts to revive the conclusions of old McTaggart by utilizing a gauge theoretic interpretation of general relativity. Earman is dismissive of A-theories; he claims that they are not part of the scientific image, though he does at least pay lip-service to Shimony's attempt to account for 'transience'. Earman sets up the problem of time as a McTaggartesque consequence of the Hamiltonian formulation of general relativity. The problem targets B-series change—different properties at different times—in that "no genuine physical magnitude takes on different values at different times" ([2002], pp. 2–3). The problem is that given the gauge theoretic conception of observables as gauge invariants, and given that time evolution is a gauge transformation (being generated by a first class constraint), the observables mustn't change from one time to the next.[214] On the assumption that time requires change,

[213] This line of response grounds the moments of time mentioned above contextually by supplying a Now, and combining this with the B-series (see Russell [1940] and Reichenbach [1947]). Thus, one provides a notion of presentness with, say, the time of utterance of a sentence, S_t, and then analyzes the tenses in terms of this 'present' and a string of earlier and later times. 'The Battle of Waterloo is Past' is parsed as 'The Battle of Waterloo is earlier than S_t'.

[214] Again, I feel that Earman limits the problem too much by focusing on the *removal* of the gauge freedom as the source of the problems, rather than the gauge freedom itself. Thus, he writes that for that "class of gauge theories where the very dynamics is implemented by a gauge transformation ... [w]hat such a theory describes when the gauge freedom ... is killed is a world without B-series change" (*loc. cit.*, 7). I say the B-series change is ruled out regardless of whether the gauge freedom is removed. The difference is subtle: when the gauge freedom is removed, time itself is removed, for time evolution is along a gauge orbit; when the gauge freedom is retained, there is time evolution of a sort, but it is gauge,

and assuming that the observables exhausts what might undergo change (that is, assuming that the set is complete), it appears that a version of McTaggart's conclusion follows: time is unreal according to Hamiltonian general relativity! If we deny that time requires change, then though there might be temporal evolution, because it is along a gauge orbit, there will be no qualitative, B-series change; just fixed values. Earman's response is to argue that general relativity is nonetheless compatible with change, though in neither the B- nor the A-series senses. Instead he introduces a 'D-series' ontology consisting of a "time ordered series of occurrences or events, with different occurrences or events occupying different positions in the series" (*ibid.*, p. 3). These are events formed from the coincidence quantities that we met in Chapter 6. Earman writes that "[t]he occurrence or non-occurrence of a coincidence event is an observable matter [in the technical sense of *observable*]...and that one such event occurs earlier than another such event events is also an observable matter ...Change now consists in the fact that different positions in the D-series are occupied by different coincidence events" (*ibid.*, p. 14). Thus, Earman maintains that his D-series is temporally ordered. But this is simply McTaggart's C-series; and, according to McTaggart, that ordering was not temporal. Earman owes us, but not give us—claiming that those who demand that Becoming is required for change "are stuck in the manifest image" (*ibid.*, p. 5)—, an explanation of how this *is* a temporal series, and in what sense change can be said to occur. He does tell us in what sense it does *not* occur, for according to Earman, "common sense B-series property change is not to be found in physical events themselves but only in the mode in which we represent these events to ourselves" (*ibid.*). However, as I later demonstrate, Earman cannot make do with a single D-series; if he must have it, then he must have many. Let us next see what Maudlin makes of Earman's account.

7.3.2 Maudlin versus Earman

Maudlin is responding to the aforementioned paper of Earman, wherein the latter upholds the seriousness of the problem of frozen dynamics, and defends a response to the problem based on the idea that there is, at a fundamental level, no B-series type change according to general relativity. As we saw, Earman argues that changes in the magnitudes of things are, at best, an artifact of the local representations (a particular chart, for example) we might choose to use to describe the world. What is real is a series of events, a D-series. Earman is very much taken with the Hamiltonian formalism, and believes that the frozen dynamics is something that must be accommodated by any sound interpretation of general relativity. I agree. Maudlin does not; rather, he thinks the frozen formalism involves "some Alice-in-Wonderland logic" (*ibid.*, p. 13)! Maudlin distinguishes two separate arguments in Earman's paper that appear to lead to the frozen formalism: the "Hamiltonian Argument" and the "Observables Argument"—corresponding, more or less, to my "states" and "observables" arguments. He takes the crux of the Hamiltonian Argument to consist in the following observation:

therefore there is no B-series change. Earman is clearly clinging to the idea that without change there is no time; but the formalism does not force this.

> Applying this standard method ["quotenting out"] to the GTR does indeed restore the determinism of the theory-but at a price. The price is that the dynamics of the theory becomes "pure gauge"; that is, states of the mathematical model which we had originally taken to represent physically different conditions occurring at different times are now deemed equivalent since they are related by a "gauge transformation". We find that what we took to be an "earlier" state of the universe is "gauge equivalent" to what we took to be a "later" state. If gauge equivalent states are taken to be physically equivalent, it follows that *there is no physical difference between the "earlier" and the "later" states:* there is no real physical change. ([Maudlin, 2002], p. 2)

Maudlin's claim is that "the key to the Hamiltonian Argument" is based "in the freedom to foliate" (*ibid.*, p. 7). A specific foliation is an essential ingredient of any Hamiltonian formulation, for we need an initial data set on a hypersurface. However, in relativistic theories there are many ways to slice up the spacetime manifold $\mathcal{M}$. Given an arbitrary foliation, a phase space can be constructed so that points of this space represent instantaneous states (in this case 3-geometries). The complete four-dimensional solution (*i.e.*, a model of general relativity) is given by a trajectory through the phase space. One and the same solution can be represented by many different trajectories depending upon the foliation that one chooses. He then claims that this yields an indeterminism of the kind that the quotienting procedure is used to eradicate; one can make foliations that agree up to some point, and then diverge thereafter. But, he claims that it is a *faux* indeterminism. The quotienting is unnecessary, and not only is it unnecessary it leads to "silly" claims such as "change is not real, but merely apparent" (*ibid.*, p. 11). Claims, says Maudlin, that Earman thinks are revealed about the deep structure of general relativity by the constrained Hamiltonian formalism. For Maudlin, any such interpretation is absurd. As he explains:

> Any interpretation which claims that the deep structure of the theory says that there is no change at all—and that leaves completely mysterious why there *seems* to be change and why the merely apparent changes are correctly predicted by the theory—so separates our experience from physical reality as to render meaningless the evidence that constitutes our grounds for believing the theory. So the only real question is not *that* the constrained Hamiltonian formalism is yielding nonsense in this case, but *why* it is yielding nonsense. And the freedom to foliate provides the perfectly comprehensible answer. ([Maudlin, 2002], p. 12)

Maudlin's opening line here is facetious. Firstly, the canonical approach is a formal framework *not* an interpretation. *Prima facie*, on a surface reading of the formalism there appears to be no scope for change, therefore, given the apparent existence of change, something is wrong. However, there *is* scope for interpreting the formalism so as to introduce change, as I show in §7.4.2. We can, on these interpretations, say why the surface reading is "yielding nonsense". The answer is related to the gauge invariant response to the hole argument (the response that is supposed to cause the problems of change in the first place): only change with respect to the manifold is ruled out, if we focus on those quantities that are independent of the manifold we can restore change by considering the 'evolving' relationships between these quantities.[215] As regards the observables argument, he rather oddly simply regurgitates what is the gauge theoretical lesson of general relativity, that local quantities cannot be observables:

> the Observables Argument gets any traction only by considering candidates for observables (values at points of the bare manifold) which are neither the sorts of things one actually uses the GTR to predict nor the sorts of things one would expect—quite apart from diffeomorphism invariance—to be observables. ([Maudlin, 2002], p. 18)

Thus, Maudlin has in fact simply accepted the gauge invariance interpretation without realizing it; he mistakenly thinks that the gauge interpretation goes hand in hand with the quotienting procedure. That values at the points of the bare manifold are not the things one predicts cannot be separated from the issues of diffeomorphism invariance, for it precisely this that results in the problems for local field quantities that we have seen in the hole argument. Thus, we can agree, and Earman will agree, that the observables argument gets off the ground by considering the 'wrong' type of observables, but this is to adopt a substantive response that buys into the gauge theoretical interpretation! (I return to this point below, for it backfires on his account of the Hamiltonian argument.)

Maudlin concludes from this double debunking that the frozen formalism problem is simply a result of either a "bad choice of formalism or a bad choice of logical form of an observable" (*ibid.*, p. 18). I proceed to attack Maudlin's account of the problem in two stages.

Firstly, putting aside his analysis of the source of the indeterminism that requires the framework of gauge theory (which I consider below), Maudlin's responses to the indeterminism are: (1) ignore it; (2) gauge fix it: and (3) quotient it. He thinks that the first two are "viable solutions" to the problem, but that the third rests on some kind of confusion (it is being applied in a domain where it should not be). Not so. The underdetermined local field quantities that gauge invariance is invoked to dispel cannot simply be ignored. To ignore it is to tend towards antirealism, for it amounts to the suggestion that we should not worry neither about how our theories represent nor what they represent. The gauge fixing response essentially sides with those who believe that the problem of change is a real problem, for it is tantamount to a resolution in terms of gauge invariant quantities: one fixes a set of coordinates, thus breaking general covariance, and defines the quantities with respect to the points of this coordinate system. Maudlin doesn't give an example of a viable gauge fixing, but the only ones I know of will involve either using the invariants of the gravitational field or else some ideal, phenomenological dust field, or some other material objects: these are wholly unrealistic idealisations. The quotienting procedure is one way sop up the indeterminism, but it is not the only way: nobody said that quotienting was the root of the problem, yet Maudlin appears convinced that it is. Quotienting would certainly eradicate

[215] In §7.5 I suggest that we should view the correlations themselves as observables, following Rovelli's interpretation; strictly speaking, on this account there is no evolution, only variation in an overall correlational structure.

any kind of evolution, since the evolution happens along a gauge orbit and the points ('temporal instants') would be identified in a quotienting strategy. But even unreduced the points represent indistinguishable states so that there will be no qualitative difference between one instant and the next: there will be no B-series (or A-series!) change. Bizarrely, Maudlin does not even consider gauge invariant observables as a viable response! Yet the observables he suggests as the kinds of things we actually measure are of just this kind: *e.g.* "the amount by which light from the sun is redshifted when it reaches the Earth" and "the position of the perihelion of Mercury *relative* to the sun" ([2002], p. 13—my italics). Or *almost* of the kind, for Maudlin assumes that the time of measurement is unproblematic. But of course, the freedom to foliate means that a time choice will be arbitrary: the time of measurement is far from unproblematic! Position relative to the sun *when*? How is the "when" of the first quantity determined? To fix matters one will need to invoke a physical clock. One then considers that above observables suggested by Maudlin when the clock reads a certain figure. Maudlin even gives an example where a physical clock is invoked: "the position of the perihelion of Mercury after some number of orbits". Here Mercury is being used as a clock, the orbits being the 'ticks'. However, now the position will need to be defined relationally, presumably by the Earth or some other useful reference object: we don't just measure the position *simpliciter* in general relativity, we always assume a frame. If we don't want the frame to be arbitrary, we had better make it physical. In case we do use an arbitrary frame, we had better be able to demonstrate that the quantity in question is independent of the specific choice. If Maudlin disagrees with this then he is talking about something other than general relativity, for he is apparently assuming that time is a fixed background structure. If he agrees, which I'm sure he must, then the observables are just my correlation observables from the previous chapter, which were, you will recall, more or less the same as Earman and Einstein's chosen coincidence observables. He gives an example of what he takes to be a good quantity for general relativity that brings this similarity to the fore: thus, he writes that "[w]hat we *can* identify by observation are the points that satisfy definite descriptions such as "the point where these geodesics which originate here meet", and against *these* sorts of [local] quantities Earman's diffeomorphism argument has exactly zero force" (*ibid.*). Indeed, but here Maudlin is essentially gauge-fixing spacetime points and then constructing gauge invariant quantities by attaching them to the physically defined points—the reasoning is that for some quantity 'ϕ', physical object 'thing', and space point $x \in \Sigma$ (or $x \in \mathcal{M}$): $\phi(thing)$ is gauge invariant but $\phi(x)$ is not. If Maudlin is willing to go this far, then why not allow that change is accounted for with just such observables: the evolution and change concerns the relations between things or quantities, rather than the having and losing of properties at times? One can form of chain of values for ϕ by using the values of 'thing' as the 'ticks' of a clock—this is essentially what Rovelli proposes (see §7.4.2). Moreover, all of this is perfectly possible in the context of the Hamiltonian formulation. Indeed, on the preferred choice of polarization, holonomies, with no local spacetime dependence, are used as the fundamental variables, thus taking on board the lesson that Maudlin has clearly internalized without being aware of it.[216]

Secondly, Maudlin diagnoses the Hamiltonian argument as the freedom to foliate a spacetime in general relativity. Different slicings of spacetime yield different trajectories through phase space, which are to represent four-dimensional solutions. But, says Maudlin, we can make a pair of foliations agree up to a point (so that their corresponding phase trajectories do likewise) and diverge thereafter (again, likewise for the trajectories). This results in an indeterminism to which the three options listed above apply. The quotienting option removes the indeterminism by declaring the solutions equivalent, and forming a reduced phase space out of the equivalence classes. Maudlin thinks this is absurd for the reasons given. But the source of the indeterminism is *not* the freedom to foliate a spacetime, it is the freedom to drag the dynamical fields around without generating a distinguishable scenario. The indeterminism concerns 'local' quantities that are attached to manifold points. Any local quantities will be altered by the dragging. We cannot even assume that we have a spacetime in the Hamiltonian formulation, for that we require a solution.

Thus, Maudlin can claim that he is willing to accept the indeterminism that follows from such gauge transformations rather than quotienting if he likes,[217] but the fact that the indeterminism is unobservable is tantamount to saying that the time-evolution is unobservable, which simply lets the problem in through the back door. As regards the observables argument it seems to me, as I hope to have demonstrated, that far from showing it to be "broken-backed", Maudlin has simply taken a stance (and a highly non-trivial one at that) with respect to the observables argument. Specifically, he opts for the view that the 'proper' observables of general relativity are relational quantities involving intersections of quantities.[218] However, what is missing from Maudlin's suggested quantities is a time of occurrence (or, in some cases, the *position* of occurrence). It is not enough to say that two things meet at a point; one must say *when* they meet, and to do this one needs a clock. Likewise, it is not enough to say when something happens, one must say *where* it happens. And the 'when' and 'where' are not given *a priori*; one can arbitrarily shift the points of the manifold around, so these cannot ground the where and the when. There is no background temporal or spatial structure in general relativity, so this will have to be a physical clock or a physical reference frame. The coincidence of the hand of the clock and the meeting of the geodesics is a diffeomorphism-invariant quantity that satisfies the constraints. It is a constant of the motion, so it does not change over time.[219] It will certainly be hard

[216] As further evidence that Maudlin misunderstands the nature of the problems of time and change, he mentions that the on the basis of his arguments, the quantum gravitational problems of time and change might be "equally chimerical" (*ibid.*, p. 18). His worry is that if local observables cause a problem in the classical theory "then we should anticipate difficulties in defining the observables in the quantum version" (*loc. cit.*, 19). But no one is suggesting that we use these kinds of observables! There are *proofs* that no such observables are available in general relativity, classical or quantum [Torre, 1993]. These quantities are not forced upon us in the Hamiltonian formulation.

[217] Something he is willing to do on the grounds that the indeterminism is "*completely phoney*" (*ibid.*, p. 9; see also p. 16).

[218] Note that Maudlin gives no account as to the nature of the 'individual' elements participating in these intersections. The standard line is to take these elements as having some physical reality independently of the relation; but this leads to serious problems as we shall see in §7.4.2.

[219] As I intimated above, and as I will discuss in §7.4.2, by stringing a sequence of such quantities together one can get a fairly robust account of change.

to write such a quantity as a phase function, but that is not of the moment for what is at stake here. As I argued, the observables Maudlin mentions sound suspiciously like Earman's coincidence quantities. This is just what many physicists take to be the 'lesson' of the hole argument and the problem of time: the proper observables are independent of the manifold and, therefore, independent of time as well as space. One way of understanding the observable content of the theory is to view the points of spacetime as relationally (dynamically) individuated in the manner Maudlin suggests; this was, of course, Stachel's position too. The problem remains: how do we reconcile this with the manifest change we seem to observe? I review some options in the next section.

7.4. CATALOGUE OF RESPONSES

Those approaches to classical and quantum gravity that attempt to understand these theories without change and time existing at a fundamental level I shall call *timeless*, and those that disagree I call *timefull*. An alternative pair of names for these views, suggested by Kuchař, is "Parmenidean" and "Heraclitean" respectively [1993]. But it is important to note that the debate here is not directly connected to the debate in the philosophy of time between 'A-theorists' and 'B-theorists' (or 'tensers' and 'detensers', if you prefer). Both of these latter camps agree that time exists, but disagree as to its nature. By contrast, the division between timefull and timeless interpretations concerns whether or not time exists *simpliciter*! I begin by reviewing several timefull responses.

7.4.1 Timefull stratagems

Recall that the observables argument required that in order to class as kosher, the relevant observables must have vanishing Poisson brackets with *all* of the constraints. This idea filtered through into the quantum version, modified appropriately. Like Maudlin, Kuchař has been a vociferous opponent of this 'liberal' gauge invariant approach to observables.[220] He agrees with the plan to the level of the diffeomorphism constraint, so that $\{\mathrm{O}, \mathcal{H}_a\} \approx 0$, $[\hat{\mathrm{O}}, \hat{\mathcal{H}}_a] \approx 0$ and $\mathcal{H}_a \Psi = 0$; but does not agree that we should apply the same reasoning to the Hamiltonian constraint. Thus, neither states nor observables should distinguish between metrics connected by Diff(Σ): only the 3-geometry $^3\mathcal{G}$ counts. But the alterations generated by the Hamiltonian constraint are a different matter says Kuchař:

> [$\mathcal{H}_\perp$] generates the dynamical change of data from one hypersurface to another. The hypersurface itself is not directly observable, just as the points $x \in \Sigma$ are not directly observables. However, the collection of the canonical data $(q_{ab}(1), p^{ab}(1))$ on the first hypersurface is clearly distinguishable from the collection $(q_{ab}(2), p^{ab}(2))$ of the evolved data on the second hypersurface.

220 Though, unlike Maudlin, he has a constructive alternative that submits to quantization. He also takes the problem of time much more seriously than Maudlin; he doesn't think that it can simply be ignored. The positions end up being very different, with Maudlin occupying a position almost identical to Earman and, as will become evident, Rovelli.

> If we could not distinguish between those two sets of data, we would never be able to observe dynamical evolution. ([Kuchař, 1993], p. 20)

Ditto for states: the Wheeler–DeWitt equation does not say that an evolved state is indistinguishable from some initial state—as the diffeomorphism constraint does—rather, it "tells us how the state evolves" (*ibid.*, p. 21). More colourfully:

> I would say that the state of the people in this room now, and their state five minute ago should not be identified. These are not merely two different descriptions of the same state. They are physically distinguishable situations. (Kuchař in [Ashtekar and Stachel, 1991], p. 139)

Thus, Kuchař concludes that "if we could observe only constants of motion, we could never observe any change" (*ibid.*). On this basis he distinguishes between two types of variable: *observables* and *perennials*. The former class are dynamical variables that remain invariant under spatial diffeomorphisms but *do not* commute with the Hamiltonian constraint; while the latter are observables that *do* commute with the Hamiltonian constraint. Kuchař's key claim is that one can observe dynamical variables that are not perennials.[221]

In their assessment of Kuchař's proposal, Belot and Earman ([1999], p. 183) claim that he "endeavours to respect the spirit of general covariance of general relativity without treating it as a principle of gauge invariance". For this reason they see his strategy as underwritten by substantivalism. I argue against the connection between the denial of gauge invariance and substantivalism in §7.6; for now I note that Kuchař *does* treat general covariance as a principle of gauge invariance as far as the diffeomorphisms of Σ are concerned (and, in the connection representation, as far as the SO(3) Gauss constraint goes). Observables are gauge invariant quantities on his approach; the crucial point is simply that the Hamiltonian constraint should not be seen as generating gauge transformations. Viewed in this light, according to Belot and Earman's own taxonomy (*ibid.*, §2), Kuchař's position should more properly be seen as underwritten by a *relationalist* interpretation of space coupled with a *substantivalist* interpretation of time! Let me spell out some more of the details of Kuchař's idea.

Kuchař's claim that observables should not have to commute with the Hamiltonian constraint leads almost inevitably to the conclusion that the observables do not act on the space of solutions; or, as he puts it "if $\Psi \in \mathcal{F}_0$ and $\hat{F}$ is an observable, $\hat{F}\Psi \notin \mathcal{F}_0$" ([1993], p. 26). This, amongst other things, motivates the *internal time* strategy, where an attempt is made to construct a time variable T from the classical phase space variables. This strategy conceives of general relativity (as described by Γ) as a parametrized field theory. The idea is to find a notion of time *before* quantization hidden amongst the phase space variables so that a time-dependent Schrödinger equation can be constructed; the quantum theory's states then evolve with respect to the background time picked out at the classical level. Kuchař's method involves finding four (scalar) fields $X^A = (\mathcal{T}(x;q,p], \mathcal{Z}^a(x;q,p])$

221 He goes further than this, arguing that perennials are in fact hard to come by. I do not deal with this aspect of his argument here. In fact, I think that relational observables show that they are not at all hard to come by. How one makes a quantum theory out of these is, of course, quite another matter. The hard task is to find quantum operators that correspond to such classical observables without facing operator ordering ambiguities, and so on.

(where $A = 0, 1, 2, 3$ and $a = 1, 2, 3$) from the full phase space Γ that when defined on $\overline{\Gamma}$ represents a spacelike embedding $X^A : \Sigma \rightarrow \mathcal{M}$ of a hypersurface Σ in the spacetime manifold $\mathcal{M}$ (without metric). These kinematical variables are to be understood as position at the manifold and the dynamical variables (separated out from the former variables within the phase space) are observables evolving along the manifold. The constraints are then understood as conditions that identify the momenta P_A conjugate to X^A with the energy-momenta of the remaining degrees of freedom: they thus determine the evolution of the true gravitational degrees of freedom between hypersurfaces.

There are two broadly 'technical' ways of dealing with Kuchař's arguments. The first involves demonstrating that general relativity is not a parametrized field theory; and the second involves showing that observing change (change that the world seems full with) is compatible with the view that all observables are constants of the motion. I deal with the second when I get to the timeless responses; the first I outline now. Clearly, we need to test whether or not the identification between the phase space Γ of general relativity and the phase space Υ of a parameterized field theory goes through. The proposal requires that there is a canonical transformation $\Phi : \Upsilon \rightarrow \Gamma$ such that $\Phi(\overline{\Upsilon}) = \overline{\Gamma}$. However, there can be no such transformation because, in general, $\overline{\Upsilon}$ is a manifold while $\overline{\Gamma}$ is not (*cf.* [Torre, 1993]). Hence, there are serious technical issues standing in the way of this approach: general relativity is not a parameterized field theory!

Along more 'philosophical' lines, one might perhaps question the line of reasoning that led Kuchař to deny that observables commute with all of the constraints in the first place. Is it an empirical input that determines the break, or is it something internal to the theory? I think that it is neither, but is instead an intuitive belief that change is a real feature of the world, and that change happens when things change by changing in the values of their observables. He takes the fact that the liberal gauge invariance position entails that observables are constants of the motion as providing a *reductio* of that view, and as providing a counterexample to Dirac's conjecture that first class constraint generate gauge transformations. But there are ways of understanding change; we can understand change as the possession of incompatible properties by things at different times (ruled out by the gauge interpretation), but we can incorporate a notion of change as variation: the rug, for example, changes from blue to green as one moves across it. Or, one can get a simulacrum of change by piecing together unchanging parts, as one finds in the old-fashioned movies. All we really need to do is explain the *appearance* of change; to assume a substantial metaphysics of time and change and then base ones physical theories on this metaphysics is a dangerous move in my opinion. Intuition strongly suggests that there is a unique notion of simultaneity; physics suggests that our intuitions need to be revised. Regardless of this, if the problem of time can be resolved in a liberalistic gauge invariant way, then we should opt for that on the grounds that violating the 'first-class constraint'–'gauge transformation' connection, that has worked so well in other gauge theories, is too high a price to pay. In keeping it we can retain a unified interpretive picture of these theories.

An alternative (internal) timefull approach uses *matter* variables coupled to spacetime geometry instead of (functionals) of the gravitational variables as above. Thus, one might consider a space filling dust field, each mote of which is considered to be a clock (*i.e.* the proper time of the motes gives a preferred time variable and, therefore, amounts to fixing a foliation). These variables are once again used to 'label' spacetime points. This includes an internal time variable against which systems can evolve, and which can function as the fixed background for the construction of the quantum theory. Another internal approach, *unimodular gravity*, amounts to a modification of general relativity, according to which the cosmological constant is taken to be a dynamical variable for which the conjugate is taken to be 'cosmological' time.[222] The upshot of this is that the Hamiltonian constraint is augmented by a cosmological constant term $\lambda + q^{-1/2}(x)$, $x \in \Sigma$, giving the super-Hamiltonian constraint $\lambda + q^{-1/2}(x)\mathcal{H}_{\perp}(x) = 0$. The presence of this extra term (or, rather, its conjugate τ) unfreezes the dynamics, thus allowing for a time-dependent Schrödinger equation describing dynamical evolution with respect to τ. The conceptual details of this approach are, however, more or less in line with gauge fixation methods like that mentioned above.[223] Another popular, but now aged approach is that which takes surfaces of constant mean curvature $\tau = q_{ab}p^{ab}/\sqrt{\det q} = \text{const}$ as providing a time coordinate by providing a privileged foliation of spacetime.[224] The basic idea underlying each of these approaches is to introduce some preferred *internal* time variable so that general relativity can be set up as a time-dependent system describing the evolution in time of a spatial geometry (possibly involving the extrinsic curvature and possibly coupled to matter or some reference fluid). With this background time parameter in hand, the quantization proceeds along the lines of other quantum field theories since there will be a non-zero Hamiltonian for the theory. Naturally, the selection of a preferred time coordinate breaks the general covariance of the theory, for it is tantamount to accepting that there is a preferred reference frame. One would have to demonstrate that the resulting quantum theory is independent of the choice.[225]

Suffice it to say that I do not think that these timefull approaches are the correct direction to go. Aside from the technical difficulties, they either represent a step backwards (*i.e.* away from general covariance) towards unphysical background structures (dust fields, and the like), or else they point to the idea that a *robust* notion of time is required to get a quantum theory up and running. The proposals in the next subsection show that this is simply false.

Before I leave the 'timefull' methods, I should first mention one more related approach: Hájiček's *perennial formalism* [1996; 1999], according to which the dynamics is constructed solely from the geometry of phase space, and no reference is

[222] The idea to use unimodularity as a response to the problem of time was originally suggested by Unruh [1989]. For a nice philosophical discussion of unimodular gravity see Earman [2003a]—§6 of his paper focuses the discussion on the problem of time. See also Isham [1992], p. 63.

[223] Isham (*ibid.*, p. 62) goes so far as to say that it is in line with reference fluid methods since it amounts to the imposition of a coordinate condition (on the metric γ_{ab}): $\det \gamma_{ab}(x^i) = 0$. See (*ibid.*, pp. 60–2) for more details on the notion of a reference fluid and how it might offer a solution to the problem of time.

[224] This approach was first suggested by York [1971]. See Beig [1994] for a nice discussion.

[225] Note that Kuchař's approach escapes this objection since it quantizes the 'multi-time' formalism according to which dynamical evolution takes place along deformations of *arbitrary* hypersurfaces embedded in $\mathcal{M}$ (see Isham, *ibid.*, p. 46).

made to spacetime. The idea is to begin with some system whose time evolution is well understood, like a Newtonian system, and transform the spacetime structure into a phase space structure so that a quantum time evolution can be reconstructed from phase space objects. Then one attempts to find similar phase spaces for systems without background spacetimes. This approach links technically to Kuchař's scheme, but conceptually it links up to the timeless approaches—especially Rovelli's evolving constants scheme. However, questions need to be asked about the way the phase space is constructed, for it is not intrinsically done, but is parasitic on what we know of phase spaces for systems with background spacetime structure (fixed metrics and connections). If the virtue of this approach is that it retains background independence, then we would surely like the formalism to reflect this property.

7.4.2 Timeless stratagems

We come now to the timeless strategies; the most radical of which is surely Barbour's. I deal with this first, and then outline the view I favour. Butterfield [2001] has written a fine account of Barbour's timelessness as outlined in the latter's book *The End of Time* [2003]; he describes the resulting position as "a curious, but coherent, position which combines aspects of modal realism *à la* Lewis and presentism *à la* Prior" (*ibid.*, p. 291). I agree that these aspects do surface; however, I disagree with his account on several key substantive points. In particular, I will argue—*contra* Butterfield—that Barbour's brand of timelessness is connected to a denial of persistence, and as such is not timeless at all; rather, it is changeless. I go further: far from denying time, Barbour has in fact *reduced* it (or, rather, the instants of time) to the points of a relative configuration space!

The central structure in Barbour's vision is the space of Riemannian metrics mod the spatial diffeomorphism group (known as "superspace"): Riem(Σ)/Diff(Σ). Choosing this space as the configuration space of the theory amounts to solving the diffeomorphism constraint; this is Barbour's *relative configuration space* that he labels "Platonia" (*ibid.*, p. 44). The Hamiltonian constraint (*i.e.* the Wheeler–DeWitt equation, Eq. (7.6)) is then understood as giving—once solved, and "once and for all" ([Barbour, 1994], p. 2875)—a *static* probability distribution over Platonia that assigns amplitudes to 3-geometries (Σ, q) in accordance with $|\Psi[q]|^2$. Each 3-geometry is taken to correspond to a "possible instant of experienced time" (*ibid.*) This much is bullet biting and doesn't get us far as it stands; there remains the problem of accounting for the appearance of change. This he does by introducing his notion of a 'time capsule,' or a 'special Now', by which he means "any fixed pattern that creates or encodes the appearance of motion, change or history" (Barbour [2003], p. 30). Barbour *conjectures* that the relative probability distribution determined by the Wheeler–DeWitt equation is peaked on time capsules; as he puts it "the timeless wavefunction of the universe concentrates the quantum mechanical probability on static configurations that are time capsules, so that the situations which have the highest probability of being experienced carry within them the appearance of time and history" (*ibid.*). What sense are we to make of this scheme?

Barbour's approach is indeed timeless in a certain sense: it contains no reference to a background temporal metric in either the classical or quantum theory. Rather, the metric is defined by the dynamics, in true Machian style. Butterfield mentions that Barbour's denial of time might sound (to a philosopher) like a simple denial of temporal becoming—*i.e.* a denial of the A-series conception of time. He rightly distances Barbour's view from this B-series conception. Strictly speaking, there is neither an A-series nor a B-series on Barbour's scheme. Barbour believes that space is fundamental, rather than spacetime.[226] This emerges from his Machian analysis of general relativity. What about Butterfield's mention of presentism and modal realism? Where do they fit in?

Presentism is the view which says that only presently existing things actually exist.[227] The view is similar in many respects to modal actualism, the view that only actually existing things exist simpliciter. Yet Butterfield claims that Barbour's view blends with modal realism. What gives? We can make sense of this apparent mismatch as follows: Barbour believes that there are *many* Nows that exist 'timelessly', even though we happen to be confined to *one*. The following passage brings the elements Butterfield mentions out to the fore[228]:

> All around NOW ... are other Nows with slightly different versions of yourself. All such nows are 'other worlds' in which there exist somewhat different but still recognizable versions of yourself. ([Barbour, 2003], p. 56)

Clearly, given the multiplicity of Nows, this cannot be presentism conceived of along Priorian lines, though we can certainly see the connection to modal realism; talk of other nows being "simultaneously present" (*ibid.*) surely separates this view from the Priorian presentist's thesis. That Barbour's approach is not a presentist approach is best brought out by the lack of temporal flow; there is no A-series change. Such a notion of change is generally tied to presentism. Indeed, the notion of many nows existing simultaneously sounds closer to eternalism than presentism; *i.e.* the view that past and future times exist with a much ontological robustness as the present time. These points also bring out analogies with the 'many-worlds' interpretation of quantum mechanics; so much so that a more appropriate characterization might be a 'many-Nows' theory.[229] Thus, I don't think that Butterfield's is an accurate diagnosis. What is the correct diagnosis?

There is a view, that has become commonplace since the advent of special relativity, that objects are four-dimensional; objects are said to 'perdure', rather than

[226] I might add that Belot writes that he does "not know of any philosopher who entertains, let alone advocates, substantivalism about space as an interpretive option for GR" ([1996], p. 83). I think that Barbour's proposal ends up looking like just such an interpretive option; a position recently defended by Pooley [in press].

[227] The consensus amongst philosophers seems to be that special and general relativity are incompatible with presentism (*cf.* Callender [2000], Saunders [2002], and Saunders [2002]). I think that special relativity allows for presentism in a certain sense—we simply need to modify what we mean by 'present' in this context, distinguishing it from what we mean in Newtonian mechanics—, and that general relativity (classical and quantum) too allows for presentism in the canonical formulation (a view recently defended by Monton [2006] in the context of timefull, 'fixed foliation' strategies). But we need to distinguish the kind of presentism that classical and quantum general relativity allows for from that which special relativity allows for, and that Newtonian mechanics allows for. But this is not the place to argue the point.

[228] Fans of Lewis' *On The Plurality of Worlds* [1986a] will notice a remarkable similarity to a certain famous passage from that work. Hence the suggested link to modal realism.

[229] Indeed, Barbour himself claims that his approach suggests what he calls a "many-instants ... interpretation of quantum mechanics" (*ibid.*). However, it seems clear that the multiplicity of Nows is as much a classical as a quantum feature.

'endure': this latter view is aligned to a three-dimensionalist account according to which objects are wholly present at each time they exist, the former view is known as 'temporal part theory'. The four-dimensionalist view is underwritten by a wide variety of concerns: for metaphysicians these concerns are to do with puzzles about change; for physics-minded philosophers they are to do with what physical theory has to say. Change over time is characterized by differences between successive temporal parts of individuals. Whichever view one chooses, the idea of *persisting* individuals plays a role; without this, the notion of change is simply incoherent, for change requires there to be a *subject* of change. Although Barbour's view is usually taken to imply a three-dimensionalist interpretation (by Butterfield for one), I think it is also perfectly compatible with a kind of temporal parts type theory. However, rather than the structure of time being linear, it is non-linear (as encoded in the relative configuration space) and the 'temporal evolution' is probabilistic (governed by a solution to the Hamiltonian constraint). We see that the parts themselves do not change or endure and they cannot perdure since they are three-dimensional items and the parts occupying distinct 3-spaces (and, indeed, the 3-spaces themselves) are not genidentical; rather, the quantum state 'jumps' around from Now to Now in accordance with the Hamiltonian constraint in such a way that the parts contain records that 'appear' to tell a story of linear evolution and persistence.[230] Properly understood, then, Barbour's views arise from a simple thesis about identity over time, *i.e.*, a denial of persistence:

> We think things persist in time because structures persist, and we mistake the structure for substance. But looking for enduring substance is like looking for time. It slips through your fingers. ([Barbour, 2003], p. 49)

In denying persisting individuals, Barbour has given a philosophical grounding for his alleged timelessness. However, as I mentioned earlier, the view that results might be seen as not at all timeless: the relative configuration space, consisting of Nows, can be seen as providing a *reduction* of time, in much the same way that Lewis' plurality of worlds provides a reduction of modal notions.[231] The space of Nows is given once and for all and does not alter, nor does the quantum state function defined over this space, and therefore the probability distribution is fixed too. But just like modality lives on in the structure of Lewis' plurality, so time lives on in the structure of Barbour's Platonia. However, also like Lewis' plurality, believing in Barbour's Platonia requires substantial imagination stretching. Of course, this isn't a knock down objection; with a proposal of this kind I think we need to assess its cogency on a cost versus benefit basis. As I show below, I think that the same result (a resolution of the problem of time) can be gotten on a tighter ontological budget. However, I think there is real value in Barbour's analysis of the problem of time, and philosophers of time would do well to further consider the

[230] Naturally, there is an obvious problem here: how are we to understand these "jumps"? If in dynamical terms, then that implies time, or at least some higher-order medium that contains the jumps. If in non-dynamical terms, then the notion becomes senseless, for the probability distribution will given once and for all.

[231] Roughly, Lewis' idea is that the notions of *necessity* and *possibility* are to be cashed out in terms of holding at all or some of a class of 'flesh and blood' worlds.

connections between Lewis' and Barbour's reductions, and the stand alone quality of the view of time that results.[232]

Not quite as radical as Barbour's are those timeless views that accept the fundamental timelessness of general relativity and quantum gravity that follows from the gauge invariant conception of observables, but attempt to introduce a notion of time and change into this picture. A standard approach along these lines is to account for time and change in terms of time-independent correlations between gauge variant quantities. The idea is that one never measures a gauge variant quantity, such as position of a particle; rather, one measures position at a time, where the time is defined by some physical clock.[233] Thus, in the general relativistic context, we might consider the spatial volume of the universe, $V(\Sigma) = \int_{\Sigma} \sqrt{-\det g}\, d^3x$; this is gauge variant (for compact Σ) and, therefore, is not an observable. Now suppose we wish to measure some quantity defined over Σ, say the total matter density $\rho(x)$, $\forall_i x_i \in \Sigma$. Of course, this too is a gauge variant quantity; but the *correlation* between V and ρ when they take on a certain value is gauge-*independent*. In this way, one can define an instant of time; one can write $\tau = \rho(V)$ or $\tau = V(\rho)$. One can then use these correlations to function as a clock giving a monotonically increasing time parameter τ against which to measure some other quantities. Unruh objects to this method along the following lines:

> one could [try to] define an instant of time by the correlation between Bryce DeWitt talking to Bill Unruh in front of a large crowd of people, and some event in the outside world one wished to measure. To do so however, one would have to express the sentence "Bryce DeWitt talking to Bill Unruh in front of a large crowd of people" in terms of physical variables of the theory which is supposed to include Bryce DeWitt, Bill Unruh, and the crowd of people. However, in the type of theory we are interested in here, those physical variables are all time independent, they cannot distinguish between "Bryce DeWitt talking to Bill Unruh in front of a large crown of people" and "Bryce DeWitt and Bill Unruh and the crowd having grown old and died and rotted in their graves". ... The subtle assumption [in the correlation view] is that the individual parts of the correlation, *e.g.* DeWitt talking, are measurable when they are not. (Unruh in [Ashtekar and Stachel, 1991], p. 267)

Belot and Earman question Unruh's interpretation of the correlation view, and suggest that it might be better understood "as a way of explaining the illusion of change in a changeless world" ([2001], p. 234). The basic idea is that one deals in quantities of the form "clock-1-reads-t_1-when-and-where-clock-2-reads-t_2". We get the illusion of change by (falsely) taking the elements of these relative (correlation) observables to be capable of being measured independently of the correlation. They suggest that Rovelli's notion of evolving constants of motion is a good way of "fleshing out" the relative observables view.

232 I expect that the view of most philosophers of time would be that Barbour has simply outlined a variation of eternalism, albeit a peculiar one.

233 See the exchange between DeWitt, Rovelli, Unruh, and Kuchař in Ashtekar & Stachel (eds.) ([1991], pp. 137–140) for a nice quick introduction to the timeless vs timefull views: Rovelli and DeWitt are firmly in favour of the correlation view, while Unruh and Kuchař are firmly against it. I outline Unruh's and Kuchař's objections below.

Rovelli's *evolving constants of motion* proposal is made within the framework of a gauge invariant interpretation. He accepts the conclusion that quantum gravity describes a fundamentally timeless reality, but argues that sense can be made of dynamics and change within such a framework. Take as a naive example of an observable m = 'the mass of the rocket'. This cannot be an observable of the theory since it changes over (coordinate) time; it fails to commute with the constraints, $\{m, \mathrm{H}\} \neq 0$, because it does not take on the same value on each Cauchy surface. Rovelli's idea is to construct a one-parameter family of observables (constants of the motion) that can represent the sorts of changing magnitudes we observe. Instead of speaking of, say, 'the mass of the rocket' or 'the mass of the rocket at t', which are both gauge dependent quantities (unless t is physical), one speaks instead of 'the mass of the rocket when it entered the asteroid belt', $m(0)$, and 'the mass of the rocket when it reached Venus', $m(1)$, and so on up until $m(n)$. These quantities are gauge invariant, and, hence, constants of the motion; but, by stringing them together in an appropriate manner, we can explain the appearance of change in a property of the rocket. The change we normally observe taking place is to be described in terms of a one-parameter family of constants of motion, $\{m(t)\}_{t\in\mathbb{R}}$, an *evolving* constant of motion.[234]

A similar criticism to Unruh's comes from Kuchař ([1993], p. 22), specifically targeting Rovelli's approach. Kuchař takes Rovelli to be advocating the view that observing "a changing dynamical variable, like Q [a particle's position, say], amounts to observing a one-parameter family $Q'(\tau_1) := Q' + P'\tau = Q - P(T - \tau)$, $\tau \in R$ of perennials" (*ibid.*, p. 22). By measuring $Q'(\tau)$ at τ_1 and τ_2 "one can infer the change of Q from $T = \tau_1$ to $T = \tau_2$" (*ibid.*). So the idea is that a changing observable can be constructed by observing correlations between two dynamical variables T and Q, so that varying τ allows one a notion of 'change of Q with respect to T'. Kuchař objects that one has no way of observing τ that doesn't smuggle in non-perennials. But this is a *gauge variant*; one doesn't need to observe τ independently of Q: we can simply stipulate that the two are a 'package deal'. In this way, I think both Unruh's and Kuchař's objections can be successfully dealt with. I outline this view further in the next section, where I attempt to strengthen the correlation solution.

Rovelli's approach has a certain appeal from a philosophical point of view. It bears similarities to temporal parts views of time and persistence. The basic idea of both of these views is that a changing individual can be constructed from unchanging parts. Change over time is conceptually no different from variation over a region of space. (I think philosophers of time might perhaps profit from a comparison of Rovelli's proposal with four-dimensionalist views.) However, technically, it is hard to construct such families of constants of motion as phase functions on the phase space of general relativity. To the extent that they can be constructed at all, they result in rather complicated functions that are hard to represent at the

[234] Rovelli, in collaboration with Connes and Rovelli [1994], has argued that the 'flow' of time can be explained as a "thermodynamical" effect, and is state dependent. The thermal time is given by the state dependent flow generated by the statistical state s over the algebra of observables: $\frac{dq}{dt} = -\{q, \log s\}$. Hence, the Hamiltonian is given by $-\log s$, so that the (statistical) state that a system occupies determines the Hamiltonian *and* the associated flow. Rovelli connects this idea up to his evolving constants proposal by identifying the thermal time flow with the one-parameter group of automorphisms of the algebra of observables (as given by the Tomita flow of a state).

quantum level (*i.e.* as quantum operators on a Hilbert space: *cf.* Hájiček [1996], p. 1369), and face the full force of the factor ordering difficulties (*cf.* Ashtekar & Stachel (eds.) [1991], p. 139).[235] For this, and other reasons, Rovelli has recently shifted to something more like the original correlation view I outlined above (see Rovelli [2002]; his earlier paper [1991] contains much the same view: elements of the evolving constants programme are retained).

As with the evolving constants of motion program, Rovelli believes that the observables of general relativity and quantum gravity are relative quantities expressing correlations between dynamical variables. The problem Rovelli sets himself in his *partial observables* program, as if in answer to Unruh's complaint, is this: "how can a correlation between two nonobservable [gauge variant] quantities be observable?" ([2002], p. 124013-1). He distinguishes between *partial* and *complete* observables, where the former is defined as a physical quantity to which we can associate a measurement leading to a number, and the latter is defined as a quantity whose value (or probability distribution) can be predicted by the relevant theory, *i.e.* a (gauge invariant) Dirac observable. Partial observables can be measured but not predicted, and complete observables are correlations between partial observables that can be both measured and predicted. The above question can then be rephrased in these terms: 'how can a pair of partial observables make a complete observable?' (see p. 124013-5). His answer is somewhat surprising, for he argues that this question is just as applicable to classical non-relativistic theories as it is to relativistic theories. However, to make sense of the answer, there is a further distinction to be made within the class of partial observables that only holds in non-general relativistic (more generally: background dependent) theories: *dependent* and *independent*. These can be understood as follows: take two partial observables, q and t (position and time); if we can write $q(t)$ but not $t(q)$ then we say that q is a dependent partial observable and t is an independent partial observable. He then traces the confusion in Unruh's objection to the notion of localization in space and time and, in particular, that this makes no sense in the context of general relativistic physics. The absolute localization admitted in non-relativistic theories means that the distinction can be disregarded in such quantum theories since "the space of observables reproduces the fixed structure of spacetime" (p. 124013-1). However, where the structure of spacetime is dynamical t and q are partial observables for which we cannot assume that an external clock or spatial reference frame exists. Going back to Unruh's example, we see that Unruh, DeWitt and the crowd of people are analogs of partial observables. Unruh assumes that the dependent/independent distinction must hold. However, this is just what Rovelli denies:

> A pre-GR theory is formulated in terms of variables (such as q) evolving as functions of certain distinguished variables (such as t). General relativistic systems are formulated in terms of variables ... that evolve with respect to each other. General relativity expresses relations between these, but in general we cannot solve for one as a function of the other. Partial observables are genuinely on the same footing. ([Rovelli, 2002], p. 124013-3)

[235] But see [Montesinos et al., 1999] for a construction of such a family for a simple $SL(2, \mathbb{R})$ model.

The theory describes relative evolution of (gauge variant) variables as functions of each other. No variable is privileged as the independent one (*cf.* [Montesinos et al., 1999], p. 5).[236] How does this resolve the problem of time? The idea is that coordinate time evolution and physical evolution are entirely different beasts. To get physical evolution, all one needs is a pair $\mathcal{C}, C$ consisting of an extended configuration space (coordinated by partial observables) and a function on $T^*\mathcal{C}$ giving the dynamics. The dynamics concerns the relations between elements of $\mathcal{C}$, and though the individual elements do not have a well defined evolution, relations between them (*i.e.* correlations) do: they are independent of coordinate time.

Let me spell this out some more, for my own response is based on Rovelli's, albeit with an interpretive twist. Consider two non-gauge invariant (*i.e.* gauge dependent) functions α and β. These are our partial observables; we can suppose that α is the volume of a compact hypersurface and that β is the matter density of a compact hypersurface. Recall that neither of these quantities is predictable, for their evolution will be gauge variant. We want to construct from this pair of partial observables a complete observable $\mathcal{E}^{\tau}_{\alpha|\beta}$ (where τ will be understood to be a 'clock' variable). To do this we consider the relational quantity that is formed by correlating the values of the two partial observables. We arbitrarily take one of the partial observables to be the 'clock', whose values will parameterize the evolution of the other. Let β be the clock. $\mathcal{E}^{\tau}_{\alpha|\beta}$ then gives the quantity that gives the value of α when, under the flow generated by the constraints, the value of β is τ. Thus, a partial observable is evolved along a gauge flow, such that the evolution is a gauge-transformation, and is to be understood as a clock 'ticking' along the gauge orbit. On its own, of course, this is an expression of the problem of change since evolution along a gauge orbit is just the problem! But when we correlate another partial observable with the values at which $\beta = \tau$ we form a time-independent observable since the value of α when $\beta = \tau$ does not change. Variation in τ allows for the formation of a 1-parameter family of complete observables that correspond to empirically observable change.[237] The evolution does not occur with respect to some background time parameter, but with respect to the values of the arbitrary clock; the complete observables will *predict* the value of α at the 'time' $\beta = \tau$. More precisely, the evolution will be a map $\mathfrak{E}^{\tau} : \mathcal{E}^{\tau_0}_{\alpha|\beta} \rightarrow \mathcal{E}^{\tau+\tau_0}_{\alpha|\beta}$, taking complete observables into complete observables. The fact that the clock β is arbitrary (since it can be chosen from $C^{\infty}(\mathcal{C}) \subset C^{\infty}(\Gamma)$) implies that the theory is a multi-fingered time formalism: there are numerous (infinitely many) choices that one can make for the clocks, and so there are numerous times—though not all choices will be 'good' clocks physically speaking.

The multi-fingered time result implies that Earman's D-series of coincidence events—which I take to be of the form of Rovelli's complete observables, albeit

236 Earman [2002] appears to endorse this view, and claims that the events (he calls the "Komar events") formed by such coincidences between gauge variant variables can be strung together to give a temporal evolution, generating a "D-series". However, I think that *coincidences* narrow the class of observables down too much. Moreover, I argue below that if Earman means to follow this kind of account—and I think it is clear that he does (see, *e.g.*, Earman [2002], p. 22)—of the evolution of observables, then the D-series cannot be formed: a unique series in incompatible with the multi-fingered time evolution that goes in tandem with the relational approach.

237 This is why the response matches temporal parts theories: temporal parts do not change in themselves, but by forming an individual from a string of such parts a persisting, changing thing emerges.

including the four invariant components of the gravitational field—(of which he speaks as if it were unique) cannot be applicable. Earman ([2002], p. 14) claims that "[t]he occurrence or non-occurrence of a coincidence event is an observable matter" and that when "one such event occurs earlier than another such event" that "is also an observable matter"; "[c]hange now consists in the fact that different positions in the D-series are occupied by different coincidence events". This is not equivalent to the B-series, consisting of a string of events which are either earlier-then, later-than, or simultaneous with each other, because, according to Earman, that "can be described in terms of the time independent correlations between gauge dependent quantities which change with time" (*ibid.*, p. 15). B-series change is an artifact of the local representations (the elements of the equivalence class of metrics) rather than a real feature of the world, that associated with the equivalence class itself (for which his D-series is supposed to apply). This is a strange way of viewing the content of B-series time, and I have never seen any philosophers of time dabbling with such concepts before: why does the B-series depend on gauge variant quantities? Perhaps it is a way to understand the B-series given an ontology that sticks by the gauge dependent quantities, but for different ontologies it needn't follow. If, for example, we adopt an ontology of events then it seems that Earman has simply constructed a B-series all over again. If we view the ontology as per Rovelli's partial/complete observables approach then the multi-fingered time makes the D-series dependent on an arbitrary choice of clock. If we are take make any sense of Earman's D-series within this framework then it will be as one series among infinitely many, each corresponding to a clock choice. But then it does nothing more than to give a name to $\mathcal{E}^{\tau_i}_{\alpha|\beta}$.

However, both Earman and Rovelli appear to want to cling to the notion that the elements of the relations (the partial observables or coinciding elements) have some independent physical reality.[238] This is most explicit is Rovelli who takes the extended configuration space (physically impossible states and all!) to have physical significance as the space of the partial observables. I agree that, without empirical evidence to the contrary, the extended space should be retained since it gives us more conceptual elbow room; but I favour a view whereby gauge invariance itself picks out the physical parts of this space. The interpretation then follows the correlation view, but with the correlates and the correlations understood as simply different aspects of one and the same basic structure. The natural interpretation of Rovelli's view is that there is no physical distinction between gauge dependent and independent quantities. This implies that there are physically real quantities that are not predictable, even though we can associate a measurement procedure with them; indeed, Rovelli claims that these variables "are the quantities with the most direct physical interpretation in the theory" (*ibid.*, p. 124013-7).

It is interesting to note how this links up to Belot and Earman's interpretive taxonomy regarding constraints and spacetime ontology. Since Belot and Earman equate the view that there are physically real quantities that do not commute

[238] Note that Rovelli reads the gauge-fixation methods involving dust variables, curvature scalars, and the like as partial observables. What occurs in these strategies is that the partial observables are taken to be independent so that they are able to function as coordinate systems. However, as Rovelli notes, since the dependent and independent players can have their roles permuted, the distinction collapses ([2002], p. 124013-4).

with the constraint with (straightforward) substantivalism, it appears that Rovelli would have to class as such, for his partial observable are just such quantities! Combined with the role reversal of Kuchař given earlier, this makes something of a mockery of their taxonomy, for they have Kuchař and Rovelli as the archetypical substantivalist and relational respectively. This, would I urge, is yet another aspect of my claim that the relationalist/substantivalist controversy doesn't get any support from those problems with their roots in the interpretation of gauge symmetries. However, I think better justice can be done to Rovelli's view if we take the measurability of the gauge variant quantities as *derived* from the more fundamental correlations of which they are a part. I explain this structuralist gloss on Rovelli's position in the next section.

7.5. ENTER STRUCTURALISM

Rovelli, and other defenders of the correlations view,[239] are of the opinion that the observables of general relativity and quantum gravity are *relative* quantities that express correlations between dynamical, and hence gauge variant, variables. The problems posed to the correlation-type timeless strategies are based upon an understanding that is couched in terms of relationalism. The fact that correlations between *material* systems are required to define instants of time (and points of space) does indeed look, superficially, to entail relationalism. I suspect that this entailment is what was motivating the objections of Unruh and Kuchař. The assumption was that if it is relations doing the work, then the relata must have some physical significance independently of these relations. This is just what I deny: the distinction between material systems and space and time simply amounts to different aspects of one and the same physical structure (*cf.* Stein [1967]). It is not that relations can be free standing; maybe they can, but in this case we have clear relata entering into the relations: DeWitt, Unruh, and a crowd of people! The question concerns the relative ontological priority of these relata over the relations. Relationalists will argue that the relations supervene upon the relata so that the relata are fundamental. Substantivalists will argue that the relata enter into their relations only in virtue of occupying a position in some underlying spatiotemporal structure that exists independently of both the relations and relata. An alternative position will see the relata as being some kind of epiphenomena or 'by-product' resulting from intersections occurring between the relations. But there is a middle way between these two extremes: neither relations nor relata have ontological priority. The relata are individuated in virtue of the relations and the relations are individuated by the relata.[240] Thus, the idea is to understand the correlation view

239 Others include DeWitt (see Ashtekar & Stachel (eds.) [1991], p. 137), Marolf [1995], Page and Wooters [1983], and, on the philosophical side, Earman ([2002]: see below). Page and Wooters' idea is that one deal with conditional probabilities for outcomes of pairs of observables. One then takes the observables as defining an instant of time (*qua* the value of a physical clock variable) at which the other observable is measured. A notion of evolution emerges in terms of the dependence of conditional probabilities on the values of the (internally defined) clock variables.

240 Thus, though admittedly similar, this should be distinguished from Teller's brand of *relational holism* (see his [1991]). Teller argues that in some cases—entanglement is the example he focuses on—we should view relations as being primitive (non-supervenient).

structurally: one cannot decompose or factor the relative observables in to their relata, since the relata have no physical significance outside (independently) of the correlations. But one need not imbue the relations themselves with ontological primacy either. Thus, one can evade the objection that gauge dependent quantities are independently measurable by taking the correlations and correlates to be interdependent.

I shall call the overall structure formed from such correlations a *correlational network*, and the correlates I shall call *correlata*. It is important to note that the correlata need not be material objects, and we can find suitable items from the vacuum case. One is able to use (any) four invariants of the metric tensor to provide an intrinsic coordinate system that one can use to set up the necessary correlational network.[241] Thus, this approach does not imply relationalism; but it does not imply substantivalism either (neither sophisticated nor straightforward). The reason is, of course, that those interpretations require a stance to be taken with regard to the primacy of some category of object (points, fields, or whatever). Each of these other positions is problematic in the context of the problem of time since they both require that some set of objects take the ontological burden to function as a clock or a field of clocks.

Earman too seems to defend a version of the correlation view. His account is based on his notion of *coincidence events*; thus, he writes:

> The occurrence or non-occurrence of a coincidence event is an observable matter ... and that one such event occurs earlier than another such event is also an observable matter. ... Call this series of coincidence events the *D-series* ... Change now consists in the fact that different positions in the D-series are occupied by different coincidence events. ([Earman, 2002], p. 14)

Earman claims that the coincidence event (represented by the functional relationship $g^{\mu\nu}(\phi^{\lambda})$: "the *Komar state*") "floats free of the points of $\mathcal{M}$" and "captures the intrinsic, gauge-independent state of the gravitational field" (*ibid.*). General covariance implies that if this state is represented by one spacetime model it is also represented by any model from a diffeomorphism class of its copies. Now, Earman's interpretation of this, and his resolution of the problem of time, is to claim that the notion of spacetime points, properties localized to points, and change couched in terms of relationships between these, is to be found "in the representations" and not "in the world" (*ibid.*). This conclusion is clearly bound to the idea that in order to have any kind of change, a *subject* is required to undergo the change and *persist* under the change. In getting rid of the notion of a subject (*i.e.* spacetime points), Earman sees the only way out as abolishing change. The idea that change is a matter of representation is one way (not a particularly endearing one, say I) of accounting for the psychological impulse to believe that the world itself contains changing things, though I think it needs spelling out in much more detail than Earman has given us. But—quite aside from the fact that I don't think the existence of

[241] This is, of course, the method developed by Bergmann & Komar [1972]. They used the four eigenvalues of the Riemann tensor. Dorato & Pauri [2006] use this method, and these 'Weyl scalars' to argue for a form of structuralism they call "spacetime structural realism". This is a far cry from what I have in mind since they retain fairly robust notions of independent object (the metric field) in their approach.

spacetime points is ruled out[242]—I don't see why Earman needs to go to this extreme; there is *variation* in the structure formed from the various correlations. True, we don't get any notion or account of time *flow* from this variation, but that is a hard enough problem outside of general relativity and quantum gravity anyway (but see Rovelli & Connes [1994]).

However, some other remarks of Earman's show that he doesn't have in mind the same view as mine. For instance, Earman makes the following observations:

> [T]he gauge interpretation of diffeomorphism invariance ... calls into question the traditional choices for conceiving the subject vs. attribute distinction. The extremal choices traditionally on offer consist of taking individuals to be nothing but bundles of properties vs. taking individuals to have a 'thisness' (*haecceitas*) that is not explained by their properties. The gauge interpretation of GTR doesn't provide any grounds for *haecceitas* of spacetime points. Nor does it fit well with taking spacetime points as bundles of properties since it denies that the properties that were supposed to make up the bundle are genuine properties. The middle way between the *haecceitas* view and the bundles-of-properties view takes individuals and properties to require each other, the slogan being that neither exists independently of the states of affairs in which individuals instantiate properties. ([Earman, 2002], pp. 16–7)

As Earman goes on to explain, in the context of general relativity this middle way fares no better than the bundle-of-properties view since the gauge interpretation of general covariance "implies that the state of affairs composed of spacetime points instantiating, say, metrical properties do not capture the literal truth about physical reality; rather, these states of affairs are best seen as representations of a reality ... that itself does not have this structure". What Earman means by "representation" in this context, is, I think, what Rovelli calls a "local universe" [1992]: a physically possible world in which properties are 'attached' to spacetime points—these will be represented by points of the extended phase space, and elements of the same gauge orbit will represent the indiscernible worlds that differ only in which spacetime points get which properties. However, as Earman and Rovelli point out, this is not how general relativity represents the world; it does so by means of an equivalence class (*i.e.* the whole of the gauge orbit) of such local universes, yielding a very 'non-local' description. However, if we extend the account Earman gives to include *relations* rather than simply properties (which clearly *do* require subjects of some sort, even if subjects are simply bundles of properties) then we can in fact get directly at the structure Earman mentions.

Instead of the view Earman outlines, I have something more along the lines of Skyrms' 'Tractarian Nominalism' [1981]. The idea here is to understand individuals, properties, and relations as 'abstractions' from the structure of the world (from facts) but not as existing independently of that structure: "We may conceive of the world not as a world of individuals or as a world of properties and relations, but as a world of facts—with individuals and relations being equally abstractions from

[242] For example, Saunders' account of identity allows that spacetime points exist as individual objects while respecting diffeomorphism symmetry.

the facts" (p. 199). Likewise, the 'totality of facts' (the structure of the world) itself is 'composed' of such facts. As regards the question of ontological priority, then, we see that relations and relata share the same status: "the Tractarian Nominalist ... takes both objects and relations quite seriously, and puts them on par. Neither is reduced to the other" (p. 202). Armstrong too defends a similar account, and it is perhaps even more applicable to the decomposition problem. Thus, speaking in terms of 'states of affairs' rather than 'facts', he writes that "while by an act of selective attention they [individuals, properties, and relations] may be *considered* apart from states of affairs in which they figure, they have no existence outside states of affairs" ([1986], p. 578). Likewise, the correlations are the fundamental things; they are things that can be measured and predicted. The correlata are measurable only in virtue of their position in the correlation, and have no independence outside of this. However, the correlata are our access point to the correlations, and this is why, I think, Rovelli imbues his partial observables with fundamental significance. If his position is to escape the interpretive troubles highlighted by Unruh and Kuchař, however, the primacy needs to be reversed and shifted to the complete observables. By taking these seriously, as an ontological, those difficulties are easily resolvable.

This *structuralist* way of understanding the correlation view avoids Unruh's and Kuchař's objections, and it sidesteps Earman's worry. Not only does it resolve these objections, and the problem of time, it also provides a suitable ontological framework for classical and quantum gravity, according to which there are neither primitive points nor objects to be individuated. Rather, one has a correlational network that fluctuates quantum mechanically as a whole. This, I suggest, is a safe and sane ontological basis from which to view time and space in both classical and quantum (canonical) gravity.

Of course, avoidance of the problem of time can hardly be said to provide an adequate *defense* of the structuralist conception of the correlation view; as we have seen, there are other alternatives that are also compatible with both the correlation view and the problem of time. For this reason, I expect to be charged with ad hocness at this point. However, the structuralist conception does allow one to sidestep difficult problems with the relationally construed correlation view, and it remains in line with the gauge invariance conception of observables, unlike the timefull responses. Furthermore, it offers a unifying perspective of the gauge invariance view of observables, since it treats the problems of space and time on an equal footing. But the charge is well taken, and I shall attempt to defend the view more directly in Chapter 9. Before I turn to this, however, let us consider how the considerations of this chapter impacts on the debate between substantivalists and relationalists.

7.6. QUANTUM GRAVITY AND SPACETIME ONTOLOGY

As with the hole argument, shift argument, and other permutation symmetry arguments, there have been many grand proclamations about of the impact of quantum gravity on the issue of spacetime ontology and the debate between substantivalists and relationalists. I think it is fair to say that the received view amongst physicists working in the field of canonical quantum gravity is that the theory supports some form of relationalism. The most explicit defender of this view has been Rovelli (most explicit in: [1992] and [1997])—Smolin [2000; 2001], Baez [2001; 2006], and Crane [1993; 1995] paint similar philosophical stances. This has been largely backed up by philosophers who have taken an interest in the subject. Belot and Earman line up gauge invariant and non-gauge invariant interpretations with relationalism and substantivalism respectively; and, as we have seen, Belot sees reduced phase space and extended phase space quantizations respectively as similarly aligned. My key point is that the methods for dealing with gauge freedom (or not, as the case may be) do not bear any relation to spacetime ontology (as charted in the substantivalism vs relationalism debate), and either side of the debate can help themselves to any of the methods. Since these methods are central to the conclusions drawn in the quantum gravity context, we see that quantum gravity does not have the bearing on spacetime ontology that is often thought to hold. My conclusion is that the methods, although indeed central to our understanding of the *structure* of space and time, cannot in fact allow us to draw deeper metaphysical morals about the nature of this structure. I apply my structuralist stance in this context and argue that recent results in loop quantum gravity can be easily accommodated by my view.

Towards the end of their review of the problem of time, Belot and Earman make the following rather metaphysically weighty claims:

> It would require considerable ingenuity to construct an (intrinsic) gauge invariant substantivalist interpretation of general relativity. And if one were to accomplish this, one's reward would be to occupy a conceptual space already occupied by relationalism. Meanwhile, one would forgo the most exciting aspect of substantivalism: it's link to approaches to quantum gravity, such as the internal time approach. To the extent that such links depend upon the traditional substantivalists' commitment to the existence of physically real quantities which do not commute with the constraints, such approaches are clearly unavailable to relationalists. ([Belot and Earman, 2001], pp. 248–9)

Their argument is based on the following line of reasoning: if spacetime points were real, then quantities like 'the curvature at point x' would be real too; but such quantities do not commute with the constraints, so spacetime points cannot be real after all. Substantivalists are then seen as being committed to the view that there are physically real quantities that do not commute with the constraints, and relationalists are committed to the denial of this. They have Karel Kuchař occupying the first position and Rovelli occupying the latter. In the next chapter I argue against the first alignment on the grounds that Kuchař *is* committed to the view that all physical quantities commute with the diffeomorphism constraint. It is true that Rovelli sees himself as occupying a relationalist position, and he sees this as following from *complete* gauge invariance. However, there are a number of reasons why this is problematic—recall for starters, from the previous chapter, that Rovelli and Kuchař can be 'permuted' over relationalist and substantivalist positions according to their taxonomy!

As I already mentioned, Belot connects the substantivalist/relationalist debate to the treatment of symmetries in Hamiltonian systems and their retention or removal respectively ([2001], p. 571).[243]

Let me first detach substantivalism from the internal time approaches. Belot ([1996], p. 241) claims that "substantivalism is ... a necessary condition for loyalty to the sort of approach to quantum gravity that Kuchař advocates"; namely an approach according to which observables commute with the diffeomorphism but not the Hamiltonian constraint. But although Kuchař might claim that his position is substantivalist (see Belot, *ibid.*, p. 238), it is quite clear that a relationalist could just as well adopt it. Indeed, given that the diffeomorphism constraint is solved, Kuchař's position will come out as relationalist according to the received view—a view that Belot elsewhere endorses (see, *e.g.*, Belot [2001]). According to Kuchař the lesson of the hole argument is that it is the geometry of a spatial manifold that has physical content: the diffeomorphism constraint should be solved for.

Next, let me disentangle the view that relationalists cannot adopt the view that there are some observables that do not commute with the constraints. I grant Belot and Earman's point that the reductive relationalist will be barred from those strategies that outlaw commutation with *all* of the constraints. However, as I hinted at above, the relationalist (even the reductive one) can help himself to Kuchař's position. The phase space there is a partially reduced one, with the gauge freedom generated by the diffeomorphisms of space modded out. This is a reasonable object for the relationalist even by Belot and Earman's lights. The fact that the observables are not to commute with the Hamiltonian constraint is no problem: the relationalist too might want to deny that the geometries relate by the Hamiltonian constraint are to be identified for exactly the reasons outlined by Kuchař. Thus, it is perfectly possible for a relationalist to deny Belot & Earman's condition.

Belot and Earman are agreed that the best (easiest) way to avoid the indeterminism that arises in the hole argument, and gauge from gauge freedom in general, is to adopt a gauge invariant interpretation. However, they make the mistake of assuming that the way to achieve this is by giving a direct interpretation of the reduced phase space. They take such interpretations as showing, in the context of general relativity, there could not "be two possible worlds with the same geometry which differ only in virtue of the way this geometry is shared out over the existent spacetime points" ([2001], p. 228). This, they say, leads to relationalism (in the absence of "an attractive form of sophisticated substantivalism"). They list several problems facing the reduced space accounts: the singular points corresponding to symmetric models; non-differentiability; and the unavailability of a set of coordinates able to separate the space's points. For these reasons they conclude that "a dark cloud hangs over the programme of providing gauge invariant interpretations of general relativity ... the present state of ignorance concerning

[243] Likewise for other philosophical stances towards the symmetry arguments considered in this book. The idea is that 'substantivalism' and 'relationalism' are linked to a certain treatment of the symmetries in *any* theory formulated in a phase space description. Thus, one could be substantivalist or relationalist about vector potentials, for example; and this would simply correspond to endorsing an extended (direct) or reduced phase space formulation respectively. Of course, I argued earlier that these links can be severed for any theory.

the structure of the reduced phase space ...—and the lingering worry that this structure may be monstrous—should give pause to advocates of gauge invariant interpretations of the theory" (*ibid.*, pp. 228–9). Perhaps this is a fair comment as far as the reduced space methods go; but such methods are not necessary for gauge invariant interpretations.

Thus, what I am denying here is that the various strategies used in responding to the problem of time and the hole argument (the analogous problem for *space* or spacetime) are related to interpretive stances regarding the nature of spacetime in general relativity. The strategies do not definitively support any such stance, nor do any such stances definitively support the strategies. Thus, what we have is an underdetermination of the various strategies and stances by each other. Whatever it is that pushes one towards a particular stance as regards the nature of spacetime, it cannot be the hole argument or the problem of time. The best these arguments can do is to tell us about the *structure* or spacetime, not its nature. However, as I argued in the previous section, for a structuralist, this is all one needs: nature just is structure!

CHAPTER 8

Symmetry and Ontology

In this and the next chapter I make good on the promise to defend a structuralist account of spacetime (and the ontology of modern, gauge-theoretical physics in general), based upon the kinds of symmetry and symmetry argument dealt with in this book. I attempt to draw the many different issues of this book together under a single, unified account. However, much of these remaining chapters is devoted to showing what kinds of ontological picture can *not* be seen as 'read off the physics', this includes standard structural realist accounts. The remainder should be seen more along the lines of a prolegomenon for future work that will fill in the finer details.

I begin with a discussion of the distinction between reductive and non-reductive moves (at the level of phase spaces), and consider the philosophical implications of each type of technical move. I consider four viewpoints that can be found in the literature: those of Belot, Saunders, Rovelli, and Redhead. I show how the anti-reductive position I have been defending differs from each of these; pulls away from the connection to the substantivalism/relationalism debate; and pushes towards a form of structuralism. In the final chapter, I then outline and defend the kind of structuralism I have in mind, and distinguish it from other (modern) views that have been given the same name: those of Stein, French & Ladyman, and Dorato & Pauri.[244]

8.1. TO REDUCE OR NOT REDUCE?

Hacking [1975] famously argued that spatially symmetric worlds could never constitute a counterexample to PII: there is always a way to redescribe the situation so that the symmetry is absent and so that there are fewer objects within the world. Belot has recently argued that the same holds for PSR with respect to those counterexamples that utilize "a multiplicity of qualitatively identical worlds related by spatiotemporal or other symmetries" ([2001], p. 2).[245] The claim is that such objects ought to be identified, and, moreover, we ought to identify as a "matter of

[244] I also consider the recent views of Stachel, Saunders and the sophisticated substantivalists which bear many similarities to the three structuralisms mentioned above: all of which are shown to be united under the banner of 'semantic universalism'. Additionally, given certain superficial similarities, I distinguish the view I develop from van Fraassen's constructive empiricism.

[245] The examples I have considered in this book have all been examples of this general type. The worlds have generally been understood as 'inhabiting' a possibility space, generally represented by points or paths in a phase space. (If you don't endorse PSR then simply redirect any reference to PSR to the possibility of indistinguishable worlds that have been responsible for the problems we have encountered). Both Hacking and Belot understand PII as operating on possibility

policy" (*ibid.*). Thus, Belot claims that PII is enforced by PSR and this is just what I will deny: PSR is certainly compatible with PII, but it is also compatible with non-reductive options too. I argue that, quite to the contrary, as a matter of policy we should *not* reduce! I connect this conclusion to the argument from Chapter 2 that spacetime ontology is not connected to possibility counting.

Belot explicitly connects his discussion to phase space descriptions of theories: the extended phase space[246] corresponds to a theory with symmetries and the reduced phase space to an empirically identical theory without the symmetries. Note that this equivalence is a purely classical affair; we can associate to any constrained classical system two phase spaces: an extended space with constraints and a reduced space obtained by solving the constraints. These spaces correspond to possibility spaces for physical systems, and as I mentioned earlier, the reduced space contains less possibilities (or possible *worlds*) than the extended space. This difference in the number of possibilities is (empirically) inert at the classical level,[247] but it can manifest itself physically in the behaviour of the respective quantum systems associated to quantizations of the two types of space. The intuitive reason is that degrees of freedom that are absent from the reduced phase space description will be present and undergo fluctuations in the extended phase space quantization.

Now, as a matter of good metaphysical practice, I agree with Belot that we should generally prefer the formulation of a theory without indistinguishable possibilities if those possibilities are redundant. But this preference is not a matter to be decided on the basis of the theory one is talking about if that theory treats the formulations with and without such possibilities as *equivalent*.[248] Put as simply

space itself, rather than on its elements (*i.e.* it is about sets of worlds rather than the *contents* of worlds). Hacking seeks to preserve PII by arguing that any symmetrical world put forward as a counterexample can be faithfully represented by a world that isn't a counterexample; while the counterexample worlds can be imagined and described, they do not constitute *genuine* possibilities (see French [1994] for details and a critique of Hacking's proposal). Belot connects this notion up to the formal possibility spaces and argues that PSR can always be protected from similar counterexamples simply by invoking PII.

246 When I speak of the 'extended phase space' I do not distinguish between constrained and unconstrained spaces; I simply mean an unreduced space of whatever kind. Generally speaking, however, I will be talking about kinematical spaces where the constraints have not yet been imposed.

247 Huggett [1999] draws on this inertness to defend the view that classical statistical mechanics is as permutation invariant as quantum statistical mechanics, and is compatible with both the reduced and extended phase spaces (or "anti-haecceitistic" and "haecceitistic" phase spaces, as he calls them). I agree with Huggett on this point, and it can be seen as applying my conclusions in the spacetime case to the case of particles—Saunders ([2003a], p. 302) endorses Huggett's line. (Note that Saunders (*loc. cit.*) claims that in §2 of our [2003], we (that is, myself and Steven French) are "clearly sympathetic to the view that Leibniz Equivalence, as applied to permutations, is incompatible with classical physics (equivalently, that classically one is committed to the use of haecceitistic phase space)". This simply isn't the case (I cannot speak for French here): there is underdetermination at both the classical and quantum levels, though the *type* of underdetermination is different in these two cases. In the classical case, the underdetermination is secured by the empirical equivalence of the reduced and extended spaces; in the quantum case, it is secured by imposing a symmetrization (initial) condition on the quantum states on the one hand and permutation invariance (reduced along the lines of Leibniz Equivalence) on the other. The view of symmetries I have been defending fully respects this conclusion; I have been at pains to uncover such underdetermination in all cases of such invariance symmetries.)

248 Clearly, in terms of the *conceptual* structure of the respective formalisms they will not be equivalent in general. The formulation without indistinguishables will not be able to accommodate counterfactual switchings of properties between individuals, for the fact that this is a symmetry (for maximal property swaps) will result in such possibilities being removed (of course, we might avail ourselves of Lewis' cheapskate haecceitism once again, so that one and the same world accommodates many possibilities). Indeed, the symmetry arguments I have examined have followed this form: properties are redistributed over individuals in such a way as to preserve observable relations. I consider these conceptual differences below and in §8.3 and the next chapter. (See Teller [1998] for a nice discussion of a related problem concerning haecceities facing constructive empiricists who are wedded to semantic universalism—on which, see §9.3.)

as this, the point I am making may seem rather obvious; yet many authors implicitly contradict this simple point. Moreover, we can work with the extended phase space and get the large or small set of possibilities from it by choosing either a direct or indirect (or selective) interpretation respectively. We do not have this elbow room if we automatically opt for the reduced space. Clearly, however, quantum theory may give us a reason to reduce or not reduce depending upon which formulation of the classical theory yields an empirically successful quantum theory. However, whether we are forced to reduce or not makes no difference to interpretive matters as they stand in the classical theory: one can still be committed to either the things represented by the symmetry operands or one can be committed to the equivalence classes of those things. The choice is purely metaphysical until a quantum theory comes along and tells us otherwise. Even if the reduced phase space is, for whatever reason, deemed to be the correct representational tool for our theories, we cannot simply see this as thereby underwriting relationalist stances, for anti-haecceitism is not a necessary part of the reductive form of relationalism, and that is something that needs arguing for independently (*cf.* Saunders [2003b] for a similar view drawn from his version of PII—more on this in §7.6).

8.2. GEOMETRIC MECHANICS AND POSSIBILITY SPACES

Let us begin by focusing our attention on the nature of the geometric spaces under consideration, diversifying and building upon the introductions presented in earlier chapters. Let (Γ, ω) be the classical *unconstrained* phase space of some constrained physical system or structure $\mathfrak{S}$. This is, of course, a symplectic manifold of dimension $2n$. A *constrained* phase space is then constructed by imposing a set of conditions $\phi_i : \Gamma \to \mathbb{R}$ $(i = 1, \ldots, n)$, known as first-class constraints. This determines a submanifold $C = \{x \in \Gamma \mid \forall_i : \phi_i(x) = 0\}$ called the *constraint surface* (of dimension $m \leqslant n$). These conditions allow us to view C as embedded in Γ, and in so doing we note that the restriction of symplectic form to the constraint surface, $\omega|_C$, giving a geometrically weaker *presymplectic* form σ, determines a foliation $F_{\omega|_C}$ of C whose leaves correspond to gauge orbits and, therefore, to phase points that represent the same physical state of $\mathfrak{S}$. A *reduced* phase space can then be constructed by forming the quotient space $\Gamma_{\text{red}} = C/F_{\omega|_C}$, resulting in a space of leaves or orbits. Crucially, Γ_{red} is a manifold[249] (with $\text{Dim}(\Gamma_{\text{red}}) = 2n - 2m$) and, given a submersion map $\pi : C \to \Gamma_{\text{red}}$, there is defined a symplectic form ω_{red} on Γ_{red}, such that $\pi^* \omega_{\text{red}} = \omega|_C$. The resulting symplectic geometry $(\Gamma_{\text{red}}, \omega_{\text{red}})$ is the phase space of the system $\mathfrak{S}$ characterized by the constraints ϕ_i *when the constraints have been solved*.

Given a simple physical system described by a theory with constraints, the above constructions have the following meaning: (1) the unconstrained phase space contains points that do not represent physically possible states for the system; that is to say, not all points of the unconstrained space are (dynamically)

[249] That the reduced phase space is a manifold does not hold generally; however, in the case of general relativity it is a disjoint union of manifolds, and so is in fact a manifold.

accessible to the system;[250] (2) the constraint surface phase space ignores these 'unphysical' points so that only points representing physically possible states remain, although these points form equivalence classes representing the same physical state; (3) the reduced phase space identifies any equivalent points on the constraint surface, so that each point represents a physically possible and physically (*i.e.* qualitatively) distinct state of the system. (In each of the chapters of this book we have studied a physical theory that admitted an analysis along such lines, and the problem has been to do with their interpretation and with which space should function as the space of physical possibilities. I have been at pains throughout to disentangle interpretive stances regarding the ontological nature of space and time from the latter issue: possibility counting simply isn't relevant.)

Distinct definitions of observables are associated with each type of space, and it is here that most of the philosophical problems we've considered have sprung from. In the unconstrained framework, an observable is simply a real-valued function on the unconstrained phase space, *i.e.* a map of the form $\mathcal{O}_\Gamma : \Gamma \rightarrow \mathbb{R}$. Of course, since some of the states of Γ are inaccessible, this characterization of the observables will produce too many; there will be observables that are in principle unmeasurable. To overcome this problem one can restrict the observables to the constraint surface, $\mathcal{O}_\mathcal{C} : \mathcal{C} \rightarrow \mathbb{R}$. However, the constraint surface is partitioned into gauge orbits by the foliation, where the usual interpretation is that the elements of such orbits are equivalent in the sense that they represent physically indistinguishable possibilities (the same physical state). In this case, with no further restriction on the form of the observables, there will be underdetermination: there will be distinct physical states that no observable can distinguish between. Of course, this is the source of the indeterminism that plagued direct interpretations of gauge theories; it is also the source of Newton's difficulties with the shift situations. To avoid this problem, the further restriction we then impose on the observables is that they be constant along gauge orbits (*i.e.* on the leaves of $F_{\omega|\mathcal{C}}$), so that observables must satisfy $\mathcal{O}_{[x]}(x) = \mathcal{O}_{[x]}(x')$ whenever $x, x' \in [x]$ (*e.g.*, when x and x' are connected by a gauge transformation). This definition is equivalent to requiring that the observables commute (weakly) with the first class constraints, $\forall_i, \{\phi_i, \mathcal{O}\} \approx 0$, where the constraints are understood as generators of gauge symmetries. Of course, this latest restriction simply amounts to a gauge invariant definition of the observables; there is no underdetermination or indeterminism because the observables are now only sensitive to differences between entire gauge orbits. Gauge-invariant observables naturally induce a function $\mathcal{O}_{[x]} : \Gamma_{\text{red}} \rightarrow \mathbb{R}$ (under the submersion map π^*), which is just to say that such functions $\mathcal{O}_{\Gamma_{\text{red}}}$ on the reduced phase space are automatically gauge invariant, corresponding as they do to gauge invariant functions on the constraint surface.

We can see two levels of surplus structure at work in the preceding descriptions: (1) the surplus associated with the *inaccessible* states of the unconstrained

[250] One option here is to claim the inaccessible states as *metaphysical possibilities* but *physical impossibilities*. A similar move is suggested by French [1989] as a response to the differences between classical and quantum statistics. One considers the principle of permutation invariance as imposing an initial condition on particle states so that state vectors are constrained to remain in one or another subspace representing particle type (boson or fermion)—see French & Rickles ([2003], pp. 222–3) for further discussion. This point fits in quite nicely with one of the main claims of this book that one can 'access' the reduced possibility set from the extended space by tweaking certain other aspects of an interpretation.

phase space (*i.e.* $\{x \mid x \in (\Gamma - \mathcal{C})\}$); and (2) the surplus associated with the gauge orbits of the constraint surface (*i.e.* $\{x \mid x \in [x] \subset \mathcal{C}\}$). I think that it is important to distinguish between these two types of surplus structure: the first type is not nearly so problematic as the second, for the latter can be taken to represent physical possibilities but the former cannot. Thus, we can write off the former 'inaccessible' type as *merely unphysical*, an artifact of representation that can be resolved by introducing a set of constraints on to the space or focusing on the space of physically accessible states represented by $\mathcal{C}$. I don't think anyone would question this. The problems concern the latter type then, and writing it off isn't as simple as it is with the unphysical surplus structure for the points of $\mathcal{C}$ are physically possible.[251] Classically, each space leads to the same physics; but quantum theory messes up this nice tidy setup. There are distinct types of quantization method associated to each of these structures and that lead to distinct (inequivalent) quantum theories.[252]

Prima facie, the reduced space looks to be the clear winner in the choice of representational space: it represents only the gauge invariant information of the system. However, in taking the quotient of a constraint surface by the gauge orbits, one loses out on certain features that are, at the very least, technically useful: manifest Lorentz invariance and locality in space, for example. It is also often hard to find a set of coordinates for the reduced space—though, on the plus side, the coordinates (and the observables) will be immediately gauge invariant if they can be found.[253] More serious are the points mentioned above, that the reduced phase space will not, in general, be a cotangent bundle of some configuration space $\mathcal{Q}_{\text{red}}$ such that $\Gamma_{\text{red}} \cong T^{*}\mathcal{Q}_{\text{red}}$. Reduced phase space is not the same as the phase space of an unconstrained phase space, and this makes quantization very difficult. Thus, non-reductive methods that quantize first and then single out the physically relevant structure from the surplus came about. Indeed, far from reducing, one often sees the opposite move being made: expansion! This idea forms the basis of BRST theory (see Henneaux & Teitelboim [1992]). Let me quickly outline the main ideas of this approach, for they may be unfamiliar to many readers.

The basic idea in BRST theory is similar in many respects to the reduced phase space methods: one wants to construct a symplectic manifold (without constraints) to function as the phase space of the gauge system one is interested in. In the BRST case, one *enlarges* the phase space of a constrained phase space (Γ, ω, ϕ_i) by adding auxiliary variables, giving the extended system $(\Gamma_{\text{ext}}, \omega_{\text{ext}})$. These auxiliary variables consist of Fermi degrees of freedom, (θ, π), called *ghosts* and their conjugate

[251] I am ignoring the kinds of selective response given by Butterfield and Maudlin here. Clearly, that idea would amount to imposing some condition on $\mathcal{C}$ such that exactly one $x \in [x] \subset \mathcal{C}$ is physically possible for each orbit of the symmetry group. Since neither party offers such a condition we can put their views aside here. Recall also that most physically interesting systems will face a Gribov obstruction. Moreover, technically (*i.e.* ignoring aspects to do with how the geometric spaces represent), such moves simply match up to the construction of a reduced phase space. Given this match, we see a possible explanation of the relationship between PII (sophisticated) and non-PII (selectivist) endorsing substantivalist options: the basic representational spaces they call upon are isomorphic.

[252] Which structure we choose to quantize on has become part of the debate between substantivalists and relationalists. I explain how below, and then assess the intrusion in §7.6.

[253] This is one of Belot and Earman's main objections to gauge invariant interpretations of general relativity (*cf.* [1999, p. 177]). The other one is the problem of time.

momenta (*anti-ghosts*) and they are chosen in such a way as to ease quantization.[254] The reason behind their introduction is to construct an operator D (the classical BRST operator) whose cohomology yields the gauge invariant functions of the theory. Quantization is carried out on *all* degrees of freedom (physical, unphysical, and ghost), and the resulting quantum system is then reduced using D. Note, however, that the BRST formalism is by no means restricted to quantum theory. It has quite respectable credentials in the classical context too: the original gauge symmetry of the classical theory is replaced by a (fermionic) rigid symmetry that acts on the expanded phase space in such a way as to encode the gauge symmetry within a simpler theory.

We thus have four available spaces (ordered according to 'size'): $(\Gamma_{\mathrm{ext}}, \omega_{\mathrm{ext}})$, (Γ, ω), $(\mathcal{C}, \sigma)$, and $(\Gamma_{\mathrm{red}}, \omega_{\mathrm{red}})$. Each of these spaces has a simple interpretation in terms of the possibility spaces I introduced in §1.1: the expanded and unconstrained phase spaces include impossible *unphysical* states (ghosts and inaccessible states in the former; inaccessible states in the latter) that, if they represent anything, represent physically impossible worlds; the constraint surface contains physically possible states (though some worlds may be multiply represented, or else there are worlds that differ haecceitistically); points of the reduced space can be put in one-to-one correspondence with physically possible worlds with no haecceitistic differences and no multiple representation. Naturally, each of these spaces has quite specific problems, and each generally produces a distinct quantum theory. However, what I am interested in this section is the question of whether any of these spaces is to be preferred on the basis of symmetries, and the symmetry arguments considered in this book. A secondary question I wish to consider is whether or not these spaces underwrite particular philosophical positions concerning spacetime ontology. Let me now deal with these issues together by isolating four views that can be seen as defending each space as a response to symmetry arguments (*i.e.* potential violations of PSR).

8.3. FOUR VIEWS ON REDUCTION

Belot believes that the existence of spacetime points is bound up with possibility counting. For example, he claims that

> [s]ubstantivalists count each possible embedding of a set of N particles into $\mathbb{R}^3$ as (being capable of) representing a distinct possibility—which is just to say that they will work with the standard $6N$ dimensional phase space when constructing mechanical theories. Relationalists about space will deny that embedding related by rigid motions can represent distinct possibilities; so they will identify points in the standard configuration space so related; thus they will employ that $3N - 6$ dimensional configuration

[254] The subject of the ontological status of ghostly variables is in need of investigation. Physicists are often ambiguous on the matter of their physical status, alternating between viewing them as a heuristic crutch and having direct physical significance (see, for example, Henneaux & Teitelboim [1992], p. 166 and Ch. 11). Unfortunately, this issue is too complicated to tackle here—see Weingard [1988] for a detailed analysis based on connections between the interpretation of ghost fields and virtual particles.

space (parameterized by the relative distances) and the $6N - 12$ dimensional phase space (parameterized by ... relative distances and velocities). ([Belot, 2000], p. 580)

Thus, according to Belot, substantivalism is bound to the extended phase space (high possibility count) and relationalism is bound to the reduced phase space (low possibility count). And his argument is that if one moves to the reduced space—as he believes one generally should—then one is committed to the non-existence of spacetime points. But we have seen that the proposed connection between possibility counting and spacetime ontology is based upon a hidden assumption about modality: the substantivalist-extended space connection requires haecceitism and the relationalism-reduced space connection requires anti-haecceitism. Thus, possibility counting has got nothing to do with spacetime ontology; it is the intrusion of modality that underwrites the supposed connection to possibility counting and particular representation spaces. The argument I presented in Chapter 2 involved showing how both the substantivalist and the relationalist could occupy both deflated and inflated possibility spaces respectively. I return to this argument again in §7.6, where I use it to dampen the expectation that quantum gravity might be decisive in matters of interpretation as regards spacetime ontology.

We need to concede, however, that the distinction between the reduced and extended spaces *is* connected to differences in possibility counting: the latter contains points that simply are not contained in the former. But the latter can, with suitable contortions, incorporate anti-haecceitistic possibility counting; so too can the reduced space incorporate the haecceitistic possibility counting of the extended space. This is simply a result of the formal and empirical equivalence of the formulations. Given this equivalence, what are the reasons for choosing one over the other? Belot argues for reduction along the following lines:

> The trick is to allow the absolutist to specify a large space of possibilities which fall into equivalence classes ... The advocate of PSR can then claim that the true space of possibilities arises by identifying equivalent absolutist possibilities, so that there is exactly one possibility corresponding to each of the absolutist's equivalence classes. ... we can always use this trick to protect PSR against refutation by Clarke's sort of examples, where indifferent possibilities are generated by the application of symmetries. ([Belot, 2001], p. 4)

Thus, as Belot notes, a direct interpretation of a theory with symmetries (of the relevant kind) will risk violating PSR and his answer is to shift to a reductive interpretation that puts orbits in to a direct correspondence with possibilities: "by always choosing interpretations ... which "factor out" symmetries ... we can ensure that our interpretations will always respect PSR" (*ibid.*, p. 7).[255] With this

[255] Belot claims that "the techniques and results of this literature [on symmetry in geometrical mechanics] promise to offer a unifying perspective on a number of classic problems in philosophy of physics (the relation between the nature of space and the nature of motion in Newtonian physics, identical particles, the nature and significance of gauge freedom and general covariance)". I agree with this statement as it stands, but Belot takes the claim too far and attempts to create alignments between philosophical stances regarding the nature of individuals and the treatment of symmetries in the areas

I don't disagree: we know—given the formal and empirical equivalence of the reduced and extended spaces—that we will always have the option of shifting to the reduced space (at least in principle) and we know that this space will have any points related by symmetries removed; since these points were responsible for the potential violations of PSR, we will indeed have resolved the difficulty. The *technical* foundations of Belot's proposal are impeccable, as one would expect. However, the question is whether this approach is necessary and, if not as I have been arguing, whether it is worth the various technical pitfalls that such approaches inevitably must face—*i.e.* the difficulties with construction mentioned above.[256] This is not to mention the chunk of possibility space that we will have jettisoned without any good physical reasons! In other words, the decision to reduce in the manner suggested by Belot is a purely *metaphysical* decision that, quite literally, makes worlds of difference.

There are two further 'technical' problems with Belot's proposal: (1) rarely do we construct 'intrinsic' reduced phase spaces for theories, generally beginning with the extended space with symmetries and then factoring them out; (2) although the extended and reduced spaces are classically equivalent, they in fact lead to distinct quantizations, and so physics *might* be decisive in choosing one over the other. The first point is simply that if Belot's PSR wielding theorist is to hold his head up high, he should be able to construct the reduced space form of a theory directly; as he points out himself, "the reduced theory knows where it came from" ([2001], p. 14).[257] The second point is more complicated and arises out of studies at the intersection of geometric mechanics and quantization (see Gotay [1984] or Plyushchay & Razumov [1995] for details). The upshot, however, is simply that the choice between extended and reduced spaces cannot simply be a matter of policy. As to Belot's underlying desire to show how PSR can always be protected by imposing PII on symmetrical worlds, there is another option that always works too: one simply views the symmetries as expressing an *indifference* concerning the states and observables of physical systems entering into them (*i.e.* along the lines of Brighouse's anti-haecceitism concerning physics). Thus, there is always a sufficient reason for the world's being where it is in a universe with a homogenous spacetime: the world is indifferent to where it placed; one position is as good as any other! Given this rather obvious possibility, Belot's account seems to be somewhat unmotivated. Further, his account faces a serious problem when one considers quantization, for certain states factored out via Belot's method might be required to fluctuate in the quantum theory.[258] Saunders offers an alternative defense of the PSR based on his idea that the individuals entering into symmetrical

of physics he mentions. The equivalences and underdetermination I have shown to hold in such contexts outlaws such alignments.

256 There are also the problems—mentioned by Belot ([2003a], p. 407)—concerning the ad hoc removal of certain points (those representing symmetrical configurations and those representing collision points) from the extended phase space in order that the reduced phase space can be constructed. Technical details involving differences between discrete and continuous symmetries are crucial here—see Belot [2003b] for more details.

257 Compare this with Earman's point that the relationalist should be able to construct his theories in relationally pure vocabulary, rather than hitching a ride on the substantivalists formulations ([1989], p. 135).

258 Belot is clearly well aware of this, of course (*cf.* [2003b], p. 221); indeed, it informs his and Earman's taxonomy of interpretations of general relativity. (See the end of this section for further details of this problem and §7.6 for a critique of Belot and Earman's taxonomy.) Note that Belot seems to shift to the view that one must await an answer from quantum theory to the question of how best to deal with symmetry [2003b]. However, if quantum statistical mechanics is anything

relations of the kind we are interested in will be weakly discernible (and absolutely indiscernible) but referentially indeterminate; the symmetries fail to get their teeth into the PSR. However, I argue that the end result, as regards the question of reduction, is the same as with Belot's proposal: there are no indiscernible possibilities.

We have already seen Saunders' view at work in the context of the hole argument. However, as Saunders points out himself, the view that emerges can be applied to "any exact symmetry in physics" ([2003a], p. 153). Saunders calls his view 'relationalism', but he sharply distinguishes this from what is usually labeled relationalism, and his main argument is that the kind of symmetry arguments I have considered in this book have "nothing to do with a *reductionist* doctrine of space or spacetime" (*ibid.*).[259] Saunders calls this latter form of relationalism "*eliminative* relationalism" and his form "non-reductive relationalism" (*ibid.*). Let us recall Leibniz's original shift argument. This was supposed to cause problems for PSR: space's being homogeneous, there was no reason why a system should be located at one part of space rather than some other. I suggested that Leibniz's use of PII can be thought of as rescuing PSR from the grip of the argument (much as Belot suggests). That does not mean PII is ad hoc, simply that Leibniz thought that commitment to PII was part and parcel of being committed to PSR. However, we cannot forget Leibniz's notion of object as given intrinsically, and its description as giving a 'complete concept'. Saunders calls this aspect of Leibniz's philosophy the "independence thesis": roughly, an object's identity is independent of anything else 'external' to that object. Saunders' claim is that Leibniz understood PII as entangled with the independence thesis: without the independence thesis, PII might allow external reasons to come into play in its protection of PSR and, in particular, in the individuation of the homogeneous parts of space, and thus bring the shift argument to a halt *without recourse to reductive* (*i.e.* eliminative) *measures*. With external reasons not playing a role, and internal reasons absent, the symmetry arguments are in clear violation of PSR. But the 'PII + independence thesis' package can be divided, and in so doing Saunders argues that a version of relationalism follows that is non-reductive precisely because it denies the independence thesis, and thus allows external factors to enter into the definition and individuation of an object. However, it remains committed to PII. I shan't go over the details again (for which, refer to §5.3.3), but simply wish to show how this scheme fits into the question of reduction vs non-reduction. The connection is clear: "relations, for Leibniz, had to be *reducible*—derivable from the monadic properties of their relata" (*ibid.* 168); when these monadic properties are equivalent so is the relational structure—the corresponding possibility space is represented by Γ_{red}.

In brief, we have the following chain of reasoning leading to Saunders' view. Leibniz's relationalism involves three components: PSR, the independence thesis, and PII. The independence thesis filters into PII, and restricts the latter principle to *internal* factors, so that relations to other things are not to be included in the

to go by, even quantum theory cannot determine the correct geometric space of the classical theory: as Huggett [1999] and French & Redhead [1988] have demonstrated, the reduced and extended formalisms are compatible with both classical and quantum theories. (Though, as I mentioned above, a lot can hang on the nature of the symmetries in question—discrete versus continuous—what goes for one type will not necessarily hold for the other.)

259 With this I am in complete agreement as my arguments from previous chapters should make clear. I differ with Saunders, however, in the way I argue for this position, and in what I think the conclusion signifies (see below).

description of an object. PSR faces trouble from the symmetry arguments, since it seems that objects related by certain symmetries count as identical in all *internal* respects, *i.e.* in all respects that matter in this case. PII, informed by the independence thesis, enters the analysis and is used to identify any such objects (points, worlds, *etc.*). This means that only internal (*i.e.* intrinsic) qualitative differences count towards numerical differences so that the differences generated by symmetries do not imply genuine physical differences. Saunders denies the independence thesis thus allowing any physical relations to individuate and, though absolute quantities—represented by *e.g.* gauge dependent variables—are eliminated in favour of relations between objects, his analysis still allows for spacetime points (and any weak discernibles) to be distinct and individuated. The upshot of this *vis-à-vis* the PSR is that the problems posed by the symmetry arguments dissolve; one uses relations to matter and events to specify points of space:

> Absolute positions disappear; under the PII points in space, considered independent of their relations with other point and with material particles, all disappear. But points in space considered independent of matter, but in relation to other points in space, are perfectly discernible (albeit only weakly), for they bear non-reflexive metrical relationships with each other. There is no problem for the PSR in consequence; there is no further question as to which spatial point underlies which pattern-position, for they are only weakly discernible. ([Saunders, 2003a], p. 174)

Saunders sees this as motivating an "even-handed approach to matter and space": things from either category can serve to individuate other members from their own and from the other category (*ibid.*, p. 176). Now, my claim was that reductive relationalism follows from the symmetry arguments considered in this book only if it is coupled to PII (construed as an anti-haecceitist principle). Saunders, however, claims that PII is not necessarily anti-haecceitist, nor is it necessarily reductive and, therefore, that the symmetry arguments do not imply reductive relationalism. His path was to deny the independence thesis and retain PII, whereas I argued both that PII wasn't a necessary part of the relationalist's position and nor was it *not* a part of the substantivalist's position. This latter point leads naturally into the sophisticated substantivalist positions; and, indeed, Saunders mentions the similarities between his own self-styled relationalist approach and these other self-styled *non-relationalist* approaches. Both approaches accept PII but the latter see reduction (*i.e.* identification of equivalent worlds) as concomitant with this, whereas Saunders does not; rather, he claims that spacetime points can have well defined identities in the absence of matter, and can be uniquely referred to in the presence of matter. One might think that this is even more substantivalist than sophisticated substantivalist positions! But Saunders agrees with me that it is ontological priority that counts when it comes to the definition of these positions, and his approach is neutral on this: each can be used to individuate the other. In this sense I would say that Saunders' position is more naturally understood as a structuralist one; indeed, he ends up in more or less the place I wish to end up, but he gets there by a different route and for different reasons.[260]

[260] I expand on the similarities and differences in the next chapter.

Thus, Saunders protects PSR by implementing but modifying PII. The result looks non-reductive, but on closer inspection the non-reductive aspect concerns objects within worlds and not worlds themselves. Since the geometric spaces correspond most closely to possibility spaces, rather than singular possible worlds and their contents, we have to inquire as to what Saunders' version of PII says about possibility space. The first thing to note is that Saunders restricts the application to worlds that have the same physical laws as our own: for different laws there may be different PIIs ([2003b], p. 297). Then, since there are no physical relations that hold between distinct possible worlds,[261] Saunders' PII reduces to Leibniz's PII and we are left with what is essentially a reductive version: "Given that possible worlds bear no physical relations to one another, it follows from the PII that numerically distinct worlds will be absolutely (and in fact strongly) discernible" (*ibid.*, p. 298). Furthermore, his discussion on the relationship between symmetry and observables shows that he in fact endorses a rather extreme reductive view. Although the PII itself does not appear anti-haecceitist or reductive at first sight, countenancing as it does physical relations, it is both of these things when applied in the context of possible worlds. Thus, although Saunders can retain spacetime points (and any weak discernibles) with his PII, he is lead to back the connection between relationalism and possibility counting that I denied in Chapter 2; according to this account, the relevant geometric space for physical theories is Γ_{red}. Hence, in the final analysis, PSR is preserved *à la* Leibniz–Belot, simply by implementing reductive PII. However, Saunders has shown us that reduction at the level of worlds does not imply eliminativism of indiscernible entities within worlds.[262] As I mentioned earlier, this is simply yet another flavour of sophisticated substantivalism, albeit one clothed in relationalist garments.

Rovelli has recently outlined an interpretation based on the full, unconstrained (extended) configuration space (along with its associated extended phase space). I presented some of the details of this program in the previous chapter; let me here simply address how it fits in with the concerns of the present section. Recall that Rovelli's claim was that a number of thorny problems from general relativity and quantum gravity can be cleaned up or resolved by utilizing his distinction between 'partial' and 'complete' observables. Partial observables are taken to coordinatize extended configuration space $\mathcal{Q}$ and complete observables coordinatize reduced phase space Γ_{red}; the "predictive content" of some dynamical theory is then given by the kernel of the map $f : \mathcal{Q} \times \Gamma_{\text{red}} \rightarrow \mathbb{R}^n$. The relevant aspect from this program for this section is captured by his claim that "the *extended* configuration space has a direct physical interpretation, as the space of the partial observables" ([2002], p. 124013-1). This space gives the *kinematics* of a theory and the *dynamics* is given by the constraints, $\phi(q^a, p_a) = 0$, on the associated extended phase space $T^*\mathcal{Q}$. *Both* are invested with physicality by Rovelli. Thus, whereas, for example, Stachel argues that the kinematical state space of a background independent theory like general

261 I think Saunders' reasoning is sound on this point. He notes that a world "is a system that is physically closed" ([2003b], p. 297), and that simply means that any physical relations that hold at that world are contained in it too.

262 Belot, on the other hand, sticks to the original PII to protect PSR. He doesn't consider Saunders' version of PII, and goes along with the idea that both absolute quantities *and* spacetime points (more generally: the things with respect to which absolute quantities are defined) are eliminated.

relativity has no physical meaning prior to a solution (so that only the dynamical state space is invested with the power to represent; kinematics being derivative), Rovelli appears to take both kinematic and dynamical spaces as equally capable.

The view Rovelli defends has some immediate philosophical interest since it is non-reductive and yet Rovelli is a self-proclaimed relationalist. Thus, prima facie, we seem to have an instance of a break between possibility counting/geometric spaces and spacetime ontology. However, it quickly becomes evident that there is a conflict between his relationalism and his choice of representational space. As regards the former, we saw that a rather naive verificationism was responsible for Rovelli's views: only measurable things are real and since spacetime location is not measurable but relations between objects are measurable, space and time are not real but are instead defined by correlations between objects. We can agree with Rovelli that the physically measurable quantities are those that are invariant under the symmetry group of a theory, *i.e.* the gauge invariant quantities. It is quite another matter to then say that these are the only physically real things, that they *exhaust* physical reality. Clearly, in Rovelli's view, however, there are plenty of physically real objects; namely, those things entering in to relations that are not themselves measurable. Any relationalism will require a definite set of material objects to generate the required relations (particles, fields, *etc.*). Rovelli's view is not that there are no objects *per se*, but that there are no objects corresponding to those that ground absolute (non-measurable, gauge variant) quantities. Thus, he moves from the fact that we never measure position in spacetime to the non-existence of spacetime points. However, his work on partial observables suggests something very different to this rather crude verificationism. Let me develop some more of the details of this latter approach.

Rovelli distinguishes between two extremes of interpretation with respect to the formal variables of a theory for a system with constraints (I have changed the notation to suit my own):

> It is sometimes claimed that the theory can only be interpreted if one finds a way to "deparameterize" the theory, namely, to select the independent variable among the variables q^a. In the opposite camp, the statement is sometimes made that only variables on the physical phase space Γ_{red} have a physical interpretation, and no interpretation should be associated with the variables of the extended configuration space Γ. ([Rovelli, 2002], p. 124013-7)

By contrast, Rovelli invests elements of Γ and $\mathcal{Q}$ (including gauge variant quantities) with physical reality; indeed, elements of the latter are taken to be "the quantities with the most direct physical interpretation" (*ibid.*). Complete observables—*i.e.* the quantities we actually measure and are able to predict uniquely (*i.e.* Bergmann/Dirac observables)—are dynamically determined *à la* Stachel:

> Such a quantity can be seen as a function on the space of solutions modulo all gauges. This space is the physical phase space of the theory Γ_{red}. ... Any complete observable can thus be expressed as a function on Γ_{red}. ([Rovelli, 2002], p. 124013-3)

Crucially, Rovelli notes that there is an equivalent description of any complete observable "as a function on the *extended* phase space having vanishing Poisson brackets with all first class constraints" (*loc. cit.*; my emphasis). Thus, we see again the formal equivalence between reduced and extended spaces even at the level of observables. In this approach, then, Rovelli distinguishes between what is observable and what there is (*i.e.* ontology), whereas before, in arguing for his relationalism, he assumed a direct connection between the two.

However, I think it is clear that Rovelli does not want to imbue what are physically *impossible* states with physical reality—that is, Γ isn't Rovelli's space of choice. That would clearly be crazy. Though he often speaks as if he means to endorse this 'crazy' metaphysic, we can best understand his view, I think, as being based upon the constraint surface $\mathcal{C}$. Thus, he speaks of a mechanical system as being completely determined by the extended, unconstrained phase space *plus* a set of constraints (if necessary). We should, therefore, view the constraints as physical 'reality conditions' and only those quantities that satisfy them as invested with reality. Nonetheless, Rovelli is still avowedly realist about non-gauge invariant quantities, quantities that do not commute with the constraints (partial observables). However, he avoids any 'hole-type' problems by defining (complete) observables as functions on the reduced space, quantities that are constant along gauge orbits of the extended space. Though nowhere near as crazy as the above metaphysics involving impossible states, the position Rovelli presents is a metaphysics nonetheless. In imbuing the gauge variant quantities with physical reality he is putting in his metaphysics by hand, for it is not being read off the physics. This is quite evident from the fact that a structuralist position can also be adopted, as I demonstrated at the end of the previous chapter. Clearly, too, one could, as Rovelli did previously, adopt the view that only the complete observables are physically real; one doesn't require the reduced space for this sort of position.

Finally, we have the expansionist option. Redhead outlines such a *liberal* view of symmetries: "forget all about gauge symmetry in the original Yang–Mills sense, and impose BRST symmetry directly as the fundamental symmetry principle" ([2003], p. 137).[263] The idea, as Redhead describes it, is to "allow non-gauge invariant quantities to enter the theory via surplus structure ... [a]nd then develop the theory by introducing still more surplus structure, such as ghost fields, antifields and so on" (*ibid.*, p. 138). He claims that this is the method that is most in line with the practice of physics. He also notes that, given the mathematical nature of the surplus structure, "this [approach] leaves us with a mysterious, even mystical, Platonist–Pythagorean role for purely mathematical considerations in theoretical physics" (*ibid.*). However, though it may be of value in the quantum gauge field theories of the electromagnetic, electroweak, and strong forces, I don't see that it is at all applicable in the context of classical and quantum general relativity in which the gauge symmetries are directly connected to the dynamics. Even if it could be shown that the BRST method is applicable, the suggested enlargement

263 He may not actually wish to be associated with this view, it isn't fully clear from the text which of the methods he endorses. However, the fact remains that this is a possible interpretive option to take with regard to gauge symmetries. If he doesn't in fact endorse this view then let us say Redhead refers to some 'other-worldly' philosopher of physics who *does* endorse it.

of phase space is a purely *classical* affair: one reduces by the BRST operator once the quantum level has been reached. Thus, the device of BRST appears to be a purely heuristic one, and cannot be seen as underwriting any *unique* interpretive stance. Indeed, in the final (quantum) analysis, the resulting picture matches, more or less, the Dirac quantization methods in that reduction is carried out at the quantum level. Classically, of course (as Redhead points out), the problem is to make sense of the auxiliary variables that are employed, and this would require considerable work. In particular, I think analysis is needed on the differences between the various senses of 'surplus structure' that come into play here: ghosts, impossible states, and indiscernible states. If it can be shown that ghosts and impossible states are of the same kind then I think this would give us grounds to reject the BRST expansionist approach.

Thus, we have four diverging views on the question of whether to reduce or not: Belot argues that we should, as a matter of general practice, reduce and get rid of the symmetries;[264] Saunders says that 'all out' reduction (*i.e.* elimination) is not necessary to get the kind of deflationary conclusion Belot wants, but nonetheless implements reduction at the level of worlds;[265] Rovelli says that we should utilize the constraint surface (or, rather, the extended space plus a set of constraints); and Redhead argues that we should expand rather than reduce or constrain. We saw above and in the previous chapter that Rovelli advocates a view whereby no reduction or gauge-fixing is carried out: the extended space and the set of constraints is sufficient to determine a sensible interpretation. With this I agree, but the interpretation I give differs from both Rovelli's and Saunders' 'non-reductive' methods. I will outline this in the next two sections. Before I get to that, let me quickly sound a warning note for hasty reduction proposals, followed by a brief summary of what I hoped to have shown thus far.

It is clear that any choice of space must come about as a result of experimental confirmation; and this can only come about at the level of quantum theory. Even then, whether or not this will choice will be possible—*i.e.* whether it could ever be shown that a certain way of counting possibilities is the correct one—is far from obvious. On a purely conceptual level I suggest that the extended space is to be preferred over the reduced space. The extended is ontologically neutral in that it allows for large and small possibility counting in a fairly unproblematic way. It leaves intact (and manifest) properties to do with symmetries, such as covariance and locality. It makes no prior assumptions about what degrees of freedom should be quantized and allowed to fluctuate (*cf.* Plyushchay & Razumov [1995], pp. 248–9). The resolutions I defended regarding the various symmetry problems considered in this book are best expressed in the extended space, where that space is minimally taken to encode a network of relationships. The symmetries mean that only the network, and correlations expressed within the network, are observable. Claims about the nature or existence of the 'nodes' of the network (*i.e.* the

264 Elena Castellani [2004] too has recently defended a similar view in her analysis of Dirac's theory of gauge systems and constrained Hamiltonian systems.

265 Essentially, Saunders' argument is that the fact that PII amounts to a reductive principle when imposed at the level of the worlds themselves does not, as if often believed (by Belot, for one), imply that PII involves elimination *within* worlds. Denying the independence thesis allows one to individuate what would have otherwise been indiscernible entities by using the relations they bear.

symmetry operands) are radically underdetermined. The reduced phase space, of course, takes a stance on what is physically relevant, and this choice is carried over to the quantum theory. Thus, there will be elements of the extended space that will not be subject to quantum fluctuations, but will be eliminated instead. Though I think the reduced phase space can be given a well-motivated structuralist defense (encoding, as it does, the supposedly physical (invariant) structure), I think that it should be a part of the *honest* structuralists manifesto that stances taken regarding to the individual elements entering into gauge-type symmetries should be avoided.[266] This, of course, includes Rovelli's partial observables realism.

I hope to have shown (or at least suggested) the following in this section:

- Theories are not bound to either reduced or extended spaces (they admit PII and non-PII-type formulations while still respecting PSR). Reduced and extended formulations are empirically equivalent.
- Reduced and extended spaces (with their associated possibility counting) are not bound to particular spacetime ontologies.
- The reduced and extended spaces *are* bound up (in a certain sense) with possibility counting: they contain different cardinalities of possibilities. But haecceitism and anti-haecceitism are nonetheless compatible with theories formulated in both spaces.
- If we choose the reduced space (without pressure from quantization) then we are cutting out possibilities in a way not dictated by the physics itself—*i.e.*, the metaphysics of possibility counting that results is not 'read off' the physics in this case.
- Since extended spaces allow 'all the options' (conceptual elbow room, as it were) we would be better off choosing such a space as the neutral base. We should, more properly, view the constraint surface as our base, for in this case the metaphysics of outlawing physically impossible states *is* easily read off the physics.

I think these points allow us to infer three more conclusions: (1) that PSR (respecting, *e.g.*, gauge invariance) does not imply PII (*i.e.* PSR does not link up to reduction, quotienting, or eliminativism); (2) that substantivalism and relationalism are not linked to the denial and endorsement of PII (reduction/elimination) respectively; and (3) that theories with gauge symmetries (or similar) cannot give decisive reasons for interpretive options concerning the nature of the individual elements connected by the symmetries. In particular, symmetry arguments like the Leibniz-shift and hole arguments, and the problem of time, cannot be used

266 Thus, I diverge quite radically from French & Ladyman's 'ontic' version of structural realism (see their [2003]—I discuss this version in §9.2). The reason: they see the underdetermination as applying to the 'individualistic' and 'non-individualist' packages only, and not as involving the eliminativist views; for this reason they drop the former package entirely and opt for an ontology of pure structure (not involving objects). These latter views are most naturally expressed in the reduced space, and it is that space that the structural realist (of the ontic brand) will wish to be aligned with: it encodes all and only the *invariant* and, they will say, physical structure. Incidentally, I think the fact the reduced space does encode this structure, and can be associated with an elimination of objects—though this is underdetermined, of course; hence my desire to stick to the extended space—, offers a quick and easy answer to the question: 'What is structure?' It is given precisely by the variables that separate the points of this space; I suggested that these should be understood as structural (*i.e.* non-reducible) correlations. Moreover, the factorization (by the gauge group) that leads to the reduced space also offers a response to Cao's objection concerning the distinction between mathematical and physical structure (on which, see French & Ladyman [2003], pp. 45–6): it is that which is invariant under this group of transformations.

to defend relationalism or substantivalism.[267] In the following sections I outline and defend these points more forcefully, and show how they impact on questions of ontology. The conclusion I draw is that if we want to read off ontologies from physical theories with symmetries then the best we can hope for is a structuralism; anything else involves auxiliary metaphysical assumptions. I turn to the connection between the findings of the present section, symmetry and structuralism in the next chapter.

267 Indeed, the symmetries give us an *explanation* of the interpretive underdetermination that occurs in theories with such symmetries. Since a non-reductive interpretation can occupy all of the conceptual spaces open to a reductive interpretation (but not always *vice versa*) on account of the nature of the symmetries, any taxonomy that aligns such ontological stances to these moves is sure to fail.

CHAPTER 9

Structuralism and Symmetry

Poincaré is generally taken to be the arch-epistemic structuralist about physics. In an oft quoted passage he claims that physical theories "teach us ... that there is such and such a relation between this thing and that", and that in naming these "things" we only name "images we substituted for the real objects which Nature will hide for ever from our eyes. The true relations between these real objects are the only reality we can attain ..." ([1905], p. 161). The crucial thing to note about this passage is that it clearly expresses an unbridgeable gap between epistemology and ontology: there *are* real objects external to our minds, but the only access we have to them comes in the form of relations between them. Since the relations express something true about the world we can gain restricted knowledge about it, but only about it's structure. The Kantian sentiment of this stance is unlikely to impress many realists, for it implies that our theories do not and cannot tell us anything about the nature of the fundamental elements of the worlds, only about their relational structure; we are forever prevented from gaining complete knowledge about the natures of objects.

A perhaps more satisfying structuralist position for the realist to adopt is an *ontological* version of structuralism, according to which the Kantian notion of 'Nature hiding real objects from our eyes' is replaced by the notion that, 'fundamentally, there just are no objects' (see Ladyman [1998]). The true relations that we *do* have access to *exhaust* the ontology. The ontology is one of pure structure, and we gain knowledge of this structure by observing relations. But the relational structure does not supervene on a set of *noumenal* objects as per Poincaré's epistemic perspective. Instead, we say that the structure is ontologically subsistent. It might be that the structure is ontologically 'prior' to or ontologically 'on all fours' with objects (relata), that are then characterized as intersections or nodes in the relational structure. Or it might be that objects simply drop out of the ontology altogether at a fundamental level. According to this stance our theories give us knowledge of reality. This is bound to not satisfy many philosophers, who will wish for a more robust characterization of this ontology of structure. This will generally amount to a desire for a *reduction* of structure to something else. Perhaps the only way to deal with these requests is to paraphrase Lewis ([1975], p. 85), who faces a similar problem with regard to his treatment of possible worlds: *Structures are what they are and not some other thing*! Fortunately, I think one can do better than this; it is symmetry that is the key to structure but the structuralism that follows is one of *extreme* deflation metaphysically speaking.

Let me quickly distinguish structuralist positions from relationalist positions; many philosophers have a strong tendency to conflate the two. Recall that relationalism was characterized by a *material dependency* thesis. The idea was that a certain structure can be given a relational description if it can be reduced to a family of relations on a primitive set of material objects. The ontological primacy is held by the set of material objects such that without them the relational structure would not exist at all. Hence, though one can be a realist about the structure, one must view it as being supervenient on (reducible to; dependent upon) some more primitive facts involving objects. Substantivalism also involves a dependency thesis and a set of primitive objects, though in this case the objects are spacetime points and it is their relations to one another that determine the overall structure. On the other hand, according to structuralism, the relational structure is not characterized by any dependency thesis at all. In fact, if anything, the dependency works the other way around (though I will argue against this below): objects are reduced to nodes in the relational structure. We might, if we wish to cling to the idea that structuralism expresses a form of relationalism, distinguish between two forms: reductive relationalism, corresponding to what I have been calling relationalism, and non-reductive relationalism, corresponding to what I have been calling structuralism. I believe that this is what Saunders [2003a] has in mind.

I have been suggesting that we should adopt a structuralist position with respect to the physics of general relativity and quantum gravity (and theories with gauge-type symmetries more generally) for two key reasons. The first is the underdetermination of metaphysics by physics—*e.g.* as brought out by French in the context of quantum statistics [1989].[268] This was seen to be a fairly ubiquitous phenomena in mathematically dense physical theories, and we found counterparts in classical mechanics, electromagnetism, gauge theory, general relativity, and quantum gravity. The second reason concerns the role that symmetries play in each of theories just mentioned. Whether we use an interpretation that retains symmetries or not, it is evident that they should play a key role in determining the ontology. We use symmetries to get to the invariant structures and these structures comprise what is physically observable. The physically invariant structures of general relativity and quantum gravity are just the diffeomorphism invariant ones (orbit constant quantities, or quantities that commute with the first class constraints in the canonical, Hamiltonian case).[269] The main interpretive debates that rage in these theories (and any other theories with symmetries on the state space) concern the status attributed to the symmetries: does one retain the symmetry or factor it out? does one quantize with the symmetries or factor them out? do the objects related by symmetries get eliminated if one factors the symmetry out? does one view the possibility set generated by the symmetry as representing physically possible worlds one-to-one, many-to-one, or not at all? I argued that in large part the issue of the reality or ontological nature of certain elements of the theory was secondary to one connected to modality and transworld identity. Once this is realised, there is

[268] French [2001] also makes use of the 'covering space'/'intrinsic space' underdetermination in the context of spacetime theories (a favourite example of conventionalists).

[269] Recall that the constraints encode in the (3+1)-dimensional, canonical context the 4-dimensional spacetime diffeomorphism invariance of the covariant theory. The diffeomorphisms comprise the gauge freedom of general relativity.

seen to be underdetermination between incompatible stances concerning ontology virtually across the board as far as theories with symmetries are concerned. The underdetermination is mathematical too, and I showed that there can be no (internal) formal basis for choosing one of these stances over any other. This leaves us without an internally well motivated metaphysical, formal, or physical basis on which to base our interpretations.[270] Hence my desire for a somewhat 'neutral' position between ontological deflationism (without objects) and inflationism (with objects—be they individuals or 'non-individuals').

In this book I have tried to bracket the issue of realism as much as possible; I took both relationalists and substantivalists to be espousing realist positions about spacetime, differing merely in how it is to be understood: spacetime points versus relational structure supervening upon material objects. Structuralism too is to be understood as a realist position differing from each of the other two positions in that the focus is shifted to the relational structure itself. All three types of position will generally agree that the structure in question is mind-independent, and that qualifies them all as espousing realism (however, minimally construed) in my book.[271] However, just as there are many quite distinct views that sit under the banners of 'relationalism' and 'substantivalism', so there are many flavours of structuralism. In this section I outline three types of structuralism (about spacetime), and show how they connect up to the issues discussed in this book. I show that each is flawed in their own quite distinctive way, and proceed to diagnose the problems. I then outline the version I think has most chance of success. The version I present focuses on the nature of symmetry and observables and the precise way that these two notions interact in the context of the problems considered in the current and previous chapters.

9.1. THREE TYPES OF STRUCTURALISM

Most flavours of structuralism are united on the point that relations are important; where they differ is on the issue of the *extent* of their importance. We have seen in our discussion of the hole argument that many responses work precisely by invoking some relations; not all of these were relationalist, strictly speaking. However, much of the division between relationalists and those who call themselves structuralists, and indeed substantivalists, turns on the question of the status of the metric field: is it spacetime or is it matter? Putting this subtle matter aside, I think it is fair to say that the three structuralist positions I consider below see themselves as advocating an alternative position to both relationalism and substantivalism. This they do in quite different ways: I take Stein as advocating an 'even handed' position with respect to matter and space, taking both to be different manifestations of the same totality; French largely follows suit, though he

270 One might suggest that Ockham's Razor—not to multiply *types* of entity beyond necessity—provides a clear cut basis. However, that is not an internal basis and is itself steeped in metaphysical presuppositions. Regardless of this, it is not clear what types of entity are necessary. Necessary how? Empirically, modally, what?

271 For example, this matches Shapiro's notion of "realism in ontology" ([1993], p. 455). Perhaps some confusion arises because of an attempted assimilation of the spacetime debate to the debate between *scientific* realists and anti-realists (see Boyd [1984] for a 'four-tier' set of conditions that have to be met in order to count as scientific realism). I have shown why I think the two should be held apart.

sees his position as involving the elimination of objects (traditionally conceived) from his ontology; Dorato *et al.* adopt a 'hybrid' view combining aspects of Hoefer's metric field substantivalism with aspects of Stachel's relationalism and that retains objects. I expand on these summaries in what follows and show where each position fails.

Stein is concerned to detach the ontology of physics from the mathematical representations used to model physical systems. Thus, although the setting up of the spacetime models requires that we distinguish levels of mathematical structure (point sets, topologies, manifolds, metrics, *etc.*), we should not read this as implying levels of ontological priority, or as implying any ontological division at all. Indeed, the models of general relativity, and spacetime theories, simply represent some spatiotemporal structure; or, rather, a substructure or "*aspect* of the structure of the world" ([1977], p. 397). That is, the fact that we model spacetime by setting up divisions between manifolds and metrics does not thereby imply that the two types of object represent some distinct *physical* objects standing in some relation to each other. This simple, and I think correct point underlies much of Stein's structuralist position. However, he is ambiguous on the *scope* of this structuralist position; for example, in his seminal paper on Newtonian spacetime, where he discusses the rough usage of the terms "ideal" and "real" by Alexander and Toulmin ([Stein, 1967], p. 276), he makes the following remarks (often wheeled as the example *par excellence* of spacetime structuralism by structural realists):

> What exactly do these authors *mean* by "ideal entities which it is helpful to consider in theory", or by a notion of theory that "has a physical application",—as opposed to entities that "exist in reality," or to "the objective existence of a cosmic substratum"? If the distinction between inertial frames of reference and those which are not inertial is a distinction that has a real application to the world; that is, if the structure [Newtonian spacetime] ... is in some sense exhibited by the world of events; and if this structure can legitimately be regarded as an explication of Newton's "absolute space and time"; then the question whether, in addition to characterizing the world in just the indicated sense, this structure of space–time also "really exists," surely *seems* to be supererogatory. [.] [T]he notion of structure of space–time cannot, in so far as it is truly applicable to the physical world, be regarded as a mere conceptual tool to be used from time to time as convenience dictates. For there is only one physical world; and if it has the postulated structure, that structure is—by hypothesis—there, once and for all. if it is not there once and for all ... then it is not there at all; although of course it may still be ... that a structure is there that approximates, in some sense, to the postulated one. ([Stein, 1967], p. 277)

Dorato reads Stein as "claiming that the traditional dispute between substantivalism and relationalism is completely analogous to that between realism and antirealism ...: neither position is tenable" ([2000], p. 1614). Realism and antirealism about spacetime are then to be cashed out as a denial that "the world of events "really exhibits" a certain geometrical structure" and the claim that "the spatiotemporal structure "really exist[s]", [in the sense of] *independent existence*" antirespectively (*ibid.*). He then takes Stein as seeing the (substantivalist) realist's claim of 'reality' as "supererogatory" and the (relationalist) denial as simply false. Dorato is unhappy with this reasoning, and well he should be, but I think that it is in fact a misreading of Stein's original paper.[272] It isn't that Stein sees the relationalist position as false; rather, he sees that position (and substantivalism) in a different light from most authors (in particular, from Alexander and Toulmin). Thus, he takes both Leibniz *qua* relationalist and Newton *qua* substantivalist as equal on the existence on the reality of space and time; or, as he puts it on "the reality of space and time as an objective framework of the phenomenal universe" ([1967], p. 277). Where they differ boils down to the nature of the relations that constitute their structure. For Leibniz the relations are grounded in "what actually is ... [and] anything that could be put it its place" (that is, actual and possible material bodies), whereas for Newton "the relations that constitute space and give its parts their individuality are ... *internal relations*"; and unlike Leibniz "Newton ... is content to postulate the entire structure of space, without attempting to derive it from or ground it in the relations of non-spatial entities" (*ibid.*, p. 278). Thus, I have to disagree with Dorato: there is *no* analogy to be made between the realism vs anti-realism and substantivalism vs relationalism debates (at least, not in this piece of Stein's opus).

French sees elements of Hoefer's metric field substantivalism in Stein's structuralism; and sees the structural characterization of spacetime points concomitant with this view as motivating a general denial of haecceitism, and so as offering a response to the hole argument. French endorses this structuralist line, but he takes such an implication as pointing to the acceptance of Leibniz equivalence ([2001], p. 27)[273]—of course, this is taken by Earman and Norton as *eo ipso* implying relationalism. Yet he claims that this does not amount to relationalism, for there is no reduction of the spacetime manifold to relations between material bodies. Rather, French sees the view that results as according "ontological status"[274] to the spacetime relations themselves and he concludes that one should be realist about "the relevant *structures*" (*ibid.*). It is these structures that "we should be realist about ... and it is to them that we should direct our philosophical attention" (*ibid.*, pp. 27–8). However, French is not forthcoming on the question of what form these structure take on: just what exactly are the "relevant structures"? Presumably they are things composed of the relations he mentions; but neither is he clear about what the "relevant relations" are. This is an important point if we wish to assess the relationship between this view and relationalism and substantivalism, and to properly distinguish it from these latter positions. I disagree too with the claim that it is relations that get assigned the ontological weight in spacetime theories; at least, I cannot see how this could be drawn from the physics in any case. Moreover, it goes clean against the basic structuralist ideal that a primitive notion

[272] Note that Dorato bases his reading on a much later paper of Stein's [Stein, 1989]. I don't think the two positions match in the way suggested by Dorato.

[273] Note, however, that Cover and O'Leary-Hawthorne [1996] argue, convincingly, that PII (underpinning Leibniz equivalence, of course) does not go hand-in-hand with anti-haecceitism. Indeed, the anti-haecceitist simply cannot make any sense out of the claims of permutations of individuals that PII and Leibniz equivalence require to function at all.

[274] Presumably by "ontological status" he means *reality*; and reality of a *non-supervenient* and *primary* sort, for many would agree that spacetime relations are elements of reality in at least *some* sense.

of individuality is not to be had, for the relations are assumed to be individuals of a sort, they can stand on their own so to speak. I think Eddington had the right idea: one should treat the relations and the relata as *mutually dependent* ontologically speaking, they come as a package deal—see, *e.g.* [Eddington, 1923; 1928]; see also [French, 2003] for a discussion of this aspect of Eddington's work (see, especially, p. 235). It was this move that enabled me to sidestep both the hole argument and Unruh and Kuchař's objections to the correlation view. More importantly, for the theme of this section, it sticks to the physics, for the symmetries do not allow us to attribute independent reality to either relations or relata; this is why we couldn't simply consider reduction options as self-evidently the correct move.

Dorato has outlined a position called "structural spacetime realism" that he claims is a "*tertium quid* ...between classical substantivalism and relationalism ...[that] sides with the latter doctrine in defending the *relational* nature of spacetime, but argues with the former that spacetime exists, at least in part, *independently* of particular physical objects and events" ([2000], pp. 1607–8). The relationalist aspect follows from Dorato's endorsement of Stachel's idea that the points of spacetime are individuated by the metric field; in his words: "spacetime points can only be identified by the relational structure provided by the gravitational field" (*ibid.*, p. 1610). The substantivalist component is captured by the thesis that "the geometrical structure used to represent them [space and time] is "really" mind-independently exemplified by the physical world" (*ibid.*, p. 1612). Finally, he claims that the position that results is a "synthesis" of substantivalism and relationalism in the sense that the "metric field is *both* matter *and* spacetime" (*ibid.*). Now, from these few passages it is quite clear that Dorato is defending something very different from both French and Stein. The details are spelt out in a more recent paper written with Pauri [Dorato and Pauri, 2006] in which they attempt to provide a more explicit defense of structural spacetime realism by essentially mixing elements of Hoefer's abandonment of primitive identity for spacetime points with Stachel's idea of the dynamical individuation of points—where, for the latter idea, they implement Bergmann and Komar's idea of using the four (scalar) eigenvalues of the Riemann tensor to define an intrinsic coordinate system.[275] They label their view "point-structuralism"; however, they note that the resulting position is "entity-realist" as regards "both the metric field and its point-events" (*ibid.*, p. 123). This suggests that their position is in fact simply Hoefer's metric field substantivalism in disguise. The similarity is indeed close, as the following passage makes clear:

> we believe that it is not at all clear whether Leibniz equivalence really grinds corn for the relationalist's mill, since the spacetime substantivalist can always ask: (1) why on earth should we identify *physical* spacetime with the bare manifold *deprived of the metric field*? (2) Why should we assume that the points of the mathematical manifold have intrinsic *physical* identity independently of the metric field? ([Dorato and Pauri, 2006], p. 128)

[275] Recall that Saunders too considered the possibility of points admitting *determinate* reference by being localized in the manner of Bergmann and Komar's suggestion.

The first question is common to Stein's and French's structuralist approaches, and we saw in §4.1 that most substantivalists about the spacetime of general relativity would distance themselves from simple manifold realism; Hoefer too concluded that the metric field was an essential part of spacetime and would have to play a part in any defensible substantivalism. The second question paves the way for sophisticated substantivalist positions, and we have seen variations in Hoefer, Saunders, and Stachel, though each sees the question as pointing in a different direction.[276] However, Dorato and Pauri are concerned to show how a coherent and robust notion of spacetime point can emerge from the formalism of general relativity. The agreement with Hoefer goes as far as metric field realism and the denial of primitive identity; the divergence of positions springs from the fact that Dorato and Pauri specify *how* the points get their individuality from the metric field and it here that the position becomes formally analogous to Stachel's. Thus, the idea is to use the metric as an individuating field for the spacetime points. Clearly, however, since the metric is not itself invariant, we need to extract the invariant information from its ten physical components. It turns out, as I have mentioned earlier, that a set of four invariant scalars[277] can be constructed and Bergmann and Komar's idea was to use these scalars to produce a kind of invariant chart for a vacuum solution of the field equations. Any quantities localized to the so defined points will be diffeomorphism-invariant, as will the points themselves, and so the hole argument problem is resolved at a stroke.

Pauri & Vallisneri also follow this Stachelian view: "it is impossible to consider the points of the space–time manifold as physically individuated without recourse to dynamical individuating fields" ([2002], p. 1). They argue that the metric field serves the latter role by separating out the "*metrical fingerprint* of point-events from the gauge variant components" (*ibid.*). Again, we find Bergmann and Komar's idea involving the intrinsic set of coordinates constructed out of the curvature invariants. However, they ground the necessity of dynamical individuating fields in the idea that "the points of a homogeneous space cannot have any intrinsic *individuality*" and support this with a quote from Weyl to the effect that such points do not have any distinguishing objective properties with respect to which they could be distinguished (*ibid.*, p. 4). They are quite clearly conflating the inability to *distinguish* and the lack of individuality. However, the two notions are quite distinct, as Saunders as brought out well in his analysis of identity.[278] The conflation is a common one in the literature on the hole argument (and, indeed, elsewhere); however, it need not detain us from the main thrust of their program. The technical details are based on Pauri & Lusanna's extension of the Bergmann–Komar approach to the Hamiltonian formalism (as developed elsewhere in this book)—see [Lusanna and Pauri, 2002]. In moving to this formalism the distinction between gauge vari-

276 In brief: Hoefer denies primitive identity for the points and claims that the metric field individuates them; Saunders claims that relations to other points individuates them and relations to matter and the metric field can allow them to be referred to; Stachel, by contrast, sees the points and the metric as entangled so that not only are the points individuated by the metric field but without a metric field there are no points.

277 These invariant scalars are eigenvalues of the Weyl tensor C_{abcd}, where the Weyl tensor is the trace free part of the Riemann tensor defined by: $R_{abcd} = C_{abcd} + \frac{2}{n-2}(g_{a[cR d]b} - g_{b[cR d]a}) - \frac{2}{(n-1)(n-2)} R g_{a[c}g_{d]b}$ (where n is the dimension of the manifold—see Wald [1984], p. 40).

278 Substrata theorists or defenders of haecceities will also clearly wish to reject this conflation.

ant and independent variables becomes apparent, and the latter are vital in the construction of the individuating fields. However, the details of their approach fall foul of my arguments from the previous section. In particular, they make essential use of the Shanmugadhasan transformation (Shanmugadhasan [1973]), which induces a gauge fixing such that the Dirac observables are restricted to the reduced space—in fact, they form a Darboux basis for the space (Pauri & Vallisneri [2002], p. 14). In projecting from the constraint surface to the reduced space they are making a decision as to which states and observables are physically real, a decision that is not read off the physics.

The first problem with these 'Italian' structuralist approaches is that, as with *any* Bergmann–Komar style approach, it is required that the spacetime admits no non-trivial symmetries—in order to work, the spacetime must be inhomogeneous and asymmetric to force the functional independence of the scalars. Only if this condition is satisfied can the four scalars be constructed.[279] A further problem is that I mentioned in relation to Saunders' non-reductive relationalism. If we consider the scalars to be 'physical' then they will surely be among the things that are quantized. If they are indeed quantized than it is difficult to make sense of the coordinate system. If they are not quantized, then it is difficult to make sense of it as a physical thing, for it will be dynamically decoupled from any quantum fields. Thus, the transition of this framework to quantum theory will most likely lead to an incoherent conceptual structure (if not in fact technical structure). The problem is that, since the metric field will become a quantum operator in quantum gravity, the scalars will undergo fluctuations (the six gauge variant degrees of freedom will not, of course: they get erased). If these scalars are the metrical fingerprint then quantization will inevitably smudge them, and the notion of a local event (with respect to these scalars) is lost. Thirdly, even if we ignore the complications that quantum theory will bring, we have the result that this approach is a 'reduced space' approach and this means that all of the interpretations will be able to access it.[280] Thus, it cannot be seen as offering unique support to structuralism.

In slightly different ways, each of the authors discussed above make the same mistake Belot makes: they assume that there is a connection between possibility counting and spacetime ontology. Since there is *some* connection between phase spaces and possibility counting, the issue of spacetime ontology is seen to be tied to phase spaces. It is true that if one desires an 'all out' anti-haecceitism, as, for example, French appears to, then the reduced space is the most obvious space to choose: the symmetries have been removed and it was just these that lead to the problems concerning individuals and indistinguishables. But this space is compatible with structuralism, relationalism, and substantivalism alike. Clearly, when mutually incompatible interpretations can occupy the same space, that space is not relevant in the characterization of these interpretations. But the underlying motivation for French is not the elimination of haecceitism *per se*, he sees this as

279 This might, of course, be met in physically realistic models; but as a constraint on interpretation, it is, I fear, too stringent. To make an interpretation dependent on these features when the theory it is an interpretation of admits symmetries unproblematically is surely missing something about that theory.

280 Note that Lusanna explicitly assumes that the gauge invariant view of observables requires the reduced space: "Leibnitz [*sic.*] equivalence is nothing else than the selection of the gauge-invariant observables" ([2003], p. 6). I hope I have said enough in the previous section to dispel this common myth.

part and parcel of his position; rather, he draws on the underdetermination of metaphysics by physics that, as we have seen, is rife in considerations of symmetry in physical theories. I go along with French's insistence on the importance of this underdetermination for the question of ontology; however, I think it is deeper than French gives it credit for. Just as both 'standard' sides of debates concerning ontology are underdetermined—*i.e.* substantivalism/relationalism; particles as individuals vs non-individuals, *etc.*—so structuralism in French's sense is implicated in this underdetermination: a fourth way is needed.[281] One's structuralism should be grounded in the *cause* of the underdetermination, rather than in the underdetermination itself; and that cause is symmetry. I the next section I attempt to pinpoint where the stumbling block is in these approaches before explicating a more honest form of structuralism that respects these points.

9.2. TO TAKE OBJECTS OR TO LEAVE THEM?

Recall that the original motivation for Worrall's structural realism was to undercut the pessimistic meta-induction[282] by arguing that "[t]here was continuity or accumulation in the shift [between theories], but the continuity is one of *form* or *structure*, not of content" ([Worrall, 1989], p. 157). French & Ladyman share Worrall's desire to provide a realist account of theory change, but do not see this alone as a sufficient basis on which to base structural realism. Thus, they argue that the ability to resolve a number of other significant problems facing realism, coming from modern physics, should also play a central role. Of the upmost importance is the underdetermination unearthed by French between interpretations of quantum particles as individuals and non-individuals. Now, since either side of this alternative requires *some* notion of object, French and Ladyman propose to simply dispense with the traditional notion of object based ontology conceived of in terms of individuality and present ontic structural realism as a "reconceptualisation of ontology ... which effects a shift from objects to structures" ([2003], p. 37).

A standard structuralist ploy, mentioned by French and Ladyman (*ibid.*), for 'constructing' objects is the *group-theoretical* approach (see, for example, Castellani [1998]). The basic idea is to look for 'objectivity' in the invariant properties as determined by the symmetry group $\mathcal{G}$ of a theory and then extract a structural notion of object from these. These properties specify the *quidditas* of the object, such that any system possessing just these properties will be a system of a certain *type*. The states of the system will then transform according to some representation $\rho(\hat{\mathcal{G}})\mathcal{S}$ of the symmetry group, so that the state space $\mathcal{S}$ concretely realizes the group with operations (functions or operators $\hat{\mathcal{G}}$) on its elements. The notion of an *elementary*

[281] French seems to acknowledge this in his paper on Hacking and PII: "Putting it bluntly, you only get as much metaphysics out of a physical theory as you put in and pulling metaphysical rabbits out of physical hats does indeed involve a certain amount of philosophical sleight of hand" ([1994], p. 466). Surely the elimination of objects from one's ontology constitutes a fairly substantial metaphysics? To quote Quine, in eliminating objects French rises above "naturalism and revert[s] to the sin of transcendental metaphysics" ([1992], p. 9). He may not care for Quinean naturalism, but this is transcendental metaphysics nonetheless.

[282] That is, the argument intended to pose a problem for realism by noting that there has been ontological discontinuity in which theories have been superseded in the past, and uses this factor to argue that the ontologies of the future will be equally fragile (see Laudan [1981] for the canonical presentation).

system (a fundamental particle, for example) is captured by choosing irreducible representations, according to which the only invariant subspaces are $\{0\}$ and $\mathcal{S}$ (where $\mathcal{S}' \subset \mathcal{S}$ is invariant if $s' \in \mathcal{S}'$ implies $\rho(\hat{\mathcal{G}})s' \in \mathcal{S}'$, $\forall\hat{\mathcal{G}}$). The notion of an *individual* object (rather than a 'mere' *type*) is grounded in the notion of *imprimitivity* systems (on which, see Varadarjan [1985] or Mackey [1978]). French and Ladyman claim that this "would leave no unknowable objects lurking in the shadows ... and it would retain a sense of objectivity understood structurally" ([2003], p. 37). However, the method of imprimitivity systems relies on the specification of a set of individuating observables, that can perform the function of uniquely determining an individual object. These will generally be observables involving spatiotemporal notions (positions, momenta, *etc.*), so that it is a background spacetime that ultimately confers the individuality and the notion of objecthood.[283] As Mackey puts it, "[a]ll we need to discuss physical events are position observables and a dynamics group" ([1978], p. 195). Clearly, in the more general case we are concerned with, involving as it does diffeomorphism invariance, such sets of properties will not do, they will not be gauge invariant and so the claim of 'objectivity' will be lost. Thus, without some gauge fixing procedure to fix such a set I am at a loss to see how the method of imprimitivity systems could even get off the ground. If one does gauge fix (perhaps *à la* Italian structuralism) then one has, as I argued above, essentially allowed a notion of object to enter anyway (not to mention the other difficulties); given this, it is not clear what work is being done by imprimitivity. But this should not be seen as lending any support to the elimination of objects view[284]; an interpretation with the concept of object is as compatible with the theories in question as one without that notion incorporated.

A similar problem faces the ontic structural realist's characterisation of a quantum field as "the structure, the whole structure, and nothing but the structure" ([French and Ladyman, 2003], p. 48). They are responding to remarks of Cao, in [2003], against *eliminativist* structural realism (without objects):

> given the rejection of particles as the basic ontology in QFT, it seems to us that the sort of developments Cao very nicely charts provide powerful support for the metaphysical SR programme. In particular, he asks about the reality of quantum fields and responds that the concept of field is used to generate the field equations which describe the structural aspects of 'these hypothetical entities' and to 'extract' the concept of particle which are the 'observable manifestations' of the same hypothetical entity. ([French and Ladyman, 2003], p. 48)

283 It is for this reason that French does not endorse Castellani's account of imprimitivity, involving, as it does, the idea that individuals are nothing but bundles of properties with spatiotemporal location conferring individuality. French prefers to view the objects (particles) as useful posits that can be understood as temporary manifestations of the structure as given by imprimitivity systems ([2000], pp. 20–1). However, for the reasons I give in what follows, I think the structuralist would be best set to forget all about imprimitivity systems.

284 Recall Quine's [1976] arguments to the effect that physics pushes towards the view that physical objects should be eliminated in favour of propertyfull regions of spacetime. As French notes ([1998], p. 94), Quine's position was based upon the permutation symmetry of quantum statistics and on the 'received view' that invariance of particles under permutations implies that there are non-individuals. However, consistency clearly requires that this result be applied to the spacetime points of general relativity too, or at least that an account be given of diffeomorphism invariance (and why it differs, if it does, from permutation invariance). However, without spacetime regions and points, Quine's position breaks down.

Now, French & Ladyman are concerned with the notion of "hypothetical entity", and argue that, inasmuch as this concept refers to something "over and above the structural aspects" it suffers from the problem of metaphysical underdetermination. In fact, I think Cao has a point when it comes to *standard* quantum field theory; what is required is an underlying substratum, an 'entity'. Quantum field theory requires a notion of locality, and this in turn requires the concept of absolute spacetime. Thus, the "structural aspects" of the quantum field are strictly parasitic on distinctly non-structural aspects of spacetime. French and Ladyman do not address this problem of background dependence. However, I think they are in luck, because when one considers background *independent* theories the non-structural rug is pulled away and one must deal in relational quantities (*i.e.* quantities not defined at spacetime points). I think they would do well to attack Cao head on in his own territory on this point: how can an underlying substratum be presumed in background independent theories?

Clearly, the ontic structuralist conception of objects is in large part motivated by French's question "Can we be realists in the face of the underdetermination of metaphysics by physics?" ([1998], p. 105). The underdetermination that French and Ladyman are concerned with is the particles as individuals vs particles as non-individuals package that appears in the context of quantum statistics. Now, as we saw above, French defends a structuralism; but it is of a distinctly eliminative flavour: "what it is about a theory that corresponds to reality are certain structural relations, while the relata themselves are regarded as ontologically eliminable" (*ibid.*, p. 107). Hence, the underdetermination is seen to provide support for structuralism and structuralism is seen to imply the elimination of individuals (objects, relata, *etc.*); or, perhaps, in more 'revisionary metaphysics' terms, as forcing the replacement of one conceptual scheme (with objects) by another (with structure). Now, there is a standard objection to this 'eliminativistic' structural realism: if we are realists about relational structure then surely this involves some further commitment to the relata that are implicated in this structure?[285] French sees no reason to be committed, and I agree that one's hand is not forced in this issue (*cf.* French & Ladyman [2003]). According to a Steinian structuralism, for example, one should not pull metaphysical conclusions from one's mathematical models too quickly, which is just what appears to be happening here. The fact that we use some domain of individuals, points, or whatever, does not thereby involve any commitment to the elements of that domain. Thus, we might find that individuals are useful in helping us construct and analyse structures, but we are at liberty to view this as purely heuristic. However, I don't think that a conclusion can be reached, from within the confines of the physics of this debate, that can break the impasse between the likes of Cao, who think that holistic structures cannot be ontologically primitive, and the likes of French and Ladyman who think

[285] See, for example, Chakravartty [1998]; French & Ladyman also attribute this objection to Redhead (see French & Ladyman [2003], p. 41), himself a self-proclaimed structuralist. We saw above that the Italian Structuralists take this objection very seriously (as do French & Ladyman—*cf. ibid.*), and as a result espouse a form of entity realism, wherein the relata are taken to have an independent existence from the relational structure, though certain elements are structurally individuated by these relations. Cao [2003; 2006] espouses a rather different view according to which the *knowledge* we can gain about unobservable entities is structural, but does not see this knowledge as being in any way ontologically exhaustive: there's more to the world than structure. Thus, Cao views physics as recommending a Poincaré style structuralism.

that they can.[286] However, the debate is freshened up somewhat by the symmetry arguments of this book: the question is, do the symmetries imply that any 'entities' have properties that are non-relational (*i.e.* do they have any intrinsic properties)? This gives us a new way of approaching the debate; but I fear that it will be no closer to being resolved because of the severe underdetermination that pervades: one's answer will depend upon one's treatment of the symmetries, and there is no internal basis for deciding which route to take in this regard.

In fact, I think with regard to the problems we have studied within this book, there is a problem here for both the eliminativist and non-eliminativist schools in that neither position is given unique support by anything in the physics. The two schools are interlocked in a debate that cannot possibly be resolved internally, so that any decision will be based upon some background metaphysical assumptions! The best one can do is to adopt a neutral position according to which there is no elimination of entities, but neither are the entities given ontological priority or independence. Rather, one views the two categories in a more even-handed way as *inter*dependent. Of course, part of the problem, and part of the motivation for this even-handedness, is the peculiar nature of the individuals (if that is what they are) that enter into the symmetries. They are identical precisely in the sense that they play the same role in the relational structure. It is just this feature that allowed for the permutation arguments (which is essentially all we have been dealing with). There were problems with positions that were committed to these individuals, as might be expected, and there was always a way of eliminating the individuals without disturbing the empirical adequacy: one can eliminate the symmetry and, in so doing, one eliminates the individuals in some sense. However, there were ways to retain the individuals *and* avoid the problems by tweaking various parts of the interpretation and the metaphysical background assumptions in particular. We are, thus, bound to have trouble here when we bring two positions together that agree about structure but disagree about individuals (entities).

Both substantivalism and relationalism require some notion of identity and individuality for the categories of entity they countenance. This individuality might be primitive, or qualitative, or else derivative from something else. Relationalism, if it contains spacetime points at all, will accept the latter option. However, what I have been pushing for is a view according to which in cases where exact symmetries operate, the laws of physics simply don't say anything about such matters. The observables are such that matters of identity and individuality are washed over. This even extends to the very existence of certain objects! This is a structuralism, I would say that it is an ontological one too; but it is not French and Ladyman's, it does not involve the elimination of objects.

Now, I am not committed to the view that *all* metaphysics is underdetermined by physics; nor even that the issue of individuality is underdetermined by physics in general. For this reason, I prefer to stay clear from the brand of ontic structural realism French and Ladyman endorse. Indeed, at present, the structuralism I have outlined can hardly said to be a realist position anymore than an anti-realist position, though I am in fact a realist. The simple point I wish to make is that the

[286] See the collection of articles in the special issue of *Synthese* (Volume **136**, 2003) for a recent snapshot of this debate.

ontic structural realist's account of physics is as underdetermined as the standard entity realist's account (be it in terms of individuals or non-individuals). Thus, concerning the status of the objects connected by exact (permutation) symmetries, there can be no empirical consequences that cannot be dealt with without them. This itself is a fairly straightforward consequence of the equivalences holding at the level of geometric spaces. Nonetheless, we *can* come up with a clear notion of structure—clearer and cleaner, I think, than French and Ladyman's notion: it is simply that which is invariant under the symmetries (as *I* have been using this term). *This* has empirical consequences, and, in the case of gauge theories like general relativity, this is the *only* aspect of the theory that has empirical consequences. Why? Simply because the gauge freedom and gauge invariance *defines* the empirical sector of the theory. But to say that this empirical sector exhausts reality, as per French and Ladyman's structural realism, is to fall in to the trap of a metaphysics that transcends the physics. That is not something I have a problem with as such; I believe that this structure is all there is, but this is not something that one can read off the physics. Moreover, the gauge invariant structure is not equivalent to the empirical structure, there are physically real quantities that we could never observe, given our limitations, and so on.

Saunders considers himself a structural realist [2003c], but given his retention of objects, this is clearly a different kettle of fish from French and Ladyman's structural realism. Firstly, recall how Saunders protects PSR from potential 'symmetry-type' violations:

> The points of space, independent of their relations to matter, unlike particles of matter, independent of their spatial relations, are in fact discernible. If, now, it were possible to refer to one point of space rather than another (without reference to matter), it would make sense to ask at which of the two points the material system is to be placed, leading to the same difficulty with the PSR. But in fact the points of space are only *weakly* discernible, so we cannot refer to any one point rather than another, and the difficulty does not arise. ([Saunders, 2003b], p. 304)

The result can be applied generally to the parts of any "highly symmetric entity" (*ibid.*). Hence, the question 'why *this* one rather than *that* one?' is ill-posed for such cases. Hoefer used a similar ploy, denying primitive identity for such parts and the result is much the same. However, Saunders' parts have their identities grounded in the PII and, as such, can be understood as well defined, numerically distinct objects (in however thin a sense); but Hoefer has no such grounding principle. Without this, however, the sense in which these parts are objects is surely rather strained. Of course, Hoefer's claim is just that this is a mistake: strip away primitive identities from a pair of objects and there remains a pair of objects, regardless of whether they are qualitatively distinguishable or not. Who is correct? I doubt one can say definitively; but we can say that Saunders' approach is the better motivated; Hoefer's rests largely upon one's philosophical intuitions. The end result is much the same though: one can have a well defined notion of (qualitatively indistinguishable) individual objects; one does not have to eliminate them; and one can escape the symmetry arguments all the same.[287]

Thus, Saunders claims that his proposal individuates objects with the PII by using the invariant properties and relations. Most of his arguments are made within the context of either classical physics or quantum mechanics, but he does mention diffeomorphism invariant field theories and claims that with the method "one obtains little more than ... relations between field values", and states that such things "are objects as events" ([2003b], p. 305). However, if one is focusing on the invariant structure then any move beyond this is making metaphysical assumptions not uniquely secured by the physics. Ontologically, any view that either retains or eliminates objects constructed as per Saunders' method or Belot's method can no longer said to be reading the metaphysics from the physics. The closest one can get to that is to treat the invariant structure itself as fundamental and even that goes too far. But structure is the best we can do, and symmetry the key to structure; as Kuchař points out: "by making a theory dependent on *more* variables, one can make it invariant with respect to a wider class of transformations" ([1993], p. 6). And by making a theory invariant with respect to a wider class of transformations, one gets to know more of the physical structure. This structure is best understood as a network of undecomposable correlations.

From what has been said so far about the similarities between the hole argument and the cases considered in the preceding chapters (Leibniz-shifts, permutation symmetry, and the gauge freedom in electromagnetism and gauge theories in general), we might expect that there is more general form of 'hole argument' that encompasses all such cases. Such a generalized hole argument would no doubt be a valuable asset; it would allow us to consider in more general, abstract terms what underlies the more specific arguments. It would, more importantly, allow us to transfer 'lessons' across the cases. Stachel offers such a generalization, and does indeed draw general conclusions that sound rather like a form of structuralism.

> The common feature of all these cases is that they result in relational theories, in the sense that there are no features of the underlying points that do not result from the ensemble of relations imposed on them by the particular model of the theory in question: aside from the underlying set and its permutation group, there are no universal, model-free structures. ([Stachel, 2002])

It is true that the "underlying points" have their *qualitative* features as result of their role in the structure; *i.e.* from the relations, as Stachel says. This does not, however, lead to relationalism. The points might, for example, possess haecceities, in which case they will have features that do not result from the relations. But Stachel has drawn attention to an important general point, that theories with gauge-type symmetries do not permit us to draw conclusions about the nature of the underlying points. Unfortunately, Stachel does just that in supposing that they are exhausted by their role in the relational structure.

The simple basic idea of structuralism in the context of general relativity is, I think, to deny any distinction between spacetime and matter as far as ontological

287 See also Saunders ([2003c], p. 131).

priority goes. Thus, substantivalism is denied, since that privileges spacetime over matter; relationalism is denied because that privileges matter over spacetime. It is just such prioritization than lands these views into serious trouble when it comes to the tribunal of quantum gravity. But they are already in some trouble in the classical context in that they go beyond the physics. The gauge symmetries allow either side of the debate to claim support from the physics, but for this reason neither side is supported at all. Arguing that there is underdetermination between the substantivalist and relationalist packages is one way to motivate a spacetime structuralism, but a more direct route is clearly preferable. I carry this out in the next section.

9.3. SURPLUS, SEMANTIC UNIVERSALISM AND MINIMAL STRUCTURALISM

Following Ismael and Van Fraassen ([2003], p. 371), let us distinguish between 'elements' and 'structure'—where the latter is defined by a set of relations on the elements, as I defined in §1.2. If we use a structure to represent a physical system or an aspect of a physical system (*i.e.*, if the structure is a mathematical model), then any surplus will, if it manifests itself at all, manifest itself through a many-to-one relationship between model and system. One and the same physical state of affairs being multiply represented is the hallmark of surplus structure. We have seen such structure in many different theories now and a (possible) many-to-one relationship is seen to occur when the structure admits a certain type of symmetry. The kinds of symmetry I have been concerned with in this book are such that they generate no change in the observable (in the technical sense) state of the system under consideration, *i.e.* global or gauge symmetries. That is to say, they do not alter the *structure* when they act on the elements—here, of course, I mean 'qualitative structure' and the observables provide the key to enable access to it. This picks out a class of symmetries that preserve all of the qualitative aspects of some model or world.

Of course, not all symmetries are like those mentioned above; most of those that occur in physics (in computations, and so on), *change* some qualitative features of the world. The world is changed but the system, taken intrinsically (*i.e.* without reference to anything *external* to the system), isn't. In their discussion of the connections between symmetries and Noether's theorems, Brading & Brown illustrate this difference with a simple example, as follows:

> One way of getting our hands on the empirical significance of a symmetry is through 'Galilean ship' type experiments. Here, we take an effectively isolated subsystem of the universe, transform it (in the case of Galileo's ship we go from the ship being at rest to the ship being in uniform motion), and observe that the two states of the subsystem are empirically indistinguishable except in relation to (parts of) the rest of the universe. Thus, in the case of Galileo's ship, no experiments carried out inside the cabin of the ship, and without reference to anything outside the ship, enable us to tell

whether the ship is at rest or moving uniformly The two states of motion are empirically indistinguishable except by looking out of the porthole. ... we apply the symmetry transformation to an effectively isolated *sub*system of the universe, yielding two *empirically distinct* scenarios across which the internal evolutions of the subsystem are empirically indistinguishable. ... However, in the case of global gauge symmetry, this approach doesn't work. Gauge transformations have no analogue to the 'Galilean ship', they have no active interpretation. While it is true that global gauge symmetry does not have the same direct empirical significance ... this does not imply that global gauge symmetry is without empirical content. *The very fact that a global gauge transformation does not lead to empirically distinct predictions is itself non-trivial.* In other words, the freedom in our descriptions is no 'mere' mathematical freedom—it is a consequence of a physically significant structural feature of the theory. ([Brading and Brown, 2003], pp. 98–9)

Let us distinguish between these two types of symmetry, calling the former variety *empirical symmetries* and the latter variety *gauge-type symmetries*.[288] Thus, the gauge-type symmetries preserve *all* qualitative structure, where 'qualitative structure' is determined by the (in principle, empirically accessible) properties and relations of a theory. The structuralism I have proffered is not general enough to include empirical symmetries; if it did include these the implication would be that there is but one unchanging structure in the world, for there would be a single element phase space representing it (after identifications have been made)! But it is the latter gauge-type symmetries that play a central role in the theories we are interested in, and these theories are the best ones we have. More to the point, within these theories, it is symmetry that plays a key role in determining what the theories are about, and therefore what the ontology of the theory is or could be. Where the symmetries of the first type are concerned, specifying an ontology (or a large part of it) will involve a significant amount of background ontology such as metrics and connections forming part of a fixed background spacetime. Without this background structure to hand, the analysis breaks down and so does the notion of particle constructed along group-theoretic lines (see the end of §9.1).

Recall the definition of a symmetry I gave in §1.2 as an automorphism of a structure, where a structure is a set of individuals with a set of properties and relations defined over them. Thus, we can view a symmetry as a permutation on the domain that has the effect of shuffling the properties and relations over the individuals without altering the structure: all relations and properties are preserved by the permutation. Now, clearly in the examples we have been considering *some* properties and relations will be altered by the permutation; if we take our individuals to possess haecceities then different properties will be assigned to the individuals in the two cases. If we fix our attention only on qualitative properties and relations then we get the desired result: invariant structure. The fact that the individuals are related in this way by a permutation implies that they are

[288] Compare this with Giulini ([2003], p. 32) who distinguishes in a similar way, though he restricts the use of the term "symmetry" to what I have called empirical symmetries and calls my gauge-type symmetries "gauge redundancies". Similarly, Ismael & van Fraassen refer to the gauge-type symmetries as "beacons of redundancy" ([2003], p. 391).

qualitatively indistinguishable, for were they distinguishable the structure would certainly not be invariant: one could call upon some property or relation to discern the structures. The individuals thus related must therefore play the same roles in the structure. The fact that many non-trivial structures of modern physics behave in just this way is quite remarkable. I would suggest that this is one possibility for the "physically significant structural feature of the theory" that Brading and Brown mentioned; it is the (qualitative) indistinguishability of the symmetry operands that is responsible. This can be understood in terms of the observables simply as an indifference to how the individuals are distributed in the relational structure.

At least this is one story; but we must be careful about stripping metaphysical conclusions from the symmetries *vis-à-vis* the status of the individuals (this includes whether or not they exist): any such conclusion will most likely have involved putting a certain amount of metaphysics in by hand. For example, if we are anti-haecceitists then the idea permuting individuals to produce distinct possibilities will not make much sense. However, this hasn't stopped many authors 'pulling metaphysical rabbits out of physical hats' (to use French's phase)! One such conclusion that has been drawn from similar arguments and reasoning is *semantic universalism*—van Fraassen's phase for the Leibnizian view that one can describe the world using only general, purely descriptive terms (*i.e.* without haecceities); or, in his own words, that "all factual description can be completely given in entirely general propositions" ([1991], p. 465).[289] Before we see how semantic universalism has been argued for, by Stachel and van Fraassen, let us get a better handle on what it says.

Here's how Cover and O'Leary-Hawthorne express the content of the semantic universalist account—or "generalist picture", as they call it:

> A general proposition is ... any proposition that is not singular, containing no individuals or haecceities; whatever determines the individual(s) that a general proposition is about, it does so indirectly via qualitative properties and relations the individual(s) happens to bear. General propositions can in a suitably rich language be expressed by sentences void of any directly referential devices such as proper names or indexicals; they correspond to sentences constructed solely from quantifiers, variables, qualitative predicates not expressing haecceities, and logical operators. ([Cover and O'Leary-Hawthorne, 1996], p. 4)

Now, it should be clear that if semantic universalism is true, then we lose out on transworld identity and *de re* modal discourse (*cf. ibid.*, p. 5). The reason is clear enough: there are not the conceptual resources to say of some particular thing in world w (with the array of properties P^i) that it is the same as the thing in world w^* with those same properties (or some different array of properties Q^j). In adopting this view, one can resolve the hole argument problem and those like it issuing from

[289] van Fraassen understands a proposition as a set of possible worlds (the 'truth value' of the proposition). If a proposition is "general" then one can permute the individuals in the worlds *salva veritae* (*ibid.*, p. 469). See Teller [1998] for an excellent presentation and critique of van Fraassen's notion of semantic universalism. I expound elements of Teller's account in which follows.

symmetries. Again, the reason is clear: those problems stem from a consideration of permuting individuals in a way that respects the overall qualitative structure (*i.e.* the qualitative properties and relations). Obviously, if one grounds all (relevant) truths in qualitative properties and relations alone (ruling out haecceities) then one avoids these problems at a stroke. Such a move is essentially made by all sophisticated substantivalists.[290] The structuralist positions outlined above follow a similar path. Thus, the notion of individuals being swapped around to give new possibilities simply does not make sense in the semantic universalist's account (*cf.* Cover and O'Leary-Hawthorne [1996], pp. 13–4). Of course, there are different ways of implementing this position. Stachel, for example, argues as follows:

> the points of the manifold are ... individuated *entirely* by the relational structure specified by some solution to the generally-covariant field equations. Remarkably enough, the elementary particles are similarly individuated by their position in a relational structure. Each particular kind of elementary particle ... may be characterized in a way that is independent of the relational structure in which its exemplars are imbricated: by their mass, spin, charge, half-life, *etc.*; but a particular elementary particle can only be individuated ... by its role in such a structure. The reason for this is ... the requirement that all relations between N of these particles be invariant under the permutation group acting on these particles. But the elementary particles and the points of space–time are the basic building blocks of our current model of the universe. ... If individuality has been lost at the level of depth to which we have currently penetrated in our physical theories, it is hard to believe that it will re-emerge if we succeed in penetrating to a deeper level in our understanding of nature. This suggests that we impose the following *generalized permutation principle* as a requirement on any candidate for a future (more) fundamental theory: Whatever the nature of the basic elements out of which it is constructed, the theory should be invariant under *all* permutations of these basic elements. ([Stachel, 2002], pp. 29–30)

Thus, Stachel is convinced that individuality (*i.e. primitive* or *intrinsic* individuality) is ruled against by the diffeomorphism and permutation invariance of general relativity and quantum mechanics. Rather, individuality is reduced to the role played by the particular parts of a relational structure, and permutations map parts to parts in such a way as to 'swap' roles. Indeed, properly understood, the idea of permutations of the parts simply do not make sense since the individuality of a 'part' is determined dynamically by the role it plays in the structure. Stachel further believes that this will be a feature of the future theory of quantum gravity, whatever it may be. Indeed, Stachel draws rather strong conclusions about the form such a theory will take, essentially ruling out loop quantum gravity and string theory at a stroke on pain of violating his generalized principle of permutation invariance (both retain a degree of background dependence) ([2002], pp. 30–1).

[290] I would place Saunders and Stachel similarly. For example, the latter speaks of a "principle of general permutation invariance" ([2002], p. 15) in a way that suggests he has the generalist picture in mind.

van Fraassen draws superficially similar conclusions in the context of quantum statistics and the question of particle identity, but draws from them deeper anti-metaphysical morals. He argues that there is an equivalence between quantum field theory (without particles; with occupation numbers) and the 'many-particle' picture:

> All models of (elementary [first quantized], non-relativistic[291]) quantum field theory can be represented by (*i.e.* are isomorphic to) ... Fock space model constructions. ... Since the latter are clearly carried out within a 'labelled particle' theory, we have a certain kind of demonstrated equivalence of the particle—and the particleless—picture. ([van Fraassen, 1991], p. 448)

He generalizes this argument using the concept of semantic universalism, as we have seen, defined by him as the thesis that all factual descriptions can be completely given by general propositions. In terms of possible worlds the view corresponds to the claim that a permutation of individuals at a world does not affect the truth values of propositions (defined as a set of such worlds). His idea is to show that whenever semantic universalism is satisfied by some theory there is an equivalence between interpretations ("pictures") with and without individuals; moreover, whenever this is the case, we can be sure that physics cannot decide which package to choose.

Now, van Fraassen's argument takes place within a highly idealized 'possible worlds'-type model $\mathrm{M} = \langle \mathrm{D}, C, R \rangle$. The models comprise a domain of individuals D and an array of "cells" C that are taken to represent qualitative differences—these are analogous to the *distinct* gauge orbits of a theory with gauge symmetry (redundancy). Thus, if two distinct individuals occupy distinct cells they will differ with respect to at least one qualitative property (each cell represents a maximally consistent set of properties). A possible world corresponds to a distribution of individuals across the cells. The standard modalities are defined by an access relation R between worlds (necessity: satisfied at *all* R-accessible worlds; possibility (contingency): satisfied at *some* R-accessible worlds). A model is then a collection of such worlds sharing the same domain, array, and access relation. If a model is closed under permutations operating on the domain (*i.e.* shuffling individuals about across the cells) then it is said to be "full". Regarding the relationship between worlds with and without individuals, van Fraassen writes:

> The models required by semantic universalism are exactly those which can be described *equally* on either view. So far we have described them in terms of individuals. But each world—a mapping of individuals into cells of a logical space—can be characterized simply as a set of occupation numbers for the cells. Closure under permutation of the access relation R entails that the R-modalities operate on fully general propositions without losing the generalities. Therefore every significant proposition can be restated entirely in terms of occupation numbers. ([van Fraassen, 1991], p. 475)

291 The qualification van Fraassen gives here is crucial since the shift to relativistic quantum field theory renders the 'particle picture' much more dubious; indeed, in the curved space quantum field theories the notion of particle doesn't seem to make any objective, *invariant* sense at all (see [Wald, 1994]).

van Fraassen claims that the occupation number representation allows one to "abstract" a model without individuals as follows: "A world is a mapping of cells into natural numbers" (*ibid.*, p. 476). This simply tells us how many instances of some properties there are. As he notes, there are many 'individual-full' models corresponding to a single 'individual-free' model; however, the equivalence class of individual-full models (*i.e.* the unique full model closed under permutations) does correspond to a single individual-free model. Thus, van Fraassen's strategy is to assume haecceitism in the initial characterization of the models—indeed, the definition of a world is tantamount to an endorsement of haecceitism. This haecceitism produces an inflated possibility set consisting of qualitatively indistinguishable worlds (*i.e.* differing solely in terms of how haecceities are distributed). The connection to my original definition of symmetry is obvious, as is the similarity to the kinds of symmetry arguments we have looked at in this book. However, van Fraassen's next move (to an occupation representation) is in effect to eliminate the haecceitism of the model by imposing a principle of permutation invariance on them. The principle of permutation invariance is taken to correspond to a principle of generality; that is, generality (of propositions) is given precisely by invariance of truth-value under permutations. Of course, this is just the notion of fullness of models defined above. Semantic universalism is then cashed out as the thesis that all models are of this kind. van Fraassen argues that there is an equivalence between these full models and the abstracted models; this is the result mentioned above: for models related by a permutation of individuals, we can mod out by the permutation (*i.e.* take the isomorphism class) to get a single abstract model. In taking the equivalence, of course, we are ignoring the underlying identity of the individuals. This leads van Fraassen to anti-haecceitism: the abstract model (without individuals) is equivalent to a unique full model (with individuals), so the notion of 'individual' on which haecceitism rests is not doing any work (*ibid.*, pp. 475–6).

This is, clearly, rather similar to what I myself have been arguing for, and something very like it formed my criticism of French and Ladyman's ontic structuralism. What are the differences? For one, I grounded my discussion more concretely in the geometric spaces used to actually describe theories; but, more importantly, I don't see that there is equivalence in the sense in which van Fraassen means. Simply by attending to the nature of the geometric spaces we can see that there is a definite *inequivalence* between the space corresponding to that with symmetries (extended: with symmetric possibilities and individuals) and that without (reduced: without symmetric possibilities or individuals).[292] Thus, the possibility structures of these interpretations are radically different; modal talk available to the latter is simply not available in the former. Teller makes a similar point with regard to van Fraassen's claim of equivalence between labelled particle $[H^n]$ and occupation number $[H^n_+]$ pictures of quantum field theory[293]:

[292] Another point to bear in mind is that the Fock space description allows for the construction of *separable* state spaces for quantum field theories. The many-particle formalism (with labels) has a state space of the form $\mathcal{H} = \bigotimes_{i=1,\ldots,\infty} H_i$ (where the H_i are individual separable Hilbert spaces); a basis in this space is given by an infinite (uncountable) sequence $|n_1, \ldots, n_i\rangle$. Perhaps this is another way to break the deadlock? Possibly, but it isn't clear to me that this demonstrates a mathematical inequivalence. For a nice discussion, see Streater & Wightman [1964].

[293] See also French ([1998], p. 107) and Butterfield ([1993] §5 and §6).

> I agree that the "pictures" are empirically equivalent, in the sense that all the facts that actually arise can, one way or another, be described in either framework. But it seems to me misleading to parley the empirical equivalence into equivalence of pictures across the board. In cutting down from H^n to H^n_+ we have lost expressive power. In H^n but not in H^n_+ we can describe cases that never occur. This fact, in turn, shows that although the cases that do occur can be described in either picture, there is an important sense in which the descriptions are not equivalent. One picture—the one using labels—describes cases that do occur in terms of a conceptual framework that facilitates saying things that cannot be said with the resources of the other picture. ([Teller, 1998], p. 131)

This corresponds to my points made within the context of the geometric spaces in the previous chapter. van Fraassen in effect uses the satisfaction of semantic universalism by certain physical theories to ontological effect. I agree that quantum statistics satisfies the principle of semantic universalism in this sense: facts about particle role do not figure in quantum statistics. This is the content—on one understanding—of permutation invariance. I argued that much the same holds for gauge invariance and diffeomorphism invariance. The observables of the theories I have considered can naturally be seen as implementing van Fraassen's idea but I don't wish to be led from this into constructive empiricism: I believe there's more to the world (that we are entitled to believe) than the empirical structure (given the empirically significant propositions). But I don't derive this from physics. Rather, all I wanted to show was that an underdetermination even more pernicious than that charted by French and Ladyman pervades our best theories. There are a number of ways to understand it; in brief: the natural space for their structuralism is the reduced space but (1) this space is (non-modally) equivalent to the extended space and (2) the reduced space allows all of the options, in particular, it is isomorphic to those spaces that result from gauge fixing most naturally aligned with robust object views. This underdetermination thus swallows up (eliminative) structural realism too. For this reason I do not wish to be lead into their brand of eliminativism. I am nonetheless a realist about our theories; certainly not an entity realist; but not a structural realist of the French & Ladyman stripe either. I believe in the invariant structure revealed by the symmetries of our theories, but I don't see how one can say that this structure is *all* there is: underdetermination forbids it.

I see two main reasons to avoid the kind of view expressed by van Fraassen in the context of the arguments considered in this book[294]—the second is, I think, stronger than the first. Firstly, it seems that, inasmuch as the position is invoked to resolve a specific type of problem in the philosophy of physics, it is liable to seem a trifle ad hoc. Certainly, if this avoidance is the only thing one bases the position on, then the charge of ad hocness is most likely vindicated. However, the general metaphysical package that the position is taken to go along with, namely anti-haecceitism, is itself rather well motivated.[295] If one happens to endorse some

[294] As a matter of fact, my metaphysical intuitions draw me towards the generalist picture and the anti-haecceitistic position it intuitively recommends. However, what I hope to have shown is simply that neither the generalist picture nor anti-haecceitism are uniquely supported by the physics in question.

[295] See, for example, Lewis [1983a; 1986a].

modal metaphysics that respects generalism, then one might have a satisfying response. However, this metaphysics will hardly be the *output* of the symmetry arguments. This brings me to the second, and more serious point and this is simply that the generalist picture is not incompatible with individuals nor with haecceitism *per se*. As regards the coexistence of individuals and generalism, I find Saunders' arguments convincing: one can stick to the idea that a purely descriptive lexicon is adequate and yet retain a notion of object, even absolutely indiscernible ones. But Saunders, of course, grants the anti-haecceitism: *de re* modal claims for such individuals simply don't make any sense (unless, perhaps, one incorporates counterpart theory)—ditto for Hoefer's primitive identity denying substantivalist.

van Fraassen argues for semantic universalism in a rather curious way; he begins by assuming haecceitism in his semantic models (toy possible worlds models) and then imposes a principle of permutation invariance on these models, concluding that the latter principle enforces anti-haecceitism. It doesn't: I have argued at many places in this book that the principles of invariance can be accommodated with and without reduction, symmetries, objects, and haecceities.[296] In other words, the principle of permutation invariance that van Fraassen's views as granting anti-haecceitism is perfectly compatible with individuals; one simply understands it as a principle of indifference concerning the structure: the (qualitative) structure is indifferent to which individuals play which roles. Thus, van Fraassen might be right that his principle of permutation invariance erases any reference to *particular* individuals, but that does not imply that one can or should *dispense* with individuals. Indeed, one can, but one need not; whatever option one chooses here is underdetermined by the physics, by the symmetries.

The observables of the theories we have considered do not distinguish between role swaps concerning individuals related by the relevant symmetries. There are a number of options, as we have seen. We can say that the objects are simply individuated by their role in the structure, as delineated by the observables (via gauge invariance). We can say that, although the swaps are not *empirically* significant, they none the less occur as distinct states and so at least have some *conceptual* significance (*e.g.* the individuals have haecceities). Or we can do away with the objects entirely, in which case the symmetries have a very different meaning.[297]

All of the above options have their plus points. None of them are dictated by the physics. We therefore need to be honest about what our theories (of the kind I have considered) can tell us about the world. In particular, they cannot tell us whether objects of the specified kind are individuals or non-individuals *e.g.* (without haecceities) and they cannot tell us if they even exist or not. Where does this leave the aspiring realist (which I claim to be myself)? I don't think panic stations need ensue for the simple reason that the symmetries, although clouding the question of objects, nonetheless provide us with definite physical

296 We have also seen how Brighouse's position allows for the irrelevance of haecceitism in considerations of physics without thereby ruling it out in principle.

297 Clearly they cannot be seen any longer as operating on objects. Indeed, I think that the question of how to understand (conceptually) symmetries is one of the most pressing problems facing advocates of the French and Ladyman ontic (eliminativist) position. This is made all the more pressing since invariance under symmetries is utilized to 'get to the structure'. Perhaps, however, the notion of using individuals as a 'heuristic crutch' can accommodate this (see, for example, French [1999])—still, the details need spelling out more than they have been thus far.

structure that is in principle measurable, invariant, and objective.[298] I espouse a view that takes this structure as an aspect of the world—treating spacetime and matter as simply different aspects of the world, *à la* Stein, thanks to the ubiquitous underdetermination—but stays silent on the extent of this structure, on the issue of whether this structure is sufficient to 'cover' what there is (call this position 'minimal ontological structuralism'[299]). I for one think that there is a whole lot more, but honesty constrains these other aspects to the box marked 'metaphysics'; beyond the structure revealed by symmetries the thread holding together physics and metaphysics snaps. Decisions concerning the latter then have to be guided by something else.[300]

Thus what we are left with is a rather deflationary stance according to which our best theories of physics cannot furnish us with any information beyond what is contained in the structure of observables, but the structure of the observables themselves forbids reading entity realist or structural realist positions from this with good conscience. Perhaps the debate has run out of steam? I hope not and, in fact, I think that Eddington's structuralist position fits this state of affairs quite well.

Eddington eschews the question of priority when it comes to relations and relata (*cf.* French [2003], p. 233):

> The relations unite the relata; the relata are the meeting points of the relations. The one is unthinkable apart from the other. I do not think that a more general starting-point of structure could be conceived. ([Eddington, 1928], pp. 230–1)

Thus, writes French, "Eddington did not regard the structure as ontologically prior ... [rather,] both relata and relations, structure and entity are ontologically entwined in that each is necessary to make sense of the other" (*ibid.*, p. 257). Now, French draws attention to a criticism of Braithwaite to the effect that the relata must be originally (*i.e.* intrinsically) distinguishable in order to be distinguishable within the relational structure; that is, they must have existence *independently* of the relational structure they find themselves enmeshed within. Of course, as French notes, Braithwaite simply misses the point: the properties conferred by the relational structure are all there is to the relata. But this does not imply any priority of the relational structure over the relata, "the two came as a package" ([2003], p. 235). Of course, Braithwaite can be seen as employing the 'no relations without relata' objection to the ontic structuralist realist approach.[301] Eddington answers

298 Not all of this structure is necessarily measurable in practice *by us*. Thus we are not restricted to the realm of the 'observable' in van Fraassen's sense. We nonetheless have 'access' to the structure through the observables (in the gauge theoretic sense).

299 This is not to be confused with van Fraassen's notion of "moderate structuralism" ([2006] 278) for that position requires an underlying substance that is the "bearer" of structure.

300 Thus, here I diverge from constructive empiricists who take empirical adequacy of theories as paramount and regard any claims that go beyond the empirical realm as out of bounds as far as belief is concerned (see van Fraassen [1980] 12). Indeed, my choice of geometric space was made to keep all of the options open where extra-empirical matters are concerned. I have no desire to wave "good-bye to metaphysics" (van Fraassen [1991], p. 480), and do not see that the underdetermination warrants it. But I also diverge from ontic structural realists who cut out any 'extra-structural' elements from their ontology as a matter of general principle.

301 Of course, the objections of Unruh and Kuchař to the (timeless) correlation responses to the problem of time are in the same vein. The idea there was that one could not have a correlation without things that are correlated. Of course, French

it in a different, and I think correct, way to the ontic brigade. The latter simply eliminate relata and argue that it is the relations that have ontological clout, not the relata. They argue for this using the underdetermination the afflicts the metaphysical positions based on objects. However, I have shown that the elimination of objects that the ontic structuralists employ is likewise underdetermined by the physics. Thus, Eddington's response is seen to offer the better stance;[302] what's more, as I showed in the previous chapters, it has the resources to deal with the problems of space and time of classical and quantum gravity when aligned with correlation-type approaches to the observables. Before we wrap things up, I first wish to further separate the position I have outlined from van Fraassen's position, thus making my own position more transparent.

9.4. MINIMAL STRUCTURALISM IS NOT CONSTRUCTIVE EMPIRICISM

Van Fraassen's constructive empiricism might look superficially similar to the position I have been calling minimal structuralism; however, as I attempt to show in this section, the positions are radically different under the surface. The superficial similarity, and ultimate difference, concerns the distinct understandings of the notion of *observable*. Firstly, and in its essentials only, what is constructive empiricism? After answering this it will be clear that my differs does indeed differ.

Constructive empiricism is founded on the claim that belief should be restricted to that which is accessible to *direct observation*—this is the source of the problems with constructive empiricism and the ultimate difference between our positions. *Belief* is not equivalent to acceptance; one can perfectly well accept a theory without believing it (the theory might have some heuristic value, for example). Science does not aim for truth—this forms the basis of van Fraassen's rejection of realism—rather, its aim is to provide us with *empirically adequate* theories. It should preserve the 'phenomena 'or the 'appearances'; these are understood to be the objects and structures of experience, experience that is not gained through 'non-sensory' technical instruments (*e.g.* telescopes; spectroscopes, *etc.*). Van Fraassen presents us with an example that is supposed to make these points clearer: the moons of Jupiter *in fact* cannot be seen with the naked eye from Earth; we use telescopes to view them. But they are nonetheless observable since "they can ... be seen without a telescope if you are close enough" ([1980], p. 16). 'Observable' thus means actually or possibly observable with the naked senses; things not satisfying this criterion are consigned to the realm of the 'unobservable'.

& Ladyman's ontic brand of structuralism would also offer a straightforward response to this problem: simple eschew the subjects of the correlations and deal with the correlations themselves. However, the elimination of objects that goes with this view—standardly signaled by a shift to a reduced space description—will be underdetermined by the physics.

302 Of course, Eddington's structuralism is far more wide-ranging this simple 'package view' suggests (see the article by French (*ibid.*) for a detailed analysis of some of the other aspects). I don't wish to be aligned with all aspects of Eddington's structuralism; only with the idea that the package view is best equipped to deal with the problems I have drawn attention to. This view can be seen, for example, in his take on the relationship between a group-structure and its elements: "the significance of a part cannot be dissociated from the system of analysis to which it belongs. As a structural concept the part is a symbol having no properties except as a constituent of the group-structure of a set of parts" ([Eddington, 1958], p. 145). In other words, structure and elements are inextricably intertwined; the correlates in a correlation exist in something like the same way a hole and its outer surface exist: they are inextricably bound together.

Though science may shift the line between 'observable' and 'unobservable', the distinction will always hold, and will always thus constrain what can be said about the link between epistemology and ontology. While van Fraassen allows that there can be perfectly meaningful theoretical statements about matters 'beyond' the line of observability, and that can be true or false, our attitude or *stance* towards these statements should be one of agnosticism. This viewpoint is able to avoid the problematic underdetermination that one faces when trying to interpret physical theories. An interpretation of a theory, you will recall, provides an ontology for that theory; it tells us how the world could be the way the theory says it is. Van Fraassen couches this idea in the semantic approach to theories, according to which a physical theory is a collection of models, the elements of which are taken to represent a possible way for the actual world to be—a mere possible representation only becomes actual if the model and physical reality are isomorphic. The problem is that there are in general many such ways, many such models, with no way, internal to the physical theory itself, of deciding between them. What we have at best, instead, is *empirical adequacy*; the models have sub-models that represent the observables, the actual appearances, and these *are* isomorphic and, therefore, good candidates for belief. Empirical adequacy for van Fraassen is just this isomorphism between empirical sub-models and phenomena. What is not captured in this isomorphism is not a candidate for rational belief. Hence, Van Fraassen's agnostic stance: one model might well be the One True Model, isomorphic to the physical world, but we can never know for there are multiple empirically adequate interpretations (models). Thus van Fraassen gives us the following example:

> If someone likes to talk in terms of spheres, I can reconstrue his every assertion ... as an assertion about points. And *vice versa*! This is not to deny that it is possible for a person to believe that points are the only real concrete individuals—what we cannot do is to say that geometry forces this view on us. ([van Fraassen, 1991], p. 451)

I agree with this, and agree that much the same holds in the context of physics. However, there are elements in both geometry and physics that we can be sure about, that we can be realist about without the shame of underdetermination; we can be realist about the *invariant* structure, and that just is the geometry. Points and spheres simply provide us with different ways of talking about the same structure. This is the point at which my position becomes distinct from constructive empiricism; the invariant structure—that which is invariant under symmetries, for example—is not necessarily observable in van Fraassen's sense, and I don't believe that there is a straightforward distinction to be made between observable and unobservable in van Fraassen's sense.

Empirical adequacy is a primary constraint on an interpretation, and I agree that a nice way of cashing the notion out is through the sub-models and isomorphisms of the semantic approach, but the aim of science goes beyond that, and can give us more than that. The minimal structuralism I outlined above goes beyond merely empirical structure. For example, let's consider the gauge invariance of general relativity. Suppose we have a simple universe with a triangle of three objects; the structure that is preserved by applying a diffeomorphism (the gauge symmetry) to the spacetime they occupy is the relative distances of the objects and the topological (and 'finer') properties of the spacetime manifold. The relative distances will count as observable according to van Fraassen, but the topological and other properties will certainly not count as such. There is no measurement—certainly not unaided!—we could perform to determine the total 4-volume of a compact universe, and yet this is an observable in my sense. We can be realist about this quantity, and it is its gauge invariance that allows us to do so with impunity; to put the point another way, we may say that the 4-volume is an *observable* but is not a *beable*, it is not something we measure. Yet this is included in my account. We don't need to devise an experiment that will provide us with some new empirical data, a reading on a computer monitor for example; even if we could, according to van Fraassen the 4-volume itself would not be observable, only the reading on the screen would be. We have a clear-cut case of disagreement, and it is a fundamental one. It goes further. Van Fraassen thinks that science is an exercise in model-building, rather than a truth-finding mission; scientists construct models that fit the observable facts, and certain models, with different ways of representing the unobservable realm, might be more useful than others. That just grinds against what I take science to be about, namely that it does aim to furnish us with TRUE stories about the world and its workings. The idea that the line between what can and cannot qualify as true belief (observable vs unobservable) is based on human physiology (*cf.* van Fraassen [1980], p. 17) and what I say is that the observables have *nothing* to do with us. I think that I have dispelled some of the problems besetting realism, that in part guided van Fraassen's rejection of it, well enough: the empirical equivalence and underdetermination that result from symmetries can be squared with realism, but not to the extent that the ontological structural realists believe, and certainly not to the extent that standard entity realists believe. The position that results can be viewed usefully as providing the safest base from which to construct ontologies in the context of modern physics; I claimed that this is best understood as a structuralist position.

To sum up this section, and indeed this book, let me state my position more plainly.[303] The underdetermination that van Fraassen is motivated by also motivated the ontological structural realists, though in very different ways! Van Fraassen argues that the underdetermination, the multiplicity of interpretations or models, should push us into retracting our belief when the interpretation goes beyond the observable (*i.e.* beyond the empirical sub-model). Thus, the underdetermination motivates van Fraassen's anti-realism; he focuses on the empirical

303 Recently, Esfeld [2004] has argued for a position superficially similar to the one I have outlined. His idea is to take from the ontic structural realists the claim that there are no intrinsic properties of things that yield structure, but to retain the things as an independent ontological category, grounding their qualitative character in the relations they bear to one another. He calls this a "moderate metaphysics of relations" in contrast to the ontic structuralists "radical metaphysics of relations". His position is based on the premise that relations do require relata—in this sense he is very close to Dorato's spacetime structural realism. I dispute this, and think that relations can be self-standing, or prior to relata. The reason I choose not to dispense with relata (objects) *tout court* is simply that the dispensation is not something that can be decided by the physics because of the equivalences holding between the pictures with and without them. I simply wish to clear the ground for the proper interpretation of physics (at least those with gauge-symmetries); physics must be supplemented by some suitable metaphysics to support the kinds of views associated with there being and not being objects. In any case, if Esfeld's position is to work, and be coherent, I think he has to adopt the kind of package deal view I prefer. Without this the relata are, in themselves, mere bare particulars and that is a metaphysics that physics can do without.

content of a theory, the bits that can be observed unaided—though this might include plates from particle collision experiments, it is the plates and not the particles that are observed. On the other hand, the same difficulty pushes the ontological structural realists into shifting their beliefs onto the structure that is common between the underdetermined interpretations, the multiply realized structure (most naturally described, as I said, by the equivalence class, within a reduced space approach). The different underdetermined interpretations, models, or solutions are simply different ways of getting to the physically real core of the world, different ways of representing the same structural facts.

Structural realists will want to say something about what happened *during* the particle collision experiment, rather than what happens when one observes the end result, the plate itself. The theory will furnish this, even though there will, or may be many apparently competing interpretations: for the structural realist there is really no competition, the true ontology is that which connects the many interpretations. The story they give will be different from that of a standard realist who will most likely wish to tell something in terms of the theoretical entities postulated by the theory, the individual particles, fields, or whatever; the structural realist story will allow for particles, but only as derivative from the 'wider' structure. This I disagree with, and I side, to a certain extent, with van Fraassen. In this chapter I have tried to show how the ontological structural realist position is subject to an underdetermination of its own; one cannot read off the physics whether there is structure that 'lives on top' of objects or is primitive, brute. The reason underlining this new underdetermination is that even with objects one can escape the traditional kinds of underdetermination that pushed van Fraassen and the structural realists; one simply looks at the observables and notes that they are insensitive to matters concerning permutations of the objects. But the structure of the observables is what I say we should focus on, this should be the object of the structural realist's devotion; it is that which is invariant under swaps of gauge related states, of certain permutations (those which do not result in a structural change) of the individual elements of the theory. But just because it is invariant under a change of the individual elements of a domain does not mean that the objects should be dispensed with; because *this* tyre is as good as *that* qualitatively identical tyre and so can be swapped, does not mean that my car can run without tyres! The view I prefer, and which sits best with general relativity and quantum gravity, was a view whereby the observables are (gauge invariant) correlations between (gauge variant) quantities (*i.e.* correlata) that cannot be viewed as independent from the whole correlation: the correlata are measurable *in virtue* of the fact that the correlation is predictable and measurable. Thus, interpretively speaking, the structure comes before the individuals since the individuals are not measurable. These observables, the correlations, are insensitive to matters concerning the individualistic ontology of spacetime points. It is in this sense that they are structural, and—given the further claim that the observables give a true account of the world—the ontology based upon them is structural realist. It is minimal because it does not say that *all* there is this structure.[304] This is the most minimal realist core that can be read

[304] I should perhaps quickly distinguish this from an *epistemic* version of structural realism. Recall that that position says that there *is* more to the world than structure, but that all we can know is the structural aspects of this "more". I say that

off the physics without falling prey to either anti-realism or metaphysical (interpretative) underdetermination.

whether there is more than the structure cannot be answered, it is underdetermined by the physics: one can be realist about the structure without holding that there is nothing but the structure. Thus, when da Costa & French say that "all that there *is*, is structure" ([2003], p. 189), they are staking out a radical metaphysical territory, they are not reading that thesis from off the formal representation of physical theories. In this sense one can shoot the ontic structural realists with their own gun; their gripe with standard entity realism is precisely that it is metaphysics not physics (*cf.* French & Ladyman [2003], p. 45). But regardless of this, we can still know that structural aspects are physically real, and knowable, despite the fact that we might not be able to observe them.

REFERENCES

[Adams, 1979] Adams, R.M. Primitive thisness and primitive identity, The Journal of Philosophy 1979, LXXVI(1), 5–26.

[Aharonov and Bohm, 1959] Aharonov, Y., Bohm, D. Significance of electromagnetic potentials in the quantum theory, Physical Review 1959, 115(3), 485–91.

[Alexander, 1956] Alexander, H.G. The Leibniz–Clarke Correspondence, Manchester University Press, 1956.

[Amelino-Camelia, 1999] Amelino-Camelia, G. Are we at the dawn of quantum-gravity phenomenology? Kowalski-Glikman, J., editor. Towards Quantum Gravity, Springer-Verlag, 1999. p. 1–44.

[Anderson, 1964] Anderson, J.L. Relativity principles and the role of coordinates in physics, in: Chiu, H.-Y., Hoffmann, W.F., editors. Gravitation and Relativity, W.A. Benjamin, Inc., 1964. p. 175–94.

[Anderson, 1967] Anderson, J.L. Principles of Relativity Physics, New York, Academic Press, 1967.

[Armstrong, 1986] Armstrong, D.M. The nature of possibility, Canadian Journal of Philosophy 1986, 16, 575–94.

[Ashtekar and Rovelli, 1992] Ashtekar, A., Rovelli, C. Connections, loops, and quantum general relativity, Classical and Quantum Gravity 1992, 9, 3–12.

[Ashtekar and Stachel, 1991] Ashtekar, A., Stachel, J. Conceptual Problems of Quantum Gravity, Basel, Birkhäuser, 1991.

[Ashtekar and Tate, 1996] Ashtekar, A., Tate, R. Lectures on Non-Perturbative Canonical Gravity, World Scientific, 1996.

[Auyang, 2001] Auyang, S.Y. Spacetime as a fundamental and inalienable structure of fields, Studies in the History and Philosophy of Modern Physics 2001, 32(2), 205–15.

[Baez and Munian, 1994] Baez, J.C., Munian, X. Gauge Fields, Knots, and Gravity, World Scientific, 1994.

[Baez, 2001] Baez, J.C. Higher-dimensional algebra and Planck scale physics, in: Callender, C., Huggett, N., editors. Physics Meets Philosophy at the Planck Scale, Cambridge University Press, 2001. p. 177–95.

[Baez, 2006] Baez, J.C. Quantum quandaries: A category-theoretic perspective, in: Rickles, D., French, S., Saatsi, J., editors. The Structural Foundations of Quantum Gravity, Oxford University Press, 2006. p. 240–65.

[Barbour and Bertotti, 1982] Barbour, J., Bertotti, B. Mach's principle and the structure of dynamical theories, Proceedings of the Royal Society of London A 1982, 382, 295–306.

[Barbour, 1994] Barbour, J. The timelessness of quantum gravity: I. the evidence from the classical theory, Classical and Quantum Gravity 1994, 11, 2853–73.

[Barbour, 2001] Barbour, J. The Discovery of Dynamics: A Study from a Machian Point of View of the Discovery and the Structure of Dynamical Theories, Oxford University Press, 2001.

[Barbour, 2003] Barbour, J. The End of Time, Weidenfeld & Nicholson, 2003.

[Bartels, 1996] Bartels, A. Modern essentialism and the problem of individuation of spacetime points, Erkenntnis 1996, 45, 25–43.

[Beall and van Fraassen, 2003] Beall, J.C., van Fraassen, B. Possibilities and Paradoxes: An Introduction to Modal and Many-Valued Logic, Oxford University Press, 2003.

[Beig, 1994] Beig, R. The classical theory of canonical general relativity, in: Ehlers, J., Friedrich, H., editors. Canonical Gravity: From Classical to Quantum, Berlin, Springer-Verlag, 1994. p. 59–80.

[Belot and Earman, 1999] Belot, G., Earman, J. From metaphysics to physics, in: Butterfield, J., Pagonis, C., editors. From Physics to Metaphysics, Cambridge University Press, 1999. p. 166–86.

[Belot and Earman, 2001] Belot, G., Earman, J. Presocratic quantum gravity, in: Callender, C., Huggett, N., editors. Physics Meets Philosophy at the Planck Scale, Cambridge University Press, 2001. p. 213–55.
[Belot, 1995] Belot, G. New work for counterpart theorists: Determinism, The British Journal for the Philosophy of Science 1995, 46, 185–95.
[Belot, 1996] Belot, G. Whatever is never and nowhere is not: Space, time, and ontology in classical and quantum gravity, PhD thesis, University of Pittsburgh, 1996.
[Belot, 1998] Belot, G. Understanding electromagnetism, The British Journal for the Philosophy of Science 1998, 49, 531–55.
[Belot, 2000] Belot, G. Geometry and motion, The British Journal for the Philosophy of Science 2000, 51, 561–95.
[Belot, 2001] Belot, G. The principle of sufficient reason, The Journal of Philosophy 2001, 99, 55–74.
[Belot, 2003a] Belot, G. Notes on symmetries, in: Brading, K., Castellani, E., editors. Symmetries in Physics: Philosophical Reflections, Cambridge University Press, 2003. p. 393–412.
[Belot, 2003b] Belot, G. Symmetry and gauge freedom, Studies in the History and Philosophy of Modern Physics 2003, 34, 189–255.
[Bergmann and Komar, 1972] Bergmann, P.G., Komar, A. The coordinate group symmetries of general relativity, International Journal of Theoretical Physics 1972, 5, 15–28.
[Bergmann and Komar, 1980] Bergmann, P.G., Komar, A. The phase space formulation of general relativity and approaches toward its canonical quantization, in: Held, A., editor. General Relativity and Gravitation: One Hundred Years After the Birth of Albert Einstein, vol. 1, Plenum Press, 1980. p. 227–54.
[Bergmann, 1961] Bergmann, P.G. Observables in general relativity, Reviews of Modern Physics 1961, 33(4), 510–4.
[Boyd, 1984] Boyd, R. The current status of scientific realism, in: Lepin, J., editor. Scientific Realism, Berkeley, University of California Press, 1984. p. 41–82.
[Brading and Brown, 2003] Brading, K., Brown, H.R. Symmetries and Noether's theorems, in: Brading, K., Castellani, E., editors. Symmetries in Physics: Philosophical Reflections, Cambridge University Press, 2003. p. 89–109.
[Brighouse, 1994] Brighouse, C. Spacetime and holes, Philosophy of Science (Proceedings) 1994, 1, 117–25.
[Brighouse, 1997] Brighouse, C. Determinism and modality, The Journal of Philosophy 1997, 48, 465–81.
[Brügmann, 1994] Brügmann, B. Loop representations, in: Ehlers, J., Friedrich, H., editors. Canonical Gravity: From Classical to Quantum, Berlin, Springer-Verlag, 1994. p. 213–53.
[Bull and Segerberg, 1984] Bull, R., Segerberg, K. Basic modal logic, in: Gabbay, D.M., Guenthner, F., editors, Handbook of Philosophical Logic, vol. 2, Dordrecht, Kluwer Academic, 1984. p. 1–88.
[Butterfield, 1984] Butterfield, J. Relationism and possible worlds, The British Journal for the Philosophy of Science 1984, 35, 101–13.
[Butterfield, 1988] Butterfield, J. Albert Einstein meets David Lewis, Philosophy of Science (Proceedings) 1988, 2, 65–81.
[Butterfield, 1989] Butterfield, J. The hole truth, The British Journal for the Philosophy of Science 1989, 40, 1–28.
[Butterfield, 1993] Butterfield, J. Identity and interpretation in quantum theory, Studies in the History and Philosophy of Science 1993, 24(3), 443–76.
[Butterfield, 2001] Butterfield, J. The end of time? The British Journal for the Philosophy of Science 2001, 53, 289–330.
[Callender, 2000] Callender, C. Shedding light on time, Philosophy of Science 2000, S587–99.
[Cao, 2001] Cao, T.Y. Prerequisites for a consistent framework of quantum gravity, Studies in the History and Philosophy of Modern Physics 2001, 32(2), 181–204.
[Cao, 2003] Cao, T.Y. Can we dissolve physical entities into mathematical structures? Synthese 2003, 136, 147–68.
[Cao, 2006] Cao, T.Y. Structural realism and quantum gravity, in: Rickles, D., French, S., Saatsi, J., editors. The Structural Foundations of Quantum Gravity, Oxford University Press, 2006. p. 40–52.
[Carlip, 1998] Carlip, S. Quantum Gravity in $2+1$ Dimensions, Cambridge University Press, 1998.

[Castellani, 1998] Castellani, E. Galilean particles: An example of constitution of objects, in: Castellani, E., editor. Interpreting Bodies: Classical and Quantum Objects in Modern Physics, Princeton University Press, 1998. p. 181–94.

[Castellani, 2004] Castellani, E. Dirac on gauges and constraints, International Journal of Theoretical Physics 2004, 43(6), 1503–14.

[Chagrov and Zakharyaschev, 1997] Chagrov, A., Zakharyaschev, M. Modal Logic, Oxford University Press, 1997.

[Chakravartty, 1998] Chakravartty, A. Semirealism, Studies in the History and Philosophy of Science 1998, 29, 391–408.

[Connes and Rovelli, 1994] Connes, A., Rovelli, C. Von Neumann algebra automorphisms and time versus thermodynamics in generally covariant quantum theories, Classical and Quantum Gravity 1994, 11, 2899–918.

[Costa and French, 2003] Da Costa, N., French, S. Science and Partial Truth, Oxford University Press, 2003.

[Cover and O'Leary-Hawthorne, 1996] Cover, J.A., O'Leary-Hawthorne, J. *Haecceitism* and anti-*Haecceitism* in Leibniz's philosophy, Noûs 1996, 30(1), 1–30.

[Crane, 1993] Crane, L. Categorical physics, http://arxiv.org/pdf/hep-th/9301061, 1993.

[Crane, 1995] Crane, L. Clocks and categories: Is quantum gravity algebraic? Journal of Mathematical Physics 1995, 36, 6180–93.

[Dicke, 1964] Dicke, R.H. The many faces of Mach, in: Chiu, H.-Y., Hoffmann, W.F., editors. Gravitation and Relativity, W.A. Benjamin, Inc., 1964. p. 121–41.

[Dirac, 1958] Dirac, P.A.M. The Principles of Quantum Mechanics, Oxford University Press, 1958.

[Dirac, 1964] Dirac, P.A.M. Lectures on Quantum Mechanics, Belfer Graduate School of Science Monographs Series, vol. 2, Yeshiva University, 1964.

[Dorato and Pauri, 2006] Dorato, M., Pauri, M. Holism and structuralism in classical and quantum GR, in: Rickles, D., French, S., Saatsi, J., editors. The Structural Foundations of Quantum Gravity, Oxford University Press, 2006. p. 121–51.

[Dorato, 2000] Dorato, M. Substantivalism, relationalism, and structural spacetime realism, Foundations of Physics 2000, 30(10), 1605–28.

[Drieschner et al., 2002] Drieschner, M., Eynck, T.O., Lyre, H. Comment on redhead: The interpretation of gauge symmetry, in: Kuhlmann, M., Lyre, H., Wayne, A., editors. Ontological Aspects of Quantum Field Theory, World Scientific, 2002. p. 303–12.

[Earman and Norton, 1987] Earman, J., Norton, J. What price substantivalism? The hole truth, The British Journal for the Philosophy of Science 1987, 38, 515–25.

[Earman, 1989] Earman, J. World Enough and Space–Time. Absolute versus Relational Theories of Space and Time, MIT Press, 1989.

[Earman, 2002] Earman, J. Throughly modern McTaggart: Or what McTaggart would have said if he had learned the general theory of relativity, Philosophers' Imprint 2002, 2(3), 1–28.

[Earman, 2003a] Earman, J. The cosmological constant, the fate of the universe, unimodular gravity, and all that, Studies in the History and Philosophy of Modern Physics 2003, 34(4), 559–77.

[Earman, 2003b] Earman, J. Ode to the constrained Hamiltonian formalism, in: Brading, K., Castellani, E., editors. Symmetries in Physics: Philosophical Reflections, Cambridge University Press, 2003. p. 140–62.

[Earman, 2006] Earman, J. Two challenges to the requirement of substantive general covariance, Synthese 2006, 148(2), 443–68.

[Eddington, 1923] Eddington, A. Mathematical Theory of Relativity, Cambridge University Press, 1923.

[Eddington, 1928] Eddington, A. The Nature of the Physical World, Cambridge University Press, 1928.

[Eddington, 1958] Eddington, A.S. The Philosophy of Physical Science, University of Michigan Press, 1958.

[Einstein, 1916] Einstein, A. The foundation of the general theory of relativity, in: Lorentz, H.A., Weyl, H., Minkowski, H., editors. The Principle of Relativity, Dover Publications, 1952. p. 111–64. 1916.

[Einstein, 1918] Einstein, A. Dialogue about objections to the theory of relativity, in: Engel, A., editor. The Berlin Years: Writings: 1918–1921, English Translation of Selected Texts. The Collected Papers of Albert Einstein, vol. 7, Princeton University Press, 2002. 1918.

[Esfeld, 2004] Esfeld, M. Quantum entanglement and a metaphysics of relations, Studies in the History and Philosophy of Modern Physics 2004, 35(4), 601–17.

[Feynman, 1962] Feynman, R., The Feynman Lectures on Physics, vol. 2, Addison–Wesley Publishing Company, 1962.

[Field, 1980] Field, H. Science Without Numbers, Princeton University Press, 1980.

[French and Krause, 1999] French, S., Krause, D. The logic of quanta, in: Conceptual Foundations of Quantum Field Theory, Cambridge University Press, 1999. p. 324–42.

[French and Ladyman, 2003] French, S., Ladyman, J. Remodeling structural realism, Synthese 2003, 136, 147–68.

[French and Redhead, 1988] French, S., Redhead, M.L.G. Quantum physics and the identity of indiscernibles, The British Journal for the Philosophy of Science 1988, 39, 233–46.

[French and Rickles, 2003] French, S., Rickles, D. Understanding permutation symmetry, in: Brading, K., Castellani, E., editors. Symmetries in Physics: Philosophical Reflections, Cambridge University Press, 2003. p. 212–38.

[French, 1989] French, S. Quantum physics and the identity of indiscernibles, Australasian Journal of Philosophy 1989, 67, 432–46.

[French, 1994] French, S. Hacking away at the identity of indiscernibles: Possible worlds and Einstein's principle of equivalence, The Journal of Philosophy 1994, 92(9), 455–66.

[French, 1998] French, S. On the withering away of physical objects, in: Castellani, E., editor. Interpreting Bodies: Classical and Quantum Objects in Modern Physics, Princeton University Press, 1998. p. 93–113.

[French, 1999] French, S. Models and mathematics in physics: The role of group theory, in: Butterfield, J., Pagonis, C., editors. From Physics to Philosophy, Cambridge University Press, 1999. p. 187–207.

[French, 2000] French, S. Symmetry, structure, and the constitution of objects, http://philsci-archive.pitt.edu/archive/00000327/, 2000.

[French, 2001] French, S. Getting out of a hole: Identity, individuality and structuralism in space–time physics, Philosophica 2001, 67, 11–29.

[French, 2003] French, S. Scribbling on the blank sheet: Eddington's structuralist conception of objects, Studies in the History and Philosophy of Modern Physics 2003, 34(2), 227–59.

[Gabbay and Guenthner, 2002] Gabbay, D.M., Guenthner, F., editors. Handbook of Philosophical Logic, Volume 2: Extensions of Classical Logic, Kluwer Academic, 2002.

[Gambini and Pullin, 1996] Gambini, R., Pullin, J. Loops, Knots, Gauge Theories, and Quantum Gravity, Cambridge University Press, 1996.

[Gaul and Rovelli, 2000] Gaul, M., Rovelli, C. Loop quantum gravity and the meaning of diffeomorphism invariance, in: Kowalski-Glikman, J., editor. Towards Quantum Gravity, Springer-Verlag, 2000. p. 277–324.

[Giulini, in press] Giulini, D. Some remarks on the notions of general covariance and background independence, in: Stamatescu, I.O., editor. Approaches to Fundamental Physics: An Assessment of Current Theoretical Ideas, Springer-Verlag, in press.

[Goldblatt, 1992a] Goldblatt, R. Logics of Time and Computation, CSLI Lecture Notes, vol. 7, Center for the Study of Language and Information, 1992.

[Goldblatt, 1992b] Goldblatt, R. Mathematics of Modality, CSLI Lecture Notes, vol. 43, Center for the Study of Language and Information, 1992.

[Gotay et al., 1998] Gotay, M.J., Isenberg, J., Marsden, J.E., Montgomery, R. Momentum maps and classical relativistic fields. Part I: Covariant field theory, http://www.citebase.org/abstract?id=oai:arXiv.org:physics/9801019, 1998.

[Gotay, 1984] Gotay, M. Constraints, reduction, and quantization, Journal of Mathematical Physics 1984, 27(8), 2051–66.

[Govaerts, 2002] Govaerts, J. The quantum geometer's universe: Particles, interactions and topology, in: Govaerts, J., Hounkonnou, M.N., Msezane, A.Z., Msezane, A.Z., editors. Contemporary Problems in Mathematical Physics, World Scientific, 2002.

[Grünbaum, 1970] Grünbaum, A. Space, time, and falsifiability. Part I, Philosophy of Science 1970, XXXVII(4), 469–588.

[Grünbaum, 1973] Grünbaum, A. Geometrodynamics and ontology, The Journal of Philosophy 1973, LXX(21), 775–800.

[Guilini, 2003] Guilini, D. That strange procedure called quantisation, in: Giulini, D., Kiefer, C., Lammerzahl, C., editors. Quantum Gravity: From Theory to Experimental Search, Springer-Verlag, 2003. p. 17–40.
[Hacking, 1975] Hacking, I. The identity of indiscernibles, The Journal of Philosophy 1975, 72, 249–56.
[Hájiček and Kijowski, 1999] Hájiček, P., Kijowski, J. Covariant gauge fixing and Kuchař decomposition, Physical Review D 1999, 61(2), 024037.
[Hájiček, 1996] Hájiček, P. Time evolution and observables in constrained systems, Classical and Quantum Gravity 1996, 13, 1353–75.
[Healey, 1997] Healey, R. Nonlocality and the Aharonov–Bohm effect, Philosophy of Science 1997, 64, 18–40.
[Healey, 1999] Healey, R. Quantum analogies: A reply to maudlin, Philosophy of Science 1999, 65, 183–203.
[Healey, 2004] Healey, R. Gauge theory and holisms, Studies in the History and Philosophy of Modern Physics 2004, 35(4), 619–42.
[Henneaux and Teitelboim, 1992] Henneaux, M., Teitelboim, C. Quantization of Gauge Systems, Princeton University Press, 1992.
[Hoefer and Cartwright, 1993] Hoefer, C., Cartwright, N. Substantivalism and the hole argument, in: Philosophical Problems of the Internal and External Worlds: Essays on the Philosophy of Adolf Grünbaum, Pittsburgh, University of Pittsburgh Press, 1993.
[Hoefer, 1996] Hoefer, C. The metaphysics of space–time substantivalism, The Journal of Philosophy 1996, 93, 5–27.
[Hoefer, 1998] Hoefer, C. Absolute versus relational spacetime: For better or worse, the debate goes on, The British Journal for the Philosophy of Science 1998, 49, 451–67.
[Horwich, 1978] Horwich, P. On the existence of time, space and space–time, Noûs 1978, 12, 397–419.
[Hu, 2001] Hu, S. Lecture Notes on Chern–Simons–Witten Theory, World Scientific, 2001.
[Huggett, 1994] Huggett, N. What are quanta, and why does it matter? Philosophy of Science (Proceedings) 1994, 2, 69–76.
[Huggett, 1999] Huggett, N. Atomic metaphysics, The Journal of Philosophy 1999, 96, 5–24.
[Isham, 1992] Isham, C.J. Canonical quantum gravity and the problem of time, in: Ibort, L.A., Rodriguez, M.A., editors. Integrable Systems, Quantum Groups, and Quantum Field Theories, Kluwer Academic, 1992. p. 157–288.
[Isham, 1993] Isham, C.J. Prima facie questions in quantum gravity, in: Ehlers, J., Friedrich, H., editors. Canonical Relativity: Classical and Quantum, Berlin, Springer-Verlag, 1993. p. 1–21.
[Ismael and van Fraassen, 2003] Ismael, J., van Fraassen, B. Symmetry as a guide to superfluous theoretical structure, in: Brading, K., Castellani, E., editors. Symmetries in Physics: Philosophical Reflections, Cambridge University Press, 2003. p. 371–92.
[Kaplan, 1975] Kaplan, D. How to Russell a Frege–Church, The Journal of Philosophy 1975, 72, 716–29.
[Kauffman, 2001] Kauffman, L., editor. Knots and Physics. 2nd edition, World Scientific, 2001.
[Krause, 1992] Krause, D. On a quasi-set theory, Notre Dame Journal of Formal Logic 1992, 33, 402–11.
[Kuchař, 1992] Kuchař, K. Time and interpretations of quantum gravity, in: Vincent, D.E., Williams, J.G., editors. Proceedings of the 4th Canadian Conference on General Relativity and Relativistic Astrophysics, World Scientific, 1992. p. 211–314.
[Kuchař, 1993] Kuchař, K. Canonical quantum gravity, http://arxiv.org/pdf/gr-qc/9304012, 1993.
[Kuhn, 1970] Kuhn, T.S. The Structure of Scientific Revolutions, University of Chicago Press, 1970.
[Ladyman, 1998] Ladyman, J. What is structural realism? Studies in the History and Philosophy of Science 1998, 29, 409–24.
[Laudan, 1981] Laudan, L. A confutation of convergent realism, Philosophy of Science 1981, 48, 19–49.
[Leeds, 1999] Leeds, S. Gauges: Aharonov, Bohm, Yang, Healey, Philosophy of Science 1999, 66, 606–27.
[Lewis, 1975] Lewis, D.K. Counterfactuals, Oxford, Basil Blackwell, 1975.
[Lewis, 1983a] Lewis, D.K. Individuation by acquaintance and by stipulation, The Philosophical Review 1983, 92, 3–32.
[Lewis, 1983b] Lewis, D.K. New work for a theory of universals, Australasian Journal of Philosophy 1983, 61, 343–77.
[Lewis, 1986a] Lewis, D.K. On The Plurality of Worlds, Oxford, Basil Blackwell, 1986.
[Lewis, 1986b] Lewis, D.K., Philosophical Papers, vol. 2, Oxford University Press, 1986.

[Loux, 1979] Loux, M., editor. The Possible and the Actual: Readings in the Metaphysics of Modality, Cornell University Press, 1979.

[Lusanna and Pauri, 2002] Lusanna, L., Pauri, M. General covariance and the objectivity of space–time point-events: The physical role of gravitational and gauge degrees of freedom, http://philsci-archive.pitt.edu/archive/00000959/, 2002.

[Lusanna, 2003] Lusanna, L. From Einstein's hole argument to Dirac and Bergmann observables, http://arxiv.org/abs/gr-qc/0302089, 2003.

[Lusanna, 2005] Lusanna, L. The chrono-geometrical structure of special and general relativity: towards a background-independent description of the gravitational field and elementary particles, in: Reimer, A., editor. General Relativity Research Trends. Horizons in World Physics, vol. 249, Nova, 2005.

[Lyre, 2001] Lyre, H. The principles of gauging, Philosophy of Science 2001, 68, S371–81.

[Lyre, 2004] Lyre, H. Holism and structuralism in $u(1)$ gauge theory, Studies in the History and Philosophy of Modern Physics 2004, 35, 597–624.

[Mackey, 1978] Mackey, G.M. Unitary Group Representations in Physics, Probability, and Number Theory, Benjamin/Cummings Publishing Co., 1978.

[Maidens, 1993] Maidens, A. Substantivalism and the hole argument, PhD thesis, University of Cambridge, 1993.

[Maidens, 1998] Maidens, A. Symmetry groups, absolute objects and action principles in general relativity, Studies in the History and Philosophy of Modern Physics 1998, 29(2), 245–72.

[Marcotty and Ledgard, 1986] Marcotty, M., Ledgard, H. The World of Programming Languages, Berlin, Springer-Verlag, 1986.

[Marolf, 1995] Marolf, D. Almost ideal clocks in quantum cosmology: A brief derivation of time, Classical and Quantum Gravity 1995, 12, 2469–86.

[Marsden and Weinstein, 1974] Marsden, J., Weinstein, A. Reduction of symplectic manifolds with symmetry, Reports in Mathematical Physics 1974, 5, 121–30.

[Marsden, 1992] Marsden, J. Lectures on Mechanics, Cambridge University Press, 1992.

[Martin, 2003] Martin, C. On continuous symmetries and the foundations of modern physics, in: Brading, K., Castellani, E., editors. Symmetries in Physics: Philosophical Reflections, Cambridge University Press, 2003. p. 29–60.

[Matschull, 1996] Matschull, H.-J. Dirac's canonical quantization programme, http://arxiv.org/pdf/quant-ph/9606031, 1996.

[Maudlin, 1988] Maudlin, T. The essence of space–time, Philosophy of Science (Proceedings) 1988, 2, 82–91.

[Maudlin, 1990] Maudlin, T. Substances and space–time: What Aristotle would have said to Einstein, Studies in the History and Philosophy of Science 1990, 21(4), 531–61.

[Maudlin, 1993] Maudlin, T. Bucket's of water and waves of space: Why space–time is probably a substance, Philosophy of Science 1993, 60, 183–203.

[Maudlin, 1998] Maudlin, T. Discussion: Healey and Aharonov–Bohm, Philosophy of Science 1998, 65, 361–8.

[Maudlin, 2002] Maudlin, T. Throughly muddled McTaggart: Or how to abuse gauge freedom to generate metaphysical monstrosities, Philosophers' Imprint 2002, 2(4), 1–19.

[McTaggart, 1908] McTaggart, J.E. The unreality of time, Mind 1908, 17(68), 457–74.

[McTaggart, 1927] McTaggart, J.E., The Nature of Existence, vol. II, Cambridge University Press, 1927.

[Melia, 1999] Melia, J. Holes, haecceitism and two conceptions of determinism, The British Journal for the Philosophy of Science 1999, 50(4), 639–64.

[Mellor, 1998] Mellor, D.H. Real Time II, Routledge, 1998.

[Misner et al., 1973] Misner, C.W., Thorne, K.S., Wheeler, J.A. Gravitation, W.H. Freeman and Company, 1973.

[Montesinos et al., 1999] Montesinos, M., Rovelli, C., Thiemann, T. An $SL(2,\mathbb{R})$ model of constrained systems with two Hamiltonian constraints, Physical Review D 1999, 60, 044009.

[Monton, 2006] Monton, B. Presentism and quantum gravity, in: Dieks, D., editor. The Ontology of Spacetime. Philosophy and Foundations of Physics, vol. 1, Elsevier, 2006. p. 263–80.

[Newton, 1962] Newton, I. De gravitatione, in: Hall, A., Hall, M., editors. Unpublished Scientific Papers of Isaac Newton, Cambridge University Press, 1962. p. 123–46.

[Norton, 1988] Norton, J. The hole argument, in: Proceedings of the Biennial Meeting of the Philosophy of Science Association: Symposia and Invited Papers, vol. 2, 1988. p. 56–64.
[Norton, 2003] Norton, J. General covariance, gauge theories, and the Kretschmann objections, in: Brading, K., Castellani, E., editors. Symmetries in Physics: Philosophical Reflections, Cambridge University Press, 2003. p. 110–23.
[Nounou, 2003] Nounou, A. A fourth way to the Aharonov–Bohm effect, in: Brading, K., Castellani, E., editors. Symmetries in Physics: Philosophical Reflections, Cambridge University Press, 2003. p. 174–99.
[Page and Wooters, 1983] Page, D.N., Wooters, W.K. Evolution without evolution: Dynamics described by stationary observables, Physical Review D 1983, 27, 2885–92.
[Pauri and Vallisneri, 2002] Pauri, M., Vallisneri, M. Ephemeral point-events: Is there a last remnant of physical objectivity? http://philsci-archive.pitt.edu/archive/00000931/, 2002.
[Plantinga, 1973] Plantinga, A. The Nature of Necessity, Oxford University Press, 1973.
[Plyuschchay and Razumov, 1995] Plyuschchay, M.S., Razumov, A.V. Dirac versus reduced phase space quantization, in: Charap, J.M., editor. Geometry of Constrained Dynamical Systems, Cambridge University Press, 1995. p. 239–50.
[Poincaré, 1905] Poincaré, H. Science and Hypothesis, Dover Publications, 1905.
[Pooley, in press] Pooley, O. The Reality of Spacetime, Oxford University Press, in press.
[Quine, 1976] Quine, W.V.O. Whither physical objects? in: Cohen, R.S., Feyerabend, P.K., Wartofsky, M.W., editors. Essays in Memory of Imre Lakatos, Dordrecht, Reidel, 1976. p. 499–502.
[Quine, 1992] Quine, W.V.O. Structure and nature, The Journal of Philosophy 1992, 89(1), 5–9.
[Redhead and Teller, 1992] Redhead, M.L.G., Teller, P. Particle labels and the theory of indistinguishable particles in quantum mechanics, The British Journal for the Philosophy of Science 1992, 42, 201–18.
[Redhead, 1975] Redhead, M.L.G. Symmetry in intertheory relations, Synthese 1975, 32, 77–112.
[Redhead, 1988] Redhead, M.L.G. A philosopher looks at quantum field theory, in: Brown, H.R., Harré, R., editors. Philosophical Foundations of Quantum Field Theory, Oxford University Press, 1988.
[Redhead, 1996] Redhead, M.L.G. From Physics to Metaphysics, Cambridge University Press, 1996.
[Redhead, 2003] Redhead, M.L.G. The interpretation of gauge symmetry, in: Brading, K., Castellani, E., editors. Symmetries in Physics: Philosophical Reflections, Cambridge University Press, 2003. p. 124–39.
[Rehren, 1990] Rehren, K.H. Quantum symmetry associated with braid group statistics, in: Numerical Methods in Fluid Dynamics. Lecture Notes in Physics, vol. 370, Springer-Verlag, 1990. p. 318–39.
[Reichenbach, 1924] Reichenbach, H. The theory of motion according to Newton, Leibniz, and Huyghens, in: Reichenbach, M., editor. Modern Philosophy of Science, London, Routledge and Kegan Paul, 1959. p. 46–66. 1924.
[Reichenbach, 1947] Reichenbach, H. Elements of Symbolic Logic, London, Macmillan, 1947.
[Reichenbach, 1958] Reichenbach, H. The Philosophy of Space and Time, Dover Publications, 1958.
[Rescher and Urquhart, 1971] Rescher, N., Urquhart, A. Temporal Logic, Berlin, Springer-Verlag, 1971.
[Rickles, 2005] Rickles, D. A new spin on the hole argument, Studies in the History and Philosophy of Modern Physics 2005, 36, 415–34.
[Rickles, 2006] Rickles, D. Time and structure in canonical gravity, in: Rickles, D., French, S., Saatsi, J., editors. The Structural Foundations of Quantum Gravity, Oxford University Press, 2006. p. 152–95.
[Rovelli, 1991] Rovelli, C. Is there incompatibility between the ways time is treated in general relativity and standard quantum mechanics, in: Ashtekarn, A., Stachel, J., editors. Conceptual Problems of Quantum Gravity, Basel, Birkhäuser, 1991. p. 126–39.
[Rovelli, 1992] Rovelli, C. What is observable in classical and quantum gravity, Classical and Quantum Gravity 1992, 8, 297–316.
[Rovelli, 1997] Rovelli, C. Halfway through the woods: Contemporary research on space and time, in: Earman, J., Norton, J., editors. The Cosmos of Science, Pittsburgh, University of Pittsburgh Press, 1997. p. 180–223.
[Rovelli, 2000] Rovelli, C. The century of the incomplete revolution: Searching for a general relativistic quantum field theory, Journal of Mathematical Physics 2000, 41(6), 3776–800.
[Rovelli, 2001] Rovelli, C. Quantum spacetime: What do we know? Callender, C., Huggett, N., editors. Physics Meets Philosophy at the Planck Scale, Cambridge University Press, 2001. p. 101–22.

[Rovelli, 2002] Rovelli, C. Partial observables, Physical Review D 2002, 65, 124013-8.

[Rovelli, 2004] Rovelli, C. Quantum Gravity, Cambridge University, 2004.

[Russell, 1940] Russell, B. An Inquiry into Meaning and Truth, New York, W.W. Norton, 1940.

[Rynasiewicz, 1994] Rynasiewicz, R. The lessons of the hole argument, The British Journal for the Philosophy of Science 1994, XCIII, 279–306.

[Rynasiewicz, 1996] Rynasiewicz, R. Absolute versus relational space–time: An outmoded debate, The Journal of Philosophy 1996, 45, 407–36.

[Saunders and Brown, 1991] Saunders, S., Brown, H.R., editors. The Philosophy of Vacuum, Oxford University Press, 1991.

[Saunders, 2002] Saunders, S. How relativity contradicts presentism, Royal Institute of Philosophy Supplement 2002, 50, 277–92.

[Saunders, 2003a] Saunders, S. Indiscernibles, general covariance, and other symmetries: The case for non-reductive relationalism, in: Ashtekar, A., Cohen, R.C., Howard, D., Renn, J., Sarkar, S., Shimony, A., editors. Revisiting the Foundations of Relativistic Physics, Festschrift in honor of John Stachel. Boston Studies in the Philosophy of Science, vol. 234, Dordrecht, Kluwer Academic, 2003. p. 151–73.

[Saunders, 2003b] Saunders, S. Physics and Leibniz's principles, in: Brading, K., Castellani, E., editors. Symmetries in Physics: Philosophical Reflections, Cambridge University Press, 2003. p. 289–307.

[Saunders, 2003c] Saunders, S. Structural realism, again, Synthese 2003, 136, 127–33.

[Shanmugadhasan, 1973] Shanmugadhasan, S. Canonical formalism for degenerate Lagrangians, Journal of Mathematical Physics 1973, 14, 677–87.

[Shapiro, 1993] Shapiro, S. Modality and ontology, Mind 1993, 102(407), 455–81.

[Shoemaker, 1969] Shoemaker, S. Time without change, The Journal of Philosophy 1969, LXVI(12), 363–81.

[Singer, 1978] Singer, I.M. Some remarks on the Gribov ambiguity, Communications in Mathematical Physics 1978, 60, 7–12.

[Sklar, 1974] Sklar, L. Space, Time, and Spacetime, University of California Press, 1974.

[Sklar, 1985] Sklar, L. Philosophy and Spacetime Physics, University of California Press, 1985.

[Skyrms, 1981] Skyrms, B. Tractarian nominalism, Philosophical Studies 1981, 40, 199–206.

[Smolin, 1991] Smolin, L. Space and time in the quantum universe, in: Ashtekar, A., Stachel, J., editors. Conceptual Problems of Quantum Gravity, Basel, Birkhäuser, 1991. p. 228–91.

[Smolin, 2000] Smolin, L. The present moment in quantum cosmology: Challenges to the arguments for the elimination of time, http://philsci-archive.pitt.edu/archive/00000153/, 2000.

[Smolin, 2001] Smolin, L. Three Roads to Quantum Gravity: A New Understanding of Space, Time and the Universe, Phoenix, 2001.

[Smolin, 2006] Smolin, L. The case for background independence, in: Rickles, D., French, S., Saatsi, J., editors. The Structural Foundations of Quantum Gravity, Oxford University Press, 2006. p. 196–239.

[Smolin, in press] Smolin, L. Generic predictions of quantum theories of gravity, in: Oriti, D., editor. Approaches to Quantum Gravity—Toward a New Understanding of Space, Time, and Matter, Cambridge University Press, in press.

[Stachel, 1989] Stachel, J. Einstein's search for general covariance, 1912–1915, in: Howard, D., Stachel, J., editors. Einstein and the History of General Relativity. Einstein Studies, vol. 1, Basel, Birkhäuser, 1989. p. 13–42.

[Stachel, 1993] Stachel, J. The meaning of general covariance: The hole story, in: Earman, J., Janis, A., Massey, G., editors. Philosophical Problems of the Internal and External Worlds: Essays on the Philosophy of Adolf Grünbaum, Pittsburgh, University of Pittsburgh Press, 1993. p. 129–60.

[Stachel, 2002] Stachel, J. 'The relations between things' and the 'things between relations': The deeper meaning of the hole argument, in: Malament, D., editor. Reading Natural Philosophy: Essays in the History and Philosophy of Science and Mathematics, Chicago and LaSalle, Open Court, 2002. p. 231–66.

[Stachel, 2003] Stachel, J. A brief history of space-time, in: Ciufolini, I., Dominici, D., Lusanna, L., editors. 2001: A Relativistic Spacetime Odyssey, Proceedings of the 25th Johns Hopkins Workshop on Current Problems in Particle Theory, New Jersey, World Scientific, 2003. p. 15–34.

[Stein, 1967] Stein, H. Newtonian space–time, Texas Quarterly 1967, 10, 174–200.

[Stein, 1977] Stein, H. On space–time and ontology: Extract from a letter to Adolf Grünbaum, in: Earman, J., Glymour, C., Stachel, J., editors. Foundations of Space–Time Theories, University of Minnesota Press, 1977. p. 374–402.
[Stein, 1989] Stein, H. Yes, but ... : Some skeptical reflections of realism and anti-realism, Dialectica 1989, 43, 47–65.
[Streater and Wightman, 1964] Streater, R.F., Wightman, A.S. PCT, Spin, Statistics, and All That, New York, Benjamin, 1964.
[Teller, 1989] Teller, P. Relativity, relational holism, and the bell inequalities, in: Cushing, J.T., McMullin, E., editors. Philosophical Consequences of Quantum Theory: Reflections on Bell's Theorem, University of Notre Dame Press, 1989. p. 208–23.
[Teller, 1990] Teller, P. What the quantum field is not, Philosophical Topics 1990, 18, 175–84.
[Teller, 1991] Teller, P. Substance, relations, and arguments about the nature of space–time, The Philosophical Review 1991, 100(3), 363–97.
[Teller, 1998] Teller, P. Quantum mechanics and haecceities, in: Castellani, E., editor. Interpreting Bodies: Classical and Quantum Objects in Modern Physics, Princeton University Press, 1998. p. 114–41.
[Toretti, 1987] Torretti, R. Relativity and Geometry, Dover Publications, 1987.
[Torre, 1993] Torre, C.G. Gravitational observables and local symmetries, Physical Review D 1993, 48, R2373–6.
[Unruh, 1989] Unruh, W. Unimodular theory of canonical quantum gravity, Physical Review D 1989, 40(4), 1048–52.
[Vallentyne, 1997] Vallentyne, P. Intrinsic properties defined, Philosophical Studies 1997, 88(2), 209–19.
[van Fraassen, 1980] van Fraassen, B. The Scientific Image, Oxford University Press, 1980.
[van Fraassen, 1990] van Fraassen, B. Laws and Symmetry, Oxford University Press, 1990.
[van Fraassen, 1991] van Fraassen, B. Quantum Mechanics: An Empiricist View, Oxford University Press, 1991.
[van Fraassen, 2006] van Fraassen, B. Structure: Its shadow and substance, The British Journal for the Philosophy of Science 2006, 57(2), 275–307.
[Varadarjan, 1985] Varadarjan, V.S. Geometry of Quantum Theory, 2nd edition, New York, Springer-Verlag, 1985.
[Wald, 1984] Wald, R. General Relativity, University of Chicago Press, 1984.
[Wald, 1994] Wald, R. Quantum Field Theory in Curved Spacetime and Black Hole Thermodynamics, University of Chicago Press, 1994.
[Weingard, 1975] Weingard, R. On the ontological status of the metric field in general relativity, The Journal of Philosophy 1975, 72(14), 426–31.
[Weingard, 1988] Weingard, R. Virtual particles and the interpretation of quantum field theory, in: Brown, H.R., Harré, R., editors. Philosophical Foundations of Quantum Field Theory, Oxford University Press, 1988. p. 43–58.
[Weinstein, 1999] Weinstein, S. Gravity and gauge theory, Philosophy of Science (Proceedings) 1999, 66, S146–55.
[Weinstein, 2001] Weinstein, S. Conceptual and foundational issues in the quantization of gravity, PhD thesis, Northwestern University, 2001.
[Wheeler, 1964] Wheeler, J.A. Gravitation as geometry—II, in: Chiu, H.-Y., Hoffmann, W.F., editors. Gravitation and Relativity, W.A. Benjamin, Inc., 1964. p. 65–89.
[Wheeler, 1968] Wheeler, J.A. Superspace and the nature of quantum geometrodynamics, in: DeWitt, C.M., Wheeler, J.A., editors. Battelle Rencontres. 1967 Lectures in Mathematics and Physics, W.A. Benjamin, Inc., 1968. p. 242–307.
[Wigner, 1967] Wigner, E. Symmetries and Reflections, Indiana University Press, 1967.
[Wilson, 1959] Wilson, N.L. Substances without substrata, Review of Metaphysics 1959, 12(4), 521–39.
[Wilson, 1993] Wilson, M. There's a hole and a bucket dear Leibniz, Midwest Studies in Philosophy 1993, XVIII, 202–41.
[Woodhouse, 1980] Woodhouse, N.M.J. Geometric Quantization, Oxford University Press, 1980.
[Worrall, 1989] Worrall, J. Structural realism: The best of both worlds? Papineau, D., editor. The Philosophy of Science, Oxford University Press, 1996. p. 139–65. 1989.
[York, 1971] York, J.W. Gravitational degrees of freedom and the initial value problem, Physical Review Letters 1971, 26, 1656–8.

SUBJECT INDEX

Kolibri 360 GmbH
Ferdinand-Jühlke-Straße 7
99095 Erfurt, Deutschland
produktsicherheit@kolibri360.de